CHEMICAL ENGINEERING FACULTY DIRECTORY

2004-2005

Volume 53

Edited by

S. Joe Qin and J. Steven Swinnea
Department of Chemical Engineering
University of Texas
Austin, Texas 78712

Revised through August 2004

American Institute of Chemical Engineers
3 Park Avenue
New York, New York 10016-5991
http://www.aiche.org

Copies may be ordered from:

AIChE Publication Sales
9 Jay Gould Ct., Waldorf, MD 20602

Phone: 800-AIChemE
800-242-4363
Fax: 301-843-0159
URL: http://www.aiche.org

PUB: #D-48
ISBN: 0-8169-0966-0

Printed in the United States of America.

Preface

The format of the present edition continues the changes set in place over the past few years. Most notably e-mail addresses for individuals appear where contributing institutions have provided them. Where possible U. S. entries begin at the top of a column. We hope that this provides a useful, readable document. We have also updated the Index of Schools (now the Contents that follows this section) and added a comprehensive index of all institutions at the end of the directory.

Production of this volume would not have been possible without the inspiration of John W. Eaton, who meticulously built the formatting and document processing scripts and gathered most of the original institution information.

Corrections and Updates

We would like to produce the most accurate and useful directory possible, but we cannot do this without your help. If you spot errors, or have suggestions for ways to improve the directory, or any other comments, we would like to hear from you. Send your corrections and comments via electronic mail to

`che-faculty-directory@che.utexas.edu`

If you do not have access to electronic mail, please use the address

ChE Faculty Directory
Attn: Dr. J. Steven Swinnea
Department of Chemical Engineering (C0400)
The University of Texas at Austin
Austin, TX 78712
USA

The most convenient method, however, is to update or add your institution's information at the URL

`http://www.che.utexas.edu/AIChE/directory/update.html`

You will need the username "directory" and the password "update" to gain access to this site. Updates will be added to the next edition. Please provide as much information as possible. We are especially interested in obtaining an e-mail address for each department.

Browsing the Directory on the WWW

Much of the information contained in this directory is also available via the World Wide Web at the URL

`http://www.che.utexas.edu/che-faculty/`

The WWW version of the directory also contains electronic mail addresses and some links to university, department, and faculty home pages. Information at this site is nominally updated at the same time that the current directory is published.

S. Joe Qin and Steve Swinnea
Austin, Texas
August 2004

Contents

United States

Alabama

University of Alabama . . . 1
University of Alabama in Huntsville . . . 1
Auburn University . . . 2
University of South Alabama . . . 3
Tuskegee University . . . 3

Arizona

Arizona State University . . . 4
University of Arizona . . . 5

Arkansas

University of Arkansas . . . 6

California

University of California, Berkeley . . . 7
University of California, Davis . . . 8
University of California, Irvine . . . 9
University of California, Los Angeles . . . 10
University of California, Riverside . . . 11
University of California, San Diego . . . 11
University of California, Santa Barbara . . . 12
California Institute of Technology . . . 13
California State University, Long Beach . . . 13
California State Polytechnic University, Pomona . . . 14
San Jose State University . . . 14
University of Southern California . . . 15
Stanford University . . . 15

Colorado

University of Colorado . . . 16
Colorado School of Mines . . . 17
Colorado State University . . . 18

Connecticut

University of Connecticut . . . 19
University of New Haven . . . 20
Yale University . . . 20

Delaware

University of Delaware . . . 21

District of Columbia

Howard University . . . 22

Florida

University of Florida . . . 23
Florida Institute of Technology . . . 24
Florida A&M University-Florida State University . . . 24
University of South Florida . . . 25

Georgia

Georgia Institute of Technology . . . 26

Idaho

University of Idaho . . . 27

Illinois

University of Illinois at Chicago . . . 28
University of Illinois at Urbana-Champaign . . . 29
Illinois Institute of Technology . . . 30
Northwestern University . . . 31

Indiana
University of Notre Dame . . . 32
Purdue University . . . 33
Rose-Hulman Institute of Technology . . . 34
Tri-State University . . . 34
Iowa
University of Iowa . . . 35
Iowa State University of Science and Technology . . . 36
Kansas
University of Kansas . . . 37
Kansas State University . . . 38
Kentucky
University of Kentucky . . . 39
University of Louisville Speed Scientific School . . . 40
Louisiana
Louisiana State University . . . 41
Louisiana Tech University . . . 41
McNeese State University . . . 42
University of Louisiana at Lafayette . . . 42
Tulane University . . . 43
Maine
University of Maine . . . 43
Maryland
Johns Hopkins University . . . 44
University of Maryland, College Park . . . 45
University of Maryland, Baltimore County . . . 46
Massachusetts
University of Massachusetts Amherst . . . 47
University of Massachusetts-Lowell . . . 48
Massachusetts Institute of Technology . . . 49
Northeastern University . . . 50
Tufts University . . . 51
Worcester Polytechnic Institute . . . 52
Michigan
University of Michigan . . . 53
Michigan State University . . . 54
Michigan Technological University . . . 55
Wayne State University . . . 56
Minnesota
University of Minnesota at Duluth . . . 56
University of Minnesota at Minneapolis St. Paul . . . 57
Mississippi
University of Mississippi . . . 58
Mississippi State University . . . 58
Missouri
University of Missouri-Columbia . . . 59
University of Missouri-Rolla . . . 60
Washington University . . . 61
Montana
Montana State University . . . 62
Nebraska
University of Nebraska . . . 62
Nevada
University of Nevada, Reno . . . 63

New Hampshire
Dartmouth College 64
University of New Hampshire 64
New Jersey
New Jersey Institute of Technology 65
Princeton University 66
Rowan University 67
Rutgers, The State University of New Jersey 68
Stevens Institute of Technology 69
New Mexico
New Mexico Institute of Mining and Technology 69
University of New Mexico 70
New Mexico State University 71
New York
The City College of The City University of New York 72
Clarkson University 72
Columbia University 73
Cooper Union 73
Cornell University 74
Manhattan College 75
Polytechnic University 75
Rensselaer Polytechnic Institute 76
University of Rochester 77
State University of New York at Buffalo 78
Syracuse University 79
North Carolina
North Carolina Agricultural and Technical State University 79
North Carolina State University 80
North Dakota
University of North Dakota 81
Ohio
University of Akron 81
Case Western Reserve University 82
University of Cincinnati 83
Cleveland State University 84
University of Dayton 84
Ohio State University 85
Ohio University 86
University of Toledo 87
Youngstown State University 87
Oklahoma
University of Oklahoma 88
Oklahoma State University 89
University of Tulsa 89
Oregon
Oregon State University 90
Pennsylvania
Bucknell University 91
Carnegie Mellon University 92
Drexel University 93
Lafayette College 93
Lehigh University 94
University of Pennsylvania 95
Pennsylvania State University 96

University of Pittsburgh . . . 97
Villanova University . . . 98
Widener University . . . 98
Puerto Rico
University of Puerto Rico . . . 99
Rhode Island
Brown University . . . 100
University of Rhode Island . . . 100
South Carolina
Clemson University . . . 101
University of South Carolina . . . 102
South Dakota
South Dakota School of Mines and Technology . . . 102
Tennessee
Christian Brothers University . . . 103
The University of Tennessee at Chattanooga . . . 103
University of Tennessee at Knoxville . . . 104
Tennessee Technological University . . . 105
Vanderbilt University . . . 105
Texas
University of Houston . . . 106
Lamar University . . . 107
Prairie View A&M University . . . 107
Rice University . . . 108
Texas A&M University . . . 109
Texas A&M University-Kingsville . . . 110
University of Texas at Austin . . . 111
Texas Tech University . . . 112
Utah
Brigham Young University . . . 113
University of Utah . . . 114
Virginia
Hampton University . . . 115
University of Virginia . . . 115
Virginia Polytechnic Institute and State University/Virginia Tech . . . 116
Washington
University of Washington . . . 117
Washington State University . . . 118
West Virginia
West Virginia University . . . 119
Wisconsin
University of Wisconsin-Madison . . . 120
Wyoming
University of Wyoming . . . 121
Tennessee
Pellissippi State Technical Community College . . . 122

Argentina
Instituto Tecnologico de Buenos Aires . . . 123
Universidad Nacional de la Plata . . . 123
Universidad Nacional del Litoral . . . 123
Universidad Nacional del Salta . . . 124
Universidad Nacional del Sur . . . 124

Australia
University of Adelaide . . . 125
Curtin University of Technology . . . 125
University of Melbourne . . . 126
Monash University . . . 126
University of New South Wales . . . 127
University of Newcastle . . . 127
University of Queensland . . . 127
Royal Melbourne Institute of Technology . . . 128
University of Sydney . . . 128

Bangladesh
Bangladesh University of Engineering and Technology . . . 129

Belgium
University of Gent . . . 129
University of Liege . . . 129

Brazil
Universidade Estadual de Campinas (UNICAMP) . . . 130
Universidade Estadual de Maringa . . . 131
Universidade Federal de Minas Gerais . . . 131
Universidade Federal Do Rio De Janeiro . . . 132
Universidade de Sao Paulo USP . . . 133
Universidade Federal de Uberlandia . . . 133

Canada
University of British Columbia . . . 134
Dalhousie University . . . 135
University of Alberta . . . 135
University of Calgary . . . 136
Ecole Polytechnique-University of Montreal . . . 137
Lakehead University . . . 137
Universite Laval . . . 137
McGill University . . . 138
McMaster University . . . 138
University of Ottawa . . . 139
University of New Brunswick . . . 139
Queen's University . . . 140
Royal Military College of Canada . . . 140
Ryerson University . . . 141
University of Saskatchewan . . . 141
Universite de Sherbrooke . . . 141
University of Toronto . . . 142
University of Waterloo . . . 143
University of Western Ontario . . . 144

Chile
Universidad de Concepcion . . . 144
Pontificia Universidad Catolica de Chile . . . 145
Universidad de Santiago de Chile . . . 145
Universidad Tecnica Federico . . . 146
Universidad Catolica de Valparaiso . . . 146

Peoples Republic of China
Beijing University of Chemical Technology . . . 146
South China University of Technology . . . 147
Tianjin University . . . 147
Tsinghua University . . . 149
Zhejiang University . . . 150

Colombia
Universidad Industrial de Santander . . . 152

Croatia
University of Zagreb . . . 152

Czech Republic
Prague Institute of Chemical Technology . . . 153
University of Pardubice . . . 153

Finland
Abo Akademi University . . . 154
Helsinki University of Technology . . . 154
Lappeenranta University of Technology . . . 154
University of Oulu . . . 154

France
Ecole Nationale Superieure des Ingenieurs en Arts Chimiques et Technologiques (ENSIACET) . . 155
Ecole Nationale Superieure des Industries Chimiques (ENSIC) . . . 156

Germany
Aachen Technical University . . . 157
Technical University of Berlin . . . 157
Technische Universitaet Braunschweig . . . 157
Technical University of Clausthal . . . 158
Darmstadt University of Technology . . . 158
University of Dortmund . . . 158
Universitat Erlangen-Nurnberg . . . 158
Technical University Hamburg-Harburg . . . 159
University of Hannover . . . 159
University of Hannover, ITC . . . 160
Universitat Kaiserslautern . . . 160
Universitat Karlsruhe . . . 160
Georg-Simon-Ohm Fachhochschule Nuernberg . . . 161
Fachhochschule Mannheim, University of Applied Sciences . . . 161

Ghana
Kwame Nkrumah University of Science and Technology . . . 161

Greece
Aristotle University . . . 161
University of Patras . . . 162

Hong Kong, SAR of China
Hong Kong University of Science and Technology . . . 163

Hungary
Budapest University of Technology and Economics (Budapesti Muszaki es Gazdasagtudomanyi Egyetem) . . . 163

University of Veszprem . . . 164

India

Anna University . . . 164
Annamalai University . . . 164
Birla Institute of Technology & Science (BITS) - Pilani . . . 165
Calcutta University . . . 165
Government Engineering College . . . 165
Indian Institute of Science . . . 166
Indian Institute of Technology, Bombay . . . 166
Indian Institute of Technology, Kanpur . . . 167
Indian Institute of Technology, Kharagpur . . . 167
Indian Institute of Technology, Madras . . . 168
Indian Institute of Technology, Delhi . . . 168
Karnataka Regional Engineering College, Surathkal . . . 168
Laxminarayan Institute of Technology, Nagpur University . . . 169
Manipal Institute of Technology . . . 169
Malaviya National Institute of Technology, Jaipur . . . 169
University of Mumbai . . . 170
National Institute of Technology Warangal . . . 170
Pravara Rural Education Society's Pravara Rural Engineering College, Loni . . . 170
R. V. College of Engineering, Affiliated with Visveswaraiah Technical University . . . 171
Regional Engineering College, Affiliated with Bharathidasan University . . . 171
Indian Institute of Technology . . . 171
Siddaganga Institute of Technology, Affiliated with Viveswaraiah University . . . 172
S. V. National Institute of Technology . . . 172
Thapar Institute of Engineering and Technology . . . 172
Vellore Institute of Technology . . . 173

Iran

Amirkabir University of Technology(Tehran Polytechnic) . . . 173
Isfahan University of Technology . . . 173
Sharif University of Technology . . . 173

Ireland

Cork Institute of Technology . . . 174
University College Dublin . . . 174

Israel

Ben-Gurion University of the Negev . . . 175
Technion-Israel Institute of Technology . . . 175

Italy

Universita di Bologna . . . 176
Universita di Cagliari . . . 176
Universita de L'Aquila . . . 177
Universita di Napoli . . . 177
Universita di Pisa . . . 178
University of Salerno . . . 178
Politecnico di Milano . . . 179

Japan

Fukuoka University . . . 180
Himeji Institute of Technology . . . 180
Hiroshima University . . . 180

Kanazawa University . . . 181
Kansai University . . . 181
Kyoto University . . . 182
Kyushu University . . . 182
Nagoya University . . . 182
Niigata University . . . 183
Osaka Prefecture University . . . 183
Osaka University . . . 184
Seikei University . . . 184
Tohoku University . . . 184
University of Tokyo . . . 185
Tokyo Institute of Technology . . . 185
Toyama University . . . 186
Yamagata University . . . 186
Yamaguchi University . . . 186
Yokohama National University . . . 187

Republic of Korea
Chonnam National University . . . 187
Keimyung University . . . 187
Korea Advanced Institute of Science and Technology . . . 188
Korea University . . . 188
Kunsan National University . . . 189
Pohang University of Science and Technology . . . 189
Seoul National University . . . 189

Kuwait
Kuwait University . . . 190

Malaysia
University of Malaya . . . 190
Universiti Sains Malaysia . . . 191

Mexico
Instituto Tecnologico y de Estudios Superiores de Monterrey . . . 191
Universidad Iberoamericana . . . 192
Universidad de las Americas, Puebla . . . 192
Universidad Autonoma Metropolitana-Iztapalapa . . . 192
Universidad Autonoma de San Luis Potosi . . . 193

Mozambique
Universidade Eduardo Mondlane . . . 193

Netherlands
University of Amsterdam . . . 194
Delft University of Technology . . . 194
Eindhoven University of Technology . . . 195
Twente University of Technology . . . 195

New Zealand
University of Auckland . . . 197
University of Canterbury . . . 197

Nigeria
University of Lagos . . . 197
Obafemi Awolowo University . . . 198

Norway
Norwegian University of Science and Technology 198

Philippines
Bicol University 199
De La Salle University 199
University of Philippines at Los Banos 199
University of Santo Tomas 200

Poland
Technical University of Lodz 200
Silesian Technical University 200
Technical University of Szczecin 201
Technical University of Szczecin, Faculty of Chemical Engineering 201
Warsaw University of Technology 201
Wroclaw Technical University 202

Portugal
Universidade de Coimbra 202
Universidade do Porto 203
Instituto Superior Técnico 204

Qatar
University of Qatar 204

Romania
Petroleum and Gas Institute 204

Saudi Arabia
King Abdulaziz University 205
King Fahd University of Petroleum & Minerals 205
King Saud University 206

Singapore
National University of Singapore 206

Slovenia
University of Ljubljana 207
University of Maribor 207

Republic of South Africa
University of Cape Town 208
University of Durban-Westville 208
University of Natal 209
North-West University, Potchefstroom Campus 209
University of Pretoria 209
University of the Witwatersrand, Johannesburg 210

Spain
University of Alcala de Henares Ingenieria Quimica 210
University of Alicante 211
Autonoma University of Barcelona 211
University of Cadiz 212
University of Cantabria 212
Universidad de Castilla-La Mancha 213
Complutense University of Madrid 214

University of Granada . . . 214
University of Illes Balears . . . 215
Universidad Politecnica de Madrid-ETSI Industriales . . . 215
University of Murcia . . . 216
University of Oviedo . . . 216
University of Pais Vasco . . . 216
Rovira i Virgili University . . . 217
University of Salamanca . . . 218
Universidad de Santiago de Compostela . . . 218
Escuela Superior de Ingenieros Industriales de Sevilla . . . 219
Universidad de Sevilla . . . 219
Universitat de Valencia . . . 220
University of Valladolid . . . 220
University of Zaragoza . . . 221

Sweden
University of Lund . . . 221
Royal Institute of Technology . . . 222

Switzerland
Swiss Federal Institute of Technology . . . 222
Swiss Federal Institute of Technology, Ecole Polytechnique Federale de Lausanne . . . 223

Taiwan
Chang Gung University . . . 223
Chung Yuan Christian University . . . 224
Feng-Chia University . . . 224
National Central University . . . 224
National Cheng Kung University . . . 225
National Chung Cheng University . . . 226
National Taiwan University of Science and Technology . . . 226
National Taiwan University . . . 227
National Tsing Hua University . . . 227
Tamkang University . . . 228
Tunghai University . . . 228
Yuan Ze University . . . 229

Thailand
Chulalongkorn University . . . 229
Kasetsart University . . . 230
Prince of Songkla University . . . 230

Turkey
Ankara University . . . 230
Bogazici University . . . 231
University of Ege . . . 231
Gazi University . . . 232
Hacettepe University . . . 232
Middle East Technical University . . . 233
Yildiz Technical University . . . 233

United Arab Emirates
United Arab Emirates University . . . 234

United Kingdom
Aston University . . . 234
University of Bath . . . 234
University of Birmingham . . . 235
University of Cambridge . . . 235
University of Edinburgh . . . 236
University of Exeter . . . 236
Heriot-Watt University . . . 237
University of Leeds . . . 237
University of London (Imperial College) . . . 237
University of London (University College London) . . . 238
Loughborough University . . . 238
University of Manchester Institute of Science and Technology (UMIST) . . . 239
University of Newcastle Upon Tyne . . . 240
University of Nottingham . . . 240
Queen's University of Belfast . . . 241
University of Sheffield . . . 241
South Bank University . . . 242
University of Strathclyde . . . 242
University of Surrey . . . 242
University of Wales Swansea . . . 243
University of Teesside . . . 243

Venezuela
Universidad Central de Venezuela . . . 243

Zimbabwe
National University of Science and Technology . . . 244

Omega Chi Epsilon 245

Index of U.S. Faculty 247

Index of Institutions 261

United States

Alabama

University of Alabama

Chemical and Biological Engineering Department
Box 870203
Tuscaloosa, AL 35487-0203
Deliveries: A127 Bevill Building

University phone .... (205) 348-6010
Department phone .... (205) 348-6450
Department FAX .... (205) 348-7558

Professors

April, Gary C. (Head) .... (205) 348-6452
gcapril@coe.eng.ua.edu
Arnold, David W. .... (205) 348-1741
darnold@coe.eng.ua.edu
Clements, Wm. C. Jr. (Emeritus) ..... (205) 348-1726
wclements@coe.eng.ua.edu
Gupta, Arunava .... (205) 348-3822
agupta@Mint.ua.edu
Lane, Alan M. .... (205) 348-1729
alane@coe.eng.ua.edu
McKinley, Marvin D. (Emeritus) ...... (205) 348-6486
mmckinley@coe.eng.ua.edu

Associate Professors

Carlson, Eric S. .... (205) 348-1581
ecarlson@coe.eng.ua.edu
Clark, Peter E. .... (205) 348-1682
pclark@coe.eng.ua.edu
Johnson, Duane T. .... (205) 348-8402
djohnson@coe.eng.ua.edu
Weaver, Mark (Adjunct) .... (205) 348-7073
mweaver@coe.eng.ua.edu
Wiest, John M. .... (205) 348-1727
jwiest@coe.eng.ua.edu

Assistant Professors

Brazel, Chris .... (205) 348-9738
cbrazel@coe.eng.ua.edu
Klein, Tonya M. .... (205) 348-9744
tklein@coe.eng.ua.edu
Ritchie, Stephen M. .... (205) 348-2712
sritchie@coe.eng.ua.edu
Turner, C. Heath .... (205) 348-1733
hturner@coe.eng.ua.edu

Accredited by: ABET

Degrees granted 2003-2004:

B.S.: 13 M.S.: 5 Ph.D.: 1

Graduate advisor: John M. Wiest

Undergraduate advisor: David Arnold

Student organization: AIChE
Advisor: Peter Clark, Tonya Klein, Chris Brazel

Department reports to: Dean of Engineering

Placement service: Career Center (205) 348-5848

University of Alabama in Huntsville

Department of Chemical and Materials Engineering
EB 130
Huntsville, AL 35899
Deliveries: 301 Sparkman Dr.

University phone .... (256) 824-6340
Department phone .... (256) 824-6810
Department FAX .... (256) 824-6839

Professors

Cerro, Ramon L. (Chair) .... (256) 824-7313
rlc@che.uah.edu
Chen, Chien-Pin .... (256) 824-6194
cchen@che.uah.edu
Chittur, Krishnan K. .... (256) 824-6850
kchittur@che.uah.edu
Smith, James E. Jr. .... (256) 824-6439
jesmith@che.uah.edu

Associate Professors

Banish, Michael .... (256) 824-6969
banishm@email.uah.edu
Weimer, Jeffrey .... (256) 824-6954
jjweimer@matsci.uah.edu

Assistant Professors

Taconi, Katherine .... (256) 824-6874
ktaconi@che.uah.edu

Accredited by: ABET

Degrees granted 2003-2004:

B.S.: 18 M.S.: 4 Ph.D.: 1

Graduate advisor: Jeffrey Weimer

Undergraduate advisor: Faculty

Student organization: Chemical Engineering Student's Club
Advisor: James E. Smith

Department reports to: Jorge Aunon, Dean, (256) 824-6474

Placement service: John R. Shrout (256) 824-6612, Director, Career Planning and Placement Office

Auburn University

Department of Chemical Engineering
230 Ross Hall
Auburn University, AL 36849-5127

University phone .... (334) 844-4000
Department phone .... (334) 844-4827
Department FAX .... (334) 844-2063

Professors

Chambers, Robert P. .... (334) 844-2054
chambers@eng.auburn.edu
Cullinan, Harry T. (Center Director) .. (334) 844-2016
cullinan@eng.auburn.edu
Curtis, Christine W. .... (334) 844-5872
curticw@auburn.edu
Elnashaie, Said Salah .... (334) 844-2060
nashaie@eng.auburn.edu
Guin, James A. .... (334) 844-2020
jaguin@eng.auburn.edu
Gupta, Ram .... (334) 844-2013
gupta@eng.auburn.edu
Krishnagopalan, Gopal A. (Assoc. Chair) (334) 844-2011
gopalk@eng.auburn.edu
Lee, Y. Y. .... (334) 844-2019
yylee@eng.auburn.edu
Maples, Glennon .... (334) 844-2041
maplegl@auburn.edu
Moore, O. C. (Emeritus)
Neuman, Ronald D. .... (334) 844-2017
rdneuman@eng.auburn.edu
Roberts, Christopher (Chair) .... (334) 844-2036
croberts@eng.auburn.edu
Shim, Jae-Jin (Visiting) .... (334) 844-4827
Tarrer, A. Ray .... (334) 844-2018
atarrer@eng.auburn.edu
Tatarchuk, Bruce J. (Center Director) . (334) 844-2023
brucet@eng.auburn.edu

Associate Professors

Duke, Steve R. .... (334) 844-2087
srduke@eng.auburn.edu
Hirth, Leo J. (Emeritus) .... (334) 844-2006
lhirth@eng.auburn.edu
Vives, Donald L. (Emeritus)

Assistant Professors

Ashurst, William (Bob) .... (334) 844-4827
Byrne, Mark E. .... (334) 844-4827
mbyrne@eng.auburn.edu
Eden, Mario .... (334) 844-4827
mre@eng.auburn.edu
Placek, Timothy D. .... (334) 844-2022
tplacek@eng.auburn.edu

Instructors

Mills, David R. .... (334) 844-2015
drmills@eng.auburn.edu

Other Faculty

Cahela, Don .... (334) 844-2026
caheldr@eng.auburn.edu
Hodges, Russel .... (334) 844-2031
hodgeru@eng.auburn.edu
Huang, Xiwen .... (334) 844-2057
huangxi@eng.auburn.edu
Liu, Juncheng .... (334) 844-4827
Lu, Yong .... (334) 844-3331
yonglu@eng.auburn.edu
Park, SangKwon (Visiting) .... (334) 844-4827
Sunwoo, Changshin (Visiting) .... (334) 844-2034
Zhu, Wen-Hua .... (334) 844-2005
whzhu@eng.auburn.edu

Accredited by: ABET, SACS

Degrees granted 2003-2004:

B.S.: 58 M.S.: 10 Ph.D.: 2

Graduate advisor: Ram Gupta

Undergraduate advisor: Jennifer Harris (334) 844-2030

Student organization: AIChE; Omega Chi Epsilon; TAPPI, PIMA
Advisor: Yoon Y. Lee, Said Elnashaie; Gopal Krishnagopalan

Department reports to: Larry Benefield (334) 844-4326, Dean of Engineering

Placement service: Nancy Bernard, Coordinator (334) 844-4744

University of South Alabama

Department of Chemical Engineering
EGLB 244
Mobile, AL 36688-0002

University phone .... (251) 460-6101
Department phone .... (251) 460-6160
Department FAX .... (251) 461-1485

Professors

Askew, William C. .... (251) 460-7563
waskew@jaguar1.usouthal.edu
Dhawan, Jagdish C. .... (251) 460-7462
jdhawan@jaguar1.usouthal.edu
Harrison, B. Keith (Chair) .... (251) 460-7461
kharriso@jaguar1.usouthal.edu
Huddleston, Ted (Emeritus)
thuddles@jaguar1.usouthal.edu
Rodriguez, Harold V. (Emeritus)
Sylvester, Nicholas .... (251) 461-1432
nsylvest@usouthal.edu

Assistant Professors

Misra, Manish .... (251) 460-6160
manish@usouthal.edu
Morisani, Stephen J. .... (251) 460-7463
smorisan@jaguar1.usouthal.edu

Accredited by: ABET

Degrees granted 2003-2004:

B.S.: 17 M.S.: 12

Graduate advisor: B. K. Harrison

Undergraduate advisor: B. K. Harrison

Student organization: AIChE
Advisor: J. C. Dhawan

Department reports to: Dr. John Steadman, Dean of Engineering

Placement service: Allan McPeak, Director of Co-op and Placement

Tuskegee University

Chemical Engineering Department
Tuskegee, AL 36088
Deliveries: Engineering Building, Room 513

University phone .... (334) 727-8011
Department phone .... (334) 727-8089
Department FAX .... (334) 724-4188

Professors

Krishnagopalan, Jaya .... (334) 727-8795
krishj@tuskegee.edu
Kwon, K. C. .... (334) 727-8976
kwonk@tuskegee.edu
Nosa, Egiebor .... (334) 724-4265
egibor@tuskegee.edu
Vahdat, Nader (Head) .... (334) 727-8978
vahdatn@tuskegee.edu

Associate Professors

Josephson, William .... (334) 727-8996
josepshon@tuskegee.edu

Assistant Professors

Floyd, Tamara .... (334) 727-8975
floydt@tuskegee.edu
Naser, Jamil .... (334) 727-8050
naserj@tuskegee.edu
Park, Yoonkook .... (334) 727-4318
parky@tuskegee.edu

Accredited by: ABET

Degrees granted 2003-2004:

B.S.: 15

Undergraduate advisor: Nader Vahdat

Student organization: AIChE
Advisor: Yoonk Park

Department reports to: Legand Burge (334) 727-8356, Dean

Placement service: Sara Stringer (334) 727-8294, Director of Cooperative Education and Placement

4 United States

Arizona

Arizona State University

Department of Chemical and Materials Engineering
P.O. Box 876006
Tempe, AZ 85287-6006
Deliveries: Engineering CNTR G Wing #202

University phone .... (480) 965-9011
Department phone .... (480) 965-3313
Department FAX .... (480) 965-0037

Professors

Berman, Neil S. (Emeritus) .... (480) 965-4113
Neil.Berman@asu.edu
Kuester, James L. (Emeritus) .... (480) 965-5071
james.kuester@asu.edu
Lin, Jerry Y. S.
Mahajan, Subhash (Chair) .... (480) 965-9710
smahajan@asu.edu
Raupp, Gregory B. (Associate Dean) .. (480) 965-2828
raupp@asu.edu
Sater, Vernon E. (Emeritus) .... (480) 965-5529
icves@mainex1.asu.ede
Wang, Joseph
Zwiebel, Imre (Emeritus) .... (480) 965-2448
BIZweibel@yahoo.com

Associate Professors

Beckman, James R. (Associate Chair) .. (480) 965-4395
jim.beckman@asu.edu
Burrows, Veronica .... (480) 965-4557
burrows@asu.edu
Sierks, Michael .... (480) 965-3313
sierks@asu.edu

Assistant Professors

Allen, Jonathan .... (480) 965-4112
Jonathan.O.Allen@asu.edu
Dillner, Ann .... (480) 965-3001
Dillner@asu.edu
Heys, Jeffrey
Park, Chan Beum .... (480) 965-0362
cbpark@asu.edu

Accredited by: ABET

Degrees granted 2003-2004:

B.S.: 35 M.S.: 13 Ph.D.: 7

Graduate advisor: Marlene Bolf

Undergraduate advisor: Xavier Oaks

Student organization: AIChE
Advisor: James Beckman

Department reports to: Peter E. Crouch (480) 965-1722, Dean

Placement service: Ray Castillo, Director (480) 965-2350

University of Arizona

Department of Chemical and Environmental Engineering
P.O. Box 210011
Tucson, AZ 85721
Deliveries: 1133 E. North Campus Drive, Room 108
University phone .... (520) 621-2211
Department phone .... (520) 621-6044
Department FAX .... (520) 621-6048

Professors

Arnold, Robert (Joint) .... (520) 621-2410
rga@engr.arizona.edu
Bier, Milan (Emeritus) .... (520) 621-6763
Edwards, Richard M. (Emeritus)
Field, James .... (520) 626-5858
jimfield@email.arizona.edu
Gross, Joseph F. (Emeritus) .... (520) 626-7559
Ogden, Kimberly (Joint) .... (520) 621-9484
ogden@email.arizona.edu
Peterson, Thomas W. (Dean) .... (520) 621-6594
peterson@erc.arizona.edu
Rehm, Thomas R. (Emeritus)
Shadman, Farhang (Director) .... (520) 621-6052
shadman@erc.arizona.edu
Sierka, Raymond (Emeritus) .... (520) 621-6586
sierka@engr.arizona.edu
Wendt, Jost O. L. (Head) .... (520) 621-2591
wendt@email.arizona.edu
White, Don H. (Emeritus) .... (520) 621-6049
dwhite@engr.arizona.edu

Associate Professors

Baygents, James .... (520) 621-6043
baygents@email.arizona.edu
Cosart, William P. (Emeritus)
Ela, Wendell (Joint) .... (520) 626-9323
wela@engr.arizona.edu
Farrell, James (Joint) .... (520) 621-2465
farrellj@engr.arizona.edu
Guzman, Roberto .... (520) 621-6041
guzmanr@engr.arizona.edu
Muscat, Anthony .... (520) 626-6580
muscat@erc.arizona.edu
Philipossian, Ara .... (520) 621-6101
ara@engr.arizona.edu
Saez, Eduardo .... (520) 621-5369
esaez@engr.arizona.edu
Sierra Alvarez, Maria Reyes .... (520) 626-2896
rsierra@email.arizona.edu

Assistant Professors

Blowers, Paul .... (520) 626-5319
blowers@engr.arizona.edu

Other Faculty

Wolf, David (Visiting) .... (520) 621-6055
wolf@engr.arizona.edu

Accredited by: ABET

Degrees granted 2003-2004:

B.S.: 26 M.S.: 8 Ph.D.: 2

Student organization: AIChE
Advisor: Paul Blowers

Department reports to: Thomas W. Peterson (520) 621-6594, Dean

Placement service: Marie Rozenblit (520) 621-2719

Arkansas

University of Arkansas

Department of Chemical Engineering
3202 Bell Engineering Center
Fayetteville, AR 72701
Deliveries: 3202 Bell Engineering Center

University phone . . . . (479) 575-2000
Department phone . . . . (479) 575-4951
Department FAX . . . . (479) 575-7926

Professors

Babcock, Robert E. . . . . (479) 575-5410
rbabcoc@engr.uark.edu
Clausen, E. C. . . . . (479) 575-5412
eclause@engr.uark.edu
Couper, James R. (Emeritus) . . . . (479) 575-5976
jrc@engr.uark.edu
Cross, Robert . . . . (479) 575-5382
racross@engr.uark.edu
Gaddy, James L. (Emeritus) . . . . (479) 575-7768
Havens, Jerry A. . . . . (479) 575-3857
jhavens@engr.uark.edu
Oxford, Charles W. (Emeritus)
Penney, W. Roy . . . . (479) 575-5681
wrp@engr.uark.edu
Silano, A. A. . . . . (479) 575-4502
Spicer, Thomas O. (Head) . . . . (479) 575-6516
tos@engr.uark.edu
Springer, Charles (Emeritus) . . . . (479) 575-5979
Thatcher, Charles M. (Emeritus) . . . . . . (479) 575-5382
Turpin, Jim L. . . . . (479) 575-5975
jturpin@engr.uark.edu
Ulrich, Richard K. . . . . (479) 575-5645
rulrich@uark.edu
Welker, J. Reed (Emeritus) . . . . (479) 575-6691
rwelker@engr.uark.edu

Associate Professors

Ackerson, Michael D. . . . . (479) 575-5978
Mike.Ackerson@processdyn.com
Beitle, Robert . . . . (479) 575-7566
rbeitle@engr.uark.edu
Thoma, Greg . . . . (479) 575-7374
gthoma@uark.edu

Instructors

Myers, Wm. A. . . . . (479) 575-5977
wam@engr.uark.edu

Accredited by: ABET

Degrees granted 2003-2004:

B.S.: 22 M.S.: 6 Ph.D.: 9

Graduate advisor: R. K. Ulrich

Undergraduate advisor: William A. Myers

Student organization: AIChE
Advisor: Ed Clausen

Department reports to: Dean of Engineering

Placement service: Career Services (479) 575-2805

California

University of California, Berkeley

Chemical Engineering Department
201 Gilman Hall, MC# 1462
Berkeley, CA 94720-1462
Deliveries: B-84 Hildebrand Hall, Berkeley, CA 94720

University phone . . . . . (510) 642-6000
Department phone . . . . . (510) 642-2291
Department FAX . . . . . (510) 642-4778

Professors

Balsara, Nitash P. . . . . . (510) 643-8973
nbalsara@cchem.berkeley.edu
Bell, Alexis T. . . . . . (510) 642-1536
Bell@cchem.berkeley.edu
Blanch, Harvey W. . . . . . (510) 642-1387
Blanch@socrates.berkeley.edu
Bromley, LeRoy A. (Emeritus) . . . . . (510) 642-2291
Cairns, Elton J. (Emeritus) . . . . . (510) 642-8537
EJcairns@lbl.gov
Chakraborty, Arup K. (Chair) . . . . . (510) 642-9275
Arup@berkeley.edu
Clark, Douglas S. . . . . . (510) 642-2408
Clark@cchem.berkeley.edu
Denn, Morton M. (Emeritus) . . . . . (212) 650-7444
denn@levdec.engr.ccny.cuny.edu
Foss, Alan S. (Emeritus) . . . . . (510) 642-4526
Foss@jwgibbs.cchem.berkeley.edu
Frechet, Jean M. J. (Joint) . . . . . (510) 643-3077
frechet@cchem.berkeley.edu
Goren, Simon L. (Emeritus) . . . . . (510) 642-2291
Graves, David B. (Associate Chair) . . . . (510) 642-2214
graves@berkeley.edu
Grens, Edward A. (Emeritus) . . . . . (510) 642-2291
Hanson, Donald N. (Emeritus) . . . . . (510) 642-2291
Iglesia, Enrique . . . . . (510) 642-9673
iglesia@cchem.berkeley.edu
Keasling, Jay D. . . . . . (510) 642-4862
keasling@socrates.berkeley.edu
King, C. Judson (Emeritus) . . . . . (510) 643-3199
cjking@berkeley.edu
Lynn, Scott (Emeritus) . . . . . (510) 642-1634
Lynn@cchem.berkeley.edu
Lyon, David N. (Emeritus) . . . . . (510) 642-2291
NBLX35A@prodigy.com
Maboudian, Roya . . . . . (510) 643-7957
maboudia@socrates.berkeley.edu
Muller, Susan J. . . . . . (510) 642-4525
muller2@socrates.berkeley.edu
Newman, John S. (Associate Chair) . . . (510) 642-4063
newman@newman.cchem.berkeley.edu
Petersen, Eugene E. (Emeritus) . . . . . (510) 642-2291
eepet@berkeley.edu
Prausnitz, John M. (Emeritus) . . . . . (510) 642-3592
prausnit@cchem.berkeley.edu
Radke, Clayton J. . . . . . (510) 642-5204
radke@cchem.berkeley.edu
Reimer, Jeffrey A. (Associate Dean) . . . (510) 642-8011
reimer@socrates.berkeley.edu
Williams, Michael C. (Emeritus) . . . . . (510) 642-2291

Assistant Professors

Katz, Alexander . . . . . (510) 643-3248
katz@cchem.berkeley.edu
Schaffer, David . . . . . (510) 643-5963
schaffer@cchem.berkeley.edu
Segalman, Rachel A. . . . . . (510) 642-7998
segalman@berkeley.edu

Instructors

Plouffe, Paul . . . . . (510) 642-8715
Plouffe@cchem.berkeley.edu
Wallman, P. Henrik . . . . . (510) 642-2295
wallman@cchem.berkeley.edu

Other Faculty

Boudart, Michel (Adjunct) . . . . . (510) 642-2291
Grossberg, A. L. . . . . . (510) 642-2291
Maiorella, Brian (Adjunct) . . . . . (510) 642-2291
bmaiorella@msn.com
Soane, David S. (Adjunct) . . . . . (510) 642-2291
Sternberg, Moshe (Adjunct) . . . . . (510) 642-2291
sternber@socrates.berkeley.edu

Accredited by ABET

Degrees granted 2003-2004:

B.S.: 91 M.S.: 5 Ph.D.: 13

Graduate advisor: John S. Newman

Undergraduate advisor: Sandra Rehling

Student organization: AIChE
Advisor: Alexander Katz

Department reports to: Dean Clayton Heathcock, College of Chemistry

Placement service: L. Hernandez-Lewis (510) 642-1162

University of California, Davis

Chemical Engineering and Materials Science
One Shields Avenue
Davis, CA 95616-5294
Deliveries: 1334 Bainer Hall

University phone .... (530) 752-1011
Department phone .... (530) 752-0400
Department FAX .... (530) 752-1031

Professors

Bell, Richard L. (Emeritus) .... (530) 752-0400
Boulton, Roger .... (530) 752-0900
rbboulton@ucdavis.edu
Browning, Nigel .... (530) 754-5358
nbrowning@ucdavis.edu
Dungan, Stephanie R. .... (530) 752-5447
srdungan@ucdavis.edu
Gates, Bruce C. .... (530) 752-3953
bcgates@ucdavis.edu
Gibeling, Jeffery C. (Dean) .... (530) 752-7037
jcgibeling@ucdavis.edu
Groza, Joanna R. .... (530) 752-8825
jrgroza@ucdavis.edu
Higgins, Brian G. .... (530) 752-8780
bghiggins@ucdavis.edu
Howitt, David G. .... (530) 752-1164
dghowitt@ucdavis.edu
Jackman, Alan P. .... (530) 752-8777
apjackman@ucdavis.edu
Lavernia, Enrique (Dean) .... (530) 752-0400
ejlavernia@ucdavis.edu
McCoy, Benjamin J. (Emeritus) .... (530) 752-0400
bjmccoy@ucdavis.edu
McDonald, Karen A. .... (530) 752-8314
kamcdonald@ucdavis.edu
Mukherjee, Amiya K. .... (530) 752-1776
akmukherjee@ucdavis.edu
Munir, Zuhair A. .... (530) 752-6348
zamunir@ucdavis.edu
Navrotsky, Alexandra .... (530) 752-3292
anavrotsky@ucdavis.edu
Palazoglu, Ahmet N. .... (530) 752-8774
anpalazoglu@ucdavis.edu
Phillips, Ronald J. .... (530) 752-2803
rjphillips@ucdavis.edu
Powell, Robert L. (Chair) .... (530) 752-8779
rlpowell@ucdavis.edu
Risbud, Subhash .... (530) 752-0474
shrisbud@ucdavis.edu
Ryu, Dewey .... (530) 752-8954
ddyryu@ucdavis.edu
Shackelford, James F. .... (530) 752-4030
jfshackelford@ucdavis.edu
Smith, J.M. (Emeritus) .... (530) 752-8773
jmsmith@ucdavis.edu
Stroeve, Pieter .... (530) 752-8778
pstroeve@ucdavis.edu
Whitaker, Stephen (Emeritus) .... (530) 752-0400
swhitaker@ucdavis.edu

Associate Professors

Block, David E. .... (530) 752-6046
deblock@ucdavis.edu
Longo, Marjorie L. .... (530) 754-6348
mllongo@ucdavis.edu
Schoenung, Julie M. .... (530) 752-5840
jmschoenung@ucdavis.edu

Assistant Professors

El-Farra, Nael .... (530) 752-0400
Faller, Roland .... (530) 752-5389
rfaller@ucdavis.edu
Kuhl, Tonya L. .... (530) 754-5911
tlkuhl@ucdavis.edu

Accredited by: ABET

Degrees granted 2003-2004:

B.S.: 33 M.S.: 6 Ph.D.: 12

Graduate advisor: Pieter Stroeve

Undergraduate advisor: Bradley Horton

Student organization: AIChE; TMS
Advisor: Tonya Kuhl

Department reports to: Enrique J. Lavernia, Dean of Engineering

Placement service: Pam Swartwood, Coordinator for Engineering

University of California, Irvine

Chemical Engineering and Materials Science
916 Engineering Tower
Irvine, CA 92697-2575

Department phone .... (949) 824-3426
Department FAX .... (949) 824-2541

Professors

DaSilva, Nancy A. (Associate Chair) ... (949) 824-8288
ndasilva@uci.edu
Earthman, James C. (Associate Chair) . (949) 824-5018
earthman@uci.edu
George, Steven C. .... (949) 824-3941
scgeorge@uci.edu
Grant, Stanley (Chair) .... (949) 824-7320
sbgrant@uci.edu
Hong, Juan .... (949) 824-8278
jhong@uci.edu
Lim, Henry C. .... (949) 824-3785
hclim@uci.edu
Mecartney, Martha .... (949) 824-2919
martham@uci.edu
Mohamed, Farghalli .... (949) 824-5807
famohame@uci.edu
Shi, Frank .... (949) 824-5362
fgshi@uci.edu
Yee, Albert F. .... (949) 824-9073
afyee@uci.edu

Associate Professors

Tellkamp, Victoria (Adjunct) .... (949) 824-9010
tellkamp@uci.edu

Assistant Professors

Lu, Jia Grace .... (949) 824-8714
jglu@uci.edu
Mumm, Daniel .... (949) 824-3858
mumm@uci.edu
Putnam, Andrew .... (949) 824-1243
aputnam@uci.edu
Shapiro, Andrew (Adjunct) .... (949) 824-8086
Andrew.A.Shapiro@jpl.nasa.gov
Venugopalan, Vasan .... (949) 824-5802
vvenugop@uci.edu
Wang, Szu-Wen .... (949) 824-2383
wangsw@uci.edu

Other Faculty

Chou, Russell (Adjunct) .... (949) 824-2368
ytchou@uci.edu

Accredited by: ABET

Degrees granted 2003-2004:

B.S.: 31 M.S.: 9 Ph.D.: 5

Graduate advisor: Vasan Venugopalan

Undergraduate advisor: Juan Hong

Student organization: AIChE

Department reports to: Nicolaos Alexopoulos, Dean

Placement service: Kathryn Van Ness, Director

University of California, Los Angeles

Department of Chemical Engineering
5531 Boelter Hall
Box 951592
Los Angeles, CA 90095-1592
Deliveries: 420 Westwood Plaza

University phone .... (310) 825-4321
Department phone .... (310) 825-2046
Department FAX .... (310) 206-4107

Professors

Cohen, Yoram .... (310) 825-8766
yoram@ucla.edu
Davis, James .... (310) 206-0011
jdavis@conet.ucla.edu
Friedlander, Sheldon K. .... (310) 825-2206
skf@ucla.edu
Hicks, Robert F. .... (310) 206-6865
rhicks@ucla.edu
Ignarro, L. J. .... (310) 825-5159
lignarro@mednet.ucla.edu
Knuth, Eldon L. (Emeritus) .... (310) 825-8485
elknuth@ucla.edu
Liao, James C. .... (310) 825-1656
liaoj@seas.ucla.edu
Manousiouthakis, Vasilios (Chair) ..... (310) 825-9385
vasilios@ucla.edu
Monbouquette, Harold G. .... (310) 825-8946
harold@seas.ucla.edu
Nobe, Ken (Emeritus) .... (310) 825-2447
nobe@seas.ucla.edu
Robinson, L. B. (Emeritus) .... (310) 825-5372
lbr@seas.ucla.edu
Senkan, Selim M. .... (310) 206-4106
senkan@seas.ucla.edu
Van Vorst, W. D. (Emeritus) .... (310) 825-2816
wvanvors@ucla.edu
Wazzan, A. R. (Emeritus) .... (310) 206-8245
wazzan@seas.ucla.edu

Associate Professors

Chang, Jane P. .... (310) 206-7980
jpchang@seas.ucla.edu
Christofides, Panagiotis D. .... (310) 794-1015
pdc@seas.ucla.edu

Assistant Professors

Orkoulas, Gerassimos .... (310) 267-0169
makis@seas.ucla.edu
Segura, Tatiana .... (310) 825-2046
Tang, Yi .... (310) 825-2046
yitang@seas.ucla.edu

Accredited by: ABET

Degrees granted 2003-2004:

B.S.: 68 M.S.: 6 Ph.D.: 4

Graduate advisor: Yoram Cohen

Undergraduate advisor: Harold G. Monbouquette

Student organization: AIChE
Advisor: Jane P. Chang

Department reports to: Dean, School of Engineering and Applied Sci.

Placement service: William Beard

University of California, Riverside

Department of Chemical and Environmental Engineering
University of California
Riverside CA 92521

University phone .... (951) 827-5185
University FAX .... (951) 827-5696
Department phone .... (951) 827-6460
Department FAX .... (951) 827-5696

Professors

Chen, Wilfred .... (951) 827-2473
wilfred@engr.ucr.edu
Deshusses, Marc A. (Chair) .... (951) 827-2477
mdeshuss@engr.ucr.edu
Haddon, Robert C. (Joint) .... (951) 827-2044
haddon@engr.ucr.edu
Matsumoto, Mark R. (Dean) .... (951) 827-5318
matsumot@engr.ucr.edu
Mulchandani, Ashok .... (951) 827-6419
adani@engr.ucr.edu
Norbeck, Joseph M. .... (951) 827-2262
Joe_Norbeck@cert.ucr.edu
Schultz, Jerome S. .... (951) 827-2111
jssbio@engr.ucr.edu

Associate Professors

Yan, Yushan .... (951) 827-2068
yushan@engr.ucr.edu

Assistant Professors

Cocker, David R. .... (951) 827-2408
dcocker@cert.ucr.edu
Kauffman, Ken J. .... (951) 827-6049
kkauffma@engr.ucr.edu
Myung, Nosang V. .... (951) 827-7710
Myung@engr.ucr.edu
Ozkan, Mihri (Joint) .... (951) 827-2900
mihri@ee.ucr.edu
Walker, Sharon L. .... (951) 827-6094
swalker@engr.ucr.edu
Wu, Jianzhong .... (951) 827-2413
jwu@engr.ucr.edu

Other Faculty

Morikis, Dimitrios .... (951) 827-2696
dmorikis@engr.ucr.edu

Accredited by: ABET

Degrees granted 2003-2004:

B.S.: 16 M.S.: 3 Ph.D.: 1

Graduate advisor: Jianzhong Wu

Undergraduate advisor: David Cocker

Student organization: AIChE
Advisor: Nosang Myung

Department reports to: Mark R. Matsumoto

University of California, San Diego

Chemical Enginnering Program
Mechanical and Aerospace Engineering Department
9500 Gilman Drive
La Jolla, CA 92093-0411
Deliveries: 371 Engineering Building II

University phone .... (858) 534-2230
Department phone .... (858) 534-3174
Department FAX .... (858) 534-4543

Professors

Chau, Pao C. .... (858) 534-6935
Miller, David R. .... (858) 534-3182
dmiller@ucsd.edu
Pozrikidis, Constantine .... (858) 534-6530
costas@ucsd.edu
Seshadri, K. .... (858) 534-4876
seshadri@ucsd.edu
Talbot, Jan B. (Director) .... (858) 534-3176
jtalbot@ucsd.edu

Associate Professors

Herz, Richard K. .... (858) 534-6540
herz@ucsd.edu

Accredited by: ABET

Degrees granted 2003-2004:

B.S.: 27 M.S.: 1

Graduate advisor: Beverley Walton

Undergraduate advisor: Gerri Johnson

Student organization: AIChE
Advisor: Pao Chau

Department reports to: Dean of Engineering (858) 534-4575

Placement service: Marci Swain (858) 534-0141

University of California, Santa Barbara

Department of Chemical Engineering
Santa Barbara, CA 93106-5080
Deliveries: Engineering II, Room 3357

University phone . . . . (805) 893-8000
Department phone . . . . (805) 893-3412
Department FAX . . . . (805) 893-4731

Professors

Aydil, Eray S. (Associate Chair) . . . . . . (805) 893-8205
aydil@engineering.ucsb.edu
Banerjee, Sanjoy . . . . (805) 893-3456
banerjee@anemone.ucsb.edu
Chmelka, Bradley F. . . . . (805) 893-3673
bradc@engineering.ucsb.edu
Doherty, Michael F. . . . . (805) 893-5309
mfd@engineering.ucsb.edu
Doyle, Frank . . . . (805) 893-8133
doyle@engineering.ucsb.edu
Fredrickson, Glenn H. . . . . (805) 893-8308
ghf@engineering.ucsb.edu
Hanna, Owen T. (Emeritus) . . . . (805) 893-3243
hanna@engineering.ucsb.edu
Israelachvili, Jacob . . . . (805) 893-8407
jacob@engineering.ucsb.edu
Kramer, Edward J. . . . . (805) 893-4999
edkramer@mrl.ucsb.edu
Lange, Fred . . . . (805) 893-8248
flange@engineering.ucsb.edu
Leal, L. Gary . . . . (805) 893-8510
lgl20@engineering.ucsb.edu
Lucas, Glenn E. . . . . (805) 893-4069
gene@engineering.ucsb.edu
McFarland, Eric . . . . (805) 893-4343
mcfar@engineering.ucsb.edu
Mellichamp, Duncan A. (Emeritus) . . . . (805) 893-2821
dmell@engineering.ucsb.edu
Pine, David (Chair) . . . . (805) 893-7383
pine@engineering.ucsb.edu
Rinker, Robert G. (Emeritus) . . . . (805) 893-2610
rinker@engineering.ucsb.edu
Sandall, Orville C. . . . . (805) 893-2908
sandall@engineering.ucsb.edu
Scott, Susannah L. . . . . (805) 893-5606
sscott@engineering.ucsb.edu
Seborg, Dale E. . . . . (805) 893-3352
seborg@engineering.ucsb.edu
Theofanous, Theofanis . . . . (805) 893-4900
theo@tcrss.ucsb.edu
Tirrell, Matthew (Dean) . . . . (805) 893-3141
tirrell@engineering.ucsb.edu
Zasadzinski, Joseph . . . . (805) 893-4769
gorilla@engineering.ucsb.edu

Assistant Professors

Daugherty, Patrick . . . . (805) 893-2610
psd@engineering.ucsb.edu
Mitragotri, Samir . . . . (805) 893-7532
samir@engineering.ucsb.edu

Accredited by: ABET

Degrees granted 2003-2004:

B.S.: 25 M.S.: 3 Ph.D.: 8

Graduate advisor: Gary Leal

Undergraduate advisor: Joe Zasadzinski

Student organization: AIChE
Advisor: Robert G. Rinker

Department reports to: Matthew Tirrell, (805) 893-3141, Dean

Placement service: Carol A. Geer (805) 893-4419

California Institute of Technology

Chemical Engineering Department
Mail Stop 210-41
Pasadena, CA 91125-4100
Deliveries: 391 S. Holliston Ave.

University phone .... (626) 395-6811
Department phone .... (626) 395-4115
Department FAX .... (626) 568-8743

Professors

Arnold, Frances H. .... (626) 395-4162
frances@cheme.caltech.edu
Brady, John F. .... (626) 395-4183
jfb@cheme.caltech.edu
Davis, Mark E. .... (626) 395-4251
mdavis@cheme.caltech.edu
Flagan, Richard C. (Head) .... (626) 395-4383
flagan@cheme.caltech.edu
Gavalas, George R. (Emeritus) .... (626) 395-4152
gavalas@cheme.caltech.edu
Kornfield, Julia A. .... (626) 395-4138
jak@cheme.caltech.edu
Seinfeld, John H. .... (626) 395-4635
seinfeld@cco.caltech.edu
Tirrell, David A. (Chair) .... (626) 395-3140
tirrell@caltech.edu
Tschoegl, Nicholas W. (Emeritus)
Wang, Zhen-Gang .... (626) 395-4647
zgw@cheme.caltech.edu

Associate Professors

Giapis, Konstantinos P. .... (626) 395-4180
giapis@cheme.caltech.edu

Assistant Professors

Asthagiri, Anand R. .... (626) 395-8130
anand@cheme.caltech.edu
Smolke, Christina D. .... (626) 395-2460
smolke@cheme.caltech.edu

Other Faculty

Wagner, Eric S. .... (626) 395-2308
ewagner@cheme.caltech.edu

Accredited by: ABET

Degrees granted 2003-2004:

B.S.: 3 M.S.: 8 Ph.D.: 9

Student organization: AIChE
Advisor: Anand R. Asthagiri

Department reports to: David A. Tirrell, Chairman, Div. of Chem. and Chemical Engineering

Placement service: Gerald Houser, Director, Career Development

California State University, Long Beach

Chemical Engineering Department
Long Beach, CA 90840
Deliveries: 1250 Bellflower Blvd.,VEC 136

University phone .... (562) 985-4111
University FAX .... (562) 985-8887
Department phone .... (562) 985-4909
Department FAX .... (562) 985-7561

Professors

Hile, Lloyd .... (562) 985-7532
lhile@csulb.edu
Jang, Larry (Chair) .... (562) 985-7533
jang@csulb.edu
Kavianian, Hamid .... (562) 985-2260
hrkche@cox.net
Lenoir, John M. (Emeritus) .... (562) 985-4909
Naimpally, Ashok .... (562) 985-1508
avnaimpa@csulb.edu
Tsai, Shirley .... (562) 985-7534
sctsai@csulb.edu

Accredited by: ABET

Degrees granted 2003-2004:

B.S.: 40 M.S.: 8

Graduate advisor: Ashok Naimpally

Undergraduate advisor: Larry Jang

Student organization: AIChE
Advisor: Larry Jang

Department reports to: Michael Mahoney, (562) 985-5121, Dean of College of Engineering

Placement service: Ed Morton (562) 985-4151, Director

California State Polytechnic University, Pomona

Chemical and Materials Engineering
3801 West Temple Avenue
Pomona, CA 91768-4069

University phone .... (909) 869-2600
Department phone .... (909) 869-2626
Department FAX .... (909) 869-6920

Professors

Aldrich, J. Winthrop (Emeritus) ...... (909) 869-2626
jwaldrich@csupomona.edu
Bray, Robert S. (Emeritus) .... (909) 869-2626
Caenepeel, Christopher L. (Chair) ..... (909) 869-2626
clcaenepeel@csupomona.edu
Feeney, Robert G. (Emeritus) .... (909) 869-2626
Hacker, Barbara .... (909) 869-2629
bahacker@csupomona.edu
Harris, William M. (Emeritus) .... (909) 869-2626
Hohmann, Edward C. (Dean) .... (909) 869-2600
echohmann@csupomona.edu
Nguyen, Thuan K. .... (909) 869-2631
tknguyen@csupomona.edu
Ontiveros, Cordelia .... (909) 869-2626
contiveros@calstate.edu
Pang, K. Hing (Emeritus) .... (909) 869-2626
khpang@csupomona.edu
Scott, Garland E. (Emeritus) .... (909) 869-2626
gescott@csupomona.edu
Sheng, Henry P. (Emeritus) .... (909) 869-2626
Stoll, A. George (Emeritus) .... (909) 869-2626
agstoll@csupomona.edu
Tassoney, Joseph P. (Emeritus) .... (909) 869-2626
Tomlinson, John L. (Emeritus) .... (909) 869-2626

Associate Professors

Ravi, Vilupanur .... (909) 869-2627

Assistant Professors

Dong, Winny .... (909) 869-2634
winnydong@csupomona.edu

Accredited by: ABET

Degrees granted 2003-2004:

B.S.: 24

Student organization: AIChE
Advisor: C.L. Caenepeel (909)-869-2626

Department reports to: Edward H. Hohmann (909) 869-2600, Dean

Placement service: Manuel Perez (909) 869-2337, Director, Career Center

San Jose State University

Department of Chemical and Materials Engineering
Engineering 385
One Washington Square
San Jose, CA 95192-0082

University phone .... (408) 924-1000
Department phone .... (408) 924-4000
Department FAX .... (408) 924-4057

Professors

Allen, Emily (Chair) .... (408) 924-4010
elallen@email.sjsu.edu
Jennings, Michael B. (Director) .... (408) 924-3926
jennimi@email.sjsu.edu
McNeil, Melanie .... (408) 924-3873
mcneil@sjsuvm1.sjsu.edu

Assistant Professors

Komives, Claire .... (408) 924-4002
ckomives@email.sjsu.edu
Young, Gregory L. .... (408) 924-3945
glyoung@email.sjsu.edu

Accredited by: ABET

Graduate advisor: M. McNeil

Student organization: AIChE
Advisor: Claire Komives

Department reports to: Belle Wei (408) 924-3800, Dean of Engineering

Placement service: Lina Melkonian (408) 924-6033, Director

University of Southern California

Chemical Engineering Department
University Park, HED 216
Los Angeles, CA 90089-1211
Deliveries: 925 Bloom Walk, HED 216

University phone .... (213) 740-2311
Department phone .... (213) 740-2225
Department FAX .... (213) 740-8053

Professors

Dougherty, Elmer L. (Emeritus) .... (213) 740-0310
Ershaghi, Iraj .... (213) 740-0321
ershaghi@usc.edu
Handy, Lyman L. (Emeritus) .... (213) 740-0315
Sahimi, Muhammad (Chair) .... (213) 740-2064
moe@iran.usc.edu
Salovey, Ronald (Emeritus) .... (213) 740-2073
salove@usc.edu
Tsotsis, Theodore T. .... (213) 740-2069
tsotsis@usc.edu
Yortsos, Yanis C. (Associate Dean) .... (213) 740-0317
yortsos@usc.edu

Associate Professors

Chang, W. Victor .... (213) 740-2072
wenji@usc.edu
Peng, Ching-An .... (213) 740-2067
capeng@usc.edu
Shing, Katherine S. .... (213) 740-2068
shing@usc.edu

Assistant Professors

Kezirian, Michael (Adjunct)
kezirian@usc.edu
Lee, C. Ted .... (213) 740-2066
tedlee@usc.edu
Wang, Pin .... (213) 740-2225

Accredited by: ABET

Degrees granted 2003-2004:

B.S.: 12 M.S.: 10 Ph.D.: 2

Graduate advisor: Dr. Katherine Shing

Student organization: AIChE
Advisor: Dr. Katherine Shing

Department reports to: Max Nikias, (213) 740-0617, Dean of Engineering

Placement service: Eileen Kohan, STU 111, Los Angeles, CA 90089-4897

Stanford University

Department of Chemical Engineering
Stauffer III, Room 113
381 North-South Axis
Stanford, CA 94305-5025

University phone .... (650) 723-2300
Department phone .... (650) 723-4906
Department FAX .... (650) 723-9780

Professors

Boudart, Michel (Emeritus) .... (650) 723-4906
Frank, Curtis W. (Chair) .... (650) 723-4573
curt@chemeng.stanford.edu
Fuller, Gerald G. .... (650) 723-9243
ggf@rio.stanford.edu
Khosla, Chaitan S. .... (650) 723-6538
ck@chemeng.stanford.edu
Madix, Robert J. .... (650) 723-2402
rjm@rio.stanford.edu
Robertson, Channing R. .... (650) 723-4986
crr@chemeng.stanford.edu
Shaqfeh, Eric S. G. (Associate Chair) .. (650) 723-3764
eric@chemeng.stanford.edu
Swartz, James .... (650) 723-5398
swartz@chemeng.stanford.edu

Associate Professors

Bao, Zhenan .... (650) 723-2419
zbao@stanford.edu
Bent, Stacey .... (650) 723-0385
stacey@chemeng.stanford.edu

Assistant Professors

Kao, Camilla .... (650) 726-0547
cammy@chemeng.stanford.edu
Musgrave, Charles .... (650) 725-9176
charles@chemeng.stanford.edu

Instructors

Pavone, Anthony .... (650) 723-4906
apavone@sric.sri.com

Accredited by ABET

Degrees granted 2003-2004:

B.S.: 29 M.S.: 26 Ph.D.: 13

Graduate advisor: Eric Shaqfeh, Gerald Fuller, James Swartz

Undergraduate advisor: Camilla Kao

Student organization: AIChE
Advisor: Camilla Kao

Department reports to: James D. Plummer, Dean

Placement service: Michelle Etter, Career Planning

Colorado

University of Colorado

Department of Chemical and Biological Engineering
Engineering Center ECCH 111
UCB 424
Boulder, CO 80309-0424

University phone .... (303) 492-0111
Department phone .... (303) 492-7471
Department FAX .... (303) 492-4341

Professors

Anseth, Kristi S. .... (303) 492-3147
kristi.anseth@colorado.edu
Bowman, Christopher N. (Chair) ...... (303) 492-3247
christopher.bowman@colorado.edu
Clough, David E. (Associate Chair) .... (303) 492-6638
david.clough@colorado.edu
Davis, Robert H. (Dean) .... (303) 492-7314
robert.davis@colorado.edu
Falconer, John L. .... (303) 492-8005
john.falconer@colorado.edu
George, Steve .... (303) 492-3398
steven.george@colorado.edu
Noble, Richard D. .... (303) 492-6100
richard.noble@colorado.edu
Ramirez, W. Fred .... (303) 492-8660
fred.ramirez@colorado.edu
Randolph, Theodore W. .... (303) 492-4776
theodore.randolph@colorado.edu
Sani, Robert L. .... (303) 492-5517
robert.sani@colorado.edu
Schwartz, Daniel .... (303) 492-5473
daniel.schwartz@colorado.edu
Stansbury, Jeff .... (303) 724-1044
jeffrey.stansbury@UCHSC.edu
Weimer, Alan W. .... (303) 492-3759
alan.weimer@colorado.edu

Associate Professors

Bryant, Stephanie
Gin, Doug .... (303) 492-7640
douglas.gin@colorado.edu
Kompala, Dhinakar S. .... (303) 492-6350
dhinakar.kompala@colorado.edu

Assistant Professors

Gill, Ryan .... (303) 492-2627
ryan.t.gill@colorado.edu
Hrenya, Christine M. .... (303) 492-7689
hrenya@colorado.edu
Mahoney, Melissa
Medlin, Will .... (303) 492-2418
Will.Medlin@colorado.edu

Instructors

deGrazia, Janet .... (303) 492-4763
deGrazia@spot.colorado.edu
Louie, Bev .... (303) 492-3687
beverly.louie@colorado.edu

Accredited by: ABET

Degrees granted 2003-2004:

B.S.: 64 M.S.: 9 Ph.D.: 12

Graduate advisor: Theodore Randolph

Undergraduate advisor: David Clough

Student organization: AIChE
Advisor: Alan Weimer

Department reports to: (303) 492-7006, Dean of Engineering and Applied Sciences

Placement service: Ann Hermann

Colorado School of Mines

Chemical Engineering Department
451 Alderson Hall
Golden, CO 80401
Deliveries: Warehouse, 900 18th Street

University phone . . . . (303) 273-3000
Department phone . . . . (303) 273-3720
Department FAX . . . . (303) 273-3730

Professors

Bunge, Annette L. . . . . (303) 273-3722
abunge@mines.edu
Dean, Anthony M. . . . . (303) 273-3643
amdean@mines.edu
Ely, James F. (Head) . . . . (303) 273-3885
jely@mines.edu
Gary, James H. (Emeritus) . . . . (303) 273-3720
Golden, John O. (Emeritus) . . . . (303) 273-3720
jgolden@mines.edu
Kidnay, Arthur J. (Emeritus) . . . . . . (303) 384-2215
akidnay@mines.edu
Miller, Ronald L. . . . . (303) 273-3892
rlmiller@mines.edu
Sloan, E. Dendy Jr. . . . . (303) 273-3723
esloan@mines.edu
Way, J. Douglas . . . . (303) 273-3519
dway@mines.edu
Yesavage, Victor F. (Emeritus) . . . . . . (303) 273-3725
vyesavag@mines.edu

Associate Professors

Dorgan, John R. . . . . (303) 273-3539
jdorgan@mines.edu
Marr, David W. M. . . . . (303) 273-3008
dmarr@mines.edu
McKinnon, J. Thomas . . . . (303) 273-3098
jmckinno@mines.edu
Wolden, Colin A. . . . . (303) 273-3544
cwolden@mines.edu
Wu, David T. . . . . (303) 384-2024
dwu@mines.edu

Other Faculty

Graboski, Michael S. . . . . (303) 273-3973
mgrabosk@mines.edu
Herring, Andy . . . . (303) 384-2082
aherring@mines.edu
Kiselev, Sergei B. . . . . (303) 273-3190
skiselev@mines.edu
Knecht, Robert D. . . . . (303) 273-3592
rknecht@mines.edu
Koh, Carolyn . . . . (303) 273-3237
ckoh@mines.edu
Miller, Kelly . . . . (303) 273-3951
ktmiller@mines.edu

Accredited by: ABET

Degrees granted 2003-2004:
B.S.: 66 M.S.: 5 Ph.D.: 6

Graduate advisor: Colin Woldin

Undergraduate advisor: James F. Ely

Student organization: AIChE
Advisor: James F. Ely

Department reports to: Nigel Middleton (303) 273-3327, Vice President for Academic Affairs

Placement service: Anna Hanley, CSM Placement Office

Colorado State University

Department of Chemical Engineering
1370 Campus Delivery
100 Glover
Fort Collins, CO 80523-1370

University phone . . . . (970) 491-1101
Department phone . . . . (970) 491-5252
Department FAX . . . . (970) 491-7369

Professors

Belfiore, Laurence A. . . . . (970) 491-5395
belfiore@engr.colostate.edu
Linden, James C. . . . . (970) 491-6122
james.linden@colostate.edu
Murphy, Vincent G. . . . . (970) 491-1791
vince@engr.colostate.edu
Reardon, Kenneth F. . . . . (970) 491-6505
reardon@engr.colostate.edu
Watson, A. Ted (Chair) . . . . (970) 491-5253
atw@engr.colostate.edu

Associate Professors

Dandy, David S. (Associate Chair) . . . . (970) 491-7437
dandy@engr.colostate.edu
Wickramasinghe, Ranil . . . . (970) 491-5276
wickram@engr.colostate.edu

Assistant Professors

Rinker, Kristina . . . . (970) 491-4648
rinker@engr.colostate.edu
Wang, Qiang (David) 970 491-2763
qwang@engr.colostate.edu

Instructors

Perkins, Tracy . . . . (970) 491-5175
tracyperkins@earthlink.net
Smith, T. Gordon
tgsmith@engr.colostate.edu

Accredited by: ABET

Degrees granted 2003-2004:

B.S.: 24 M.S.: 12 Ph.D.: 2

Graduate advisor: David S. Dandy

Undergraduate advisor: Vincent G. Murphy

Student organization: AIChE
Advisor: David S. Dandy

Department reports to: Dr. Steven Abt, Interim Dean, College of Engineering

Placement service: Ann Malen, Director, The Career Center (970) 491-5707, Career Center

Connecticut

University of Connecticut

Chemical Engineering Department
191 Auditorium Road, Unit 3222
Storrs, CT 06269-3222
Deliveries: Engineering II Building, Room 204

University phone . . . . (860) 486-2000
Department phone . . . . (860) 486-4019
Department FAX . . . . (860) 486-2959

Professors

Achenie, Luke E. K. . . . . (860) 486-2756
achenie@engr.uconn.edu
Cooper, Douglas J. (Head) . . . . (860) 486-4092
cooper@engr.uconn.edu
Fenton, James M. . . . . (860) 486-2490
jmfent@engr.uconn.edu
Helble, Joseph J. . . . . (860) 486-4602
helble@engr.uconn.edu
Shaw, Montgomery T. . . . . (860) 486-3980
shawmt@uconnvm.uconn.edu
Weiss, Robert A. . . . . (860) 486-4698
rweiss@mail.ims.uconn.edu
Wood, Thomas K. . . . . (860) 486-2483
twood@engr.uconn.edu

Associate Professors

Anderson, Thomas F. . . . . (860) 486-2473
anderson@engr.uconn.edu
Erkey, Can . . . . (860) 486-4601
cerkey@engr.uconn.edu
Mather, Patrick . . . . (860) 486-3542
mather@mail.ims.uconn.edu
Parnas, Richard S. . . . . (860) 486-9060
rparnas@mail.ims.uconn.edu

Assistant Professors

Srivastava, Ranjan . . . . (860) 486-2802
srivasta@engr.uconn.edu
Zhu, Lei . . . . (860) 486-8708
lzhu@ims.uconn.edu

Other Faculty

Bell, James P. (Emeritus) . . . . (860) 486-4019
jbell@mail.ims.uconn.edu
Bennett, Carroll O. (Emeritus) . . . . (860) 486-4019
Coughlin, Robert W. (Emeritus) . . . . . . (860) 486-4489
coughlin@engr.uconn.edu
Cutlip, Michael (Emeritus) . . . . (860) 486-0321
mcutlip@uconnvm.uconn.edu
DiBenedetto, Anthony T. (Emeritus) . . (860) 486-4019
adibened@mail.ims.uconn.edu
Howard, G. Michael (Emeritus) . . . . (860) 486-2479
howard@uconnvm.uconn.edu
Khalil, Yehia . . . . (860) 486-4020
yehiak@engr.uconn.edu
Kunz, H. Russell . . . . (860) 486-5389
russkunz@engr.uconn.edu
Sundstrom, Donald W. (Emeritus) . . . . (860) 486-4019

Accredited by: ABET

Degrees granted 2003-2004:

B.S.: 15 M.S.: 8 Ph.D.: 9

Graduate advisor: Richard S. Parnas

Student organization: AIChE
Advisor: Ranjan Srivastava

Department reports to: Amir Faghri (860) 486-2221, Dean of Engineering

Placement service: Cynthia Jones, Director (860) 486-3013

University of New Haven

Chemistry and Chemical Engineering Department
300 Boston Post Road
West Haven, CT 06516
Deliveries: Receiving Section

University phone .... (203) 932-7000
Department phone .... (203) 932-7404
Department FAX .... (203) 931-6077

Professors

Collura, Michael A. (Chair) .... (203) 932-7149
mcollura@newhaven.edu
Desio, Peter J. (Emeritus) .... (203) 932-7170
pdesio@newhaven.edu
Saliby, Michael J. .... (203) 932-7169
saliby@newhaven.edu
Wheeler, George L. .... (203) 932-7171
wheeler@snet.net

Associate Professors

Gow, Arthur S. III .... (203) 932-7173
agow@newhaven.edu
Harding, W. David .... (203) 932-7438
dharding@newhaven.edu
Schwartz, Pauline M.
pschwartz@newhaven.edu

Assistant Professors

Luzik, Eddie .... (203) 932-7006
eluzik@newhaven.edu

Other Faculty

Montagna, Peter (Adjunct)

Accredited by: ABET

Degrees granted 2003-2004:

B.S.: 4

Undergraduate advisor: W. David Harding

Student organization: AIChE
Advisor: W. David Harding

Department reports to: Dr. Zulma Toro-Ramos, Dean, School of Engineering, (203) 932-7167

Placement service: Kathryn Link (203) 932-7334

Yale University

Chemical Engineering Department
P. O. Box 208286
New Haven, CT 06520
Deliveries: 9 Hillhouse Avenue, New Haven, CT

Department phone .... (203) 432-2222
Department FAX .... (203) 432-4387

Professors

Altman, Eric .... (203) 432-4375
eric.altman@yale.edu
Crothers, Donald (Honorary) .... (203) 432-5204
donald.crothers@yale.edu
Elimelech, Menachem .... (203) 432-2789
menachem.elimelech@yale.edu
Graedel, Thomas E. (Honorary) .... (203) 432-9733
thomas.graedel@yale.edu
Haller, Gary L. .... (203) 432-4378
gary.haller@yale.edu
Loewenberg, Michael .... (203) 432-4334
michael.loewenberg@yale.edu
Pfefferle, Lisa D. .... (203) 432-4377
pfefferle@biomed.med.yale.edu
Pignatello, Joseph (Adjunct) .... (203) 432-2222
joseph.pignatello@po.state/ct.us
Rosner, Daniel E. .... (203) 432-4391
daniel.rosner@yale.edu
Saltzman, Mark (Honorary) .... (203) 432-0030
mark.saltzman@yale.edu
Wallis, James (Adjunct) .... (203) 432-4379
james.wallis@yale.edu
Walz, John Y. (Chair) .... (203) 432-4382
john.walz@yale.edu
Wikstrom, Leonard Lee (Adjunct) ..... (203) 432-6651
llwikstrom@aol.com
Zilm, Kurt (Honorary) .... (203) 432-3956
kurt.zilm@yale.edu

Associate Professors

Khalil, Yehia F. (Adjunct) .... (203) 665-3604
yehia.khalil@yale.edu
Van Tassel, Paul .... (203) 432-7983
paul.vantassel@yale.edu

Assistant Professors

Mitch, William .... (203) 432-4386
william.mitch@yale.edu

Accredited by: ABET

Degrees granted 2003-2004:

B.S.: 5 M.S.: 11 Ph.D.: 4

Undergraduate advisor: Michael Loewenberg

Department reports to: Andrew Hamilton (203) 432-4448, Deputy Provost

Placement service: Philip Jones, (203) 432-0802

Delaware

University of Delaware

Department of Chemical Engineering
Newark, DE 19716-3110
Deliveries: Colburn Laboratory, 150 Academy Street
University phone .... (302) 831-2000
Department phone .... (302) 831-2543
Department FAX .... (302) 831-1048

Professors

Barteau, Mark A. (Chair) .... (302) 831-8905
barteau@che.udel.edu
Beris, Antony N. .... (302) 831-8018
beris@che.udel.edu
Bischoff, Kenneth B. (Emeritus)
Chen, Jingguang .... (302) 831-0642
chenj@che.udel.edu
Denson, Costel D. .... (302) 831-2427
denson@che.udel.edu
Dhurjati, Prasad .... (302) 831-2879
dhurjati@che.udel.edu
Kaler, Eric W. (Dean) .... (302) 831-3553
kaler@che.udel.edu
Lenhoff, Abraham M. .... (302) 831-8989
lenhoff@che.udel.edu
Metzner, Arthur B. (Emeritus) .... (302) 831-2328
metzner@che.udel.edu
Ogunnaike, Babatunde A. .... (302) 831-4504
ogunnaik@che.udel.edu
Olson, Jon H. (Emeritus) .... (302) 831-8472
olson@che.udel.edu
Russell, T. W. Fraser .... (302) 831-1714
russell@che.udel.edu
Sandler, Stanley I. .... (302) 831-2945
sandler@che.udel.edu
Schultz, Jerold M. (Emeritus) .... (302) 831-8145
schultz@che.udel.edu
Vlachos, Dionisios .... (302) 831-2830
vlachos@che.udel.edu
Wagner, Norman J. .... (302) 831-8079
wagner@che.udel.edu
Wool, Richard P. .... (302) 831-3312
wool@che.udel.edu

Associate Professors

Buttrey, Douglas J. .... (302) 831-2034
buttrey@che.udel.edu
Lauterbach, Jochen A. .... (302) 831-6327
lauterbach@che.udel.edu
Lobo, Raul F. .... (302) 831-1261
lobo@che.udel.edu
Robinson, Anne Skaja .... (302) 831-0557
robinson@che.udel.edu
Shine, Annette D. .... (302) 831-2010
shine@che.udel.edu

Assistant Professors

Edwards, Jeremy S. .... (302) 831-8072
edwards@che.udel.edu
Furst, Eric M. .... (302) 831-0102
furst@che.udel.edu
Roberts, Christopher J. .... (302) 831-0838
roberts@che.udel.edu
Willis, Brian G. .... (302) 831-6856
willis@che.udel.edu

Other Faculty

Butera, Robert J.
Etchells, Arthur W. III
Farneth, William E.
Grant, James J. III
Grenville, Richard K.
Jones, Allan E.
LaRoche, Richard K.
Lustig, Steven R.
Lyons, James E.
Manogue, William H.
Mulholland, Kenneth L.
Schure, Mark R.
Schwaber, James S.
Short, David
Tilton, James N.
Tomb, Jean-Francois
Uebler, E. Alan
Wissbrun, Kurt F.

Accredited by ABET

Degrees granted 2003-2004:

B.S.: 49 M.S.: 9 Ph.D.: 17

Graduate advisor: Antony Beris and Annette Shine

Undergraduate advisor: Jon H. Olson

Student organization: AIChE
Advisor: Raul F. Lobo

Department reports to: Eric W. Kaler, Dean of Engineering

Placement service: David J. Berilla, Associate Director

District of Columbia

Howard University

Chemical Engineering Department
2300 6th Street, NW
Lewis K. Downing Hall, Room 1009
Washington, DC 20059

University phone .... (202) 806-6100
Department phone .... (202) 806-6624
Department FAX .... (202) 806-4635

Professors

Aluko, Mobolaji E. .... (202) 806-6617
maluko@howard.edu
Cannon, Joseph N. .... (202) 806-6626
jcannon@scs.howard.edu
Chawla, Ramesh C. .... (202) 806-4794
chawla@scs.howard.edu
Lutz, Robert J. (Visiting) .... (301) 435-1944
rlutz@Box-r.nih.gov
Rao, M. Gopala (Emeritus) .... (202) 806-4796
mgrao@comcast.com
Tharakan, John P. (Chair) .... (202) 806-6617
jtharakan@howard.edu

Associate Professors

Collins, William E. .... (202) 806-4595
wcollins@howard.edu

Assistant Professors

Ekechukwu, Kenneth (Adjunct) .... (301) 585-9569
ela@prodicy.net

Accredited by: ABET

Degrees granted 2003-2004:

B.S.: 14 M.S.: 4

Graduate advisor: Dr. Ramesh C. Chawla

Student organization: AIChE
Advisor: Dr. Joseph N. Cannon

Department reports to: James H. Johnson, Jr., Dean, College of Engineering, Architecture & Computer Science

Placement service: Samuel Hall, Office of Placement and Career Planning

Florida

University of Florida

P. O. Box 116005
Chemical Engineering Department
Gainesville, FL 32611-6005
Deliveries: Museum Road, Bldg. 723

University phone .... (352) 392-3261
Department phone .... (352) 392-0881
Department FAX .... (352) 392-9513

Professors

Anderson, Timothy J. (Associate Dean) (352) 392-0946
tim@ufl.edu

Block, Seymour S. (Emeritus) .... (352) 392-9102
block@che.ufl.edu

Crisalle, Oscar D. .... (352) 392-5120
crisalle@che.ufl.edu

Curtis, Jennifer (Chair) .... (352) 392-0882
jcurtis@che.ufl.edu

Fricke, Arthur L. (Emeritus) .... (352) 392-0884
fricke@che.ufl.edu

Hoflund, Gar B. .... (352) 392-9104
garho@hotmail.com

Johns, Lewis E. .... (352) 392-0881
johns@che.ufl.edu

Ladd, Anthony J.C. .... (352) 392-6509
ladd@che.ufl.edu

Narayanan, Ranga .... (352) 392-9103
ranga@che.ufl.edu

Orazem, Mark E. .... (352) 392-6207
meo@che.ufl.edu

Park, Chang-Won .... (352) 392-6205
park@che.ufl.edu

Rajagopalan, Raj .... (352) 392-0868
raj@eng.ufl.edu

Ren, Fan .... (352) 392-4727
ren@che.ufl.edu

Shah, Dinesh O. .... (352) 392-0877
shah@che.ufl.edu

Svoronos, Spyros A. .... (352) 392-0882
svoronos@ufl.edu

Tyner, Mack (Emeritus)

Associate Professors

Dickinson, Richard .... (352) 392-0898
dickinso@che.ufl.edu

Kirmse, Dale W. (Emeritus) .... (352) 392-0862
kirmse@che.ufl.edu

Assistant Professors

Butler, Jason .... (352) 392-2591
butler@che.ufl.edu

Chauhan, Anuj .... (352) 392-2592
chauhan@che.ufl.edu

Kopelevich, Dmitry .... (352) 392-4422
dkopelevich@che.ufl.edu

Narang, Atul .... (352) 392-0028
narang@che.ufl.edu

Weaver, Jason .... (352) 392-0869
weaver@che.ufl.edu

Accredited by: ABET

Degrees granted 2003-2004:

B.S.: 76 M.S.: 2 Ph.D.: 13

Graduate advisor: Richard Dickinson

Undergraduate advisor: Spyros A. Svoronos

Student organization: AIChE
Advisor: Gar B Hoflund

Department reports to: Pramod Khargonekar (352) 392-6000, Dean of Engineering

Placement service: Wayne Wallace, Director

Florida Institute of Technology

Chemical Engineering Department
150 West University Boulevard
Melbourne, FL 32901-6975
Deliveries: Building 540

University phone . . . . (321) 674-8000
University FAX . . . . (321) 984-8461
Department phone . . . . (321) 674-8068
Department FAX . . . . (321) 674-7565

Professors

Mason, Donald R. (Emeritus)

Associate Professors

Jennings, Paul A. (Head) . . . . (321) 674-7561
jennings@fit.edu
Tomadakis, Manolis M. . . . . (321) 674-7243
tomadaki@fit.edu
Whitlow, Jonathan E. . . . . (321) 674-7354
whitlow@fit.edu

Assistant Professors

Brenner, James R. . . . . (321) 674-7560
jbrenner@fit.edu
Pozo de Fernandez, Maria E. . . . . (321) 674-8838
mpozo@fit.edu

Other Faculty

Barile, Ron (Adjunct) . . . . (321) 674-8068

Accredited by: ABET, SACS

Degrees granted 2003-2004:

B.S.: 13 M.S.: 4

Graduate advisor: Paul A. Jennings

Undergraduate advisor: All Faculty

Student organization: AIChE
Advisor: Maria E. Pozo de Fernandez

Department reports to: J. Ronald Bailey (321) 674-8020, Dean of College of Engineering

Placement service: Dona Gaynor, Career Services and Cooperative Education

Florida A&M University-Florida State University

Department of Chemical Engineering
Room A131 FAMU-FSU College of Engineering
2525 Pottsdamer Street
Tallahassee, FL 32310-6046
Deliveries: 2525 Pottsdamer Street, A131 CEB

University phone . . . . (850) 644-2525
Department phone . . . . (850) 410-6149
Department FAX . . . . (850) 410-6150

Professors

Alamo, Rufina . . . . (850) 410-6376
alamo@eng.fsu.edu
Locke, Bruce R. . . . . (850) 410-6165
locke@eng.fsu.edu
Palanki, Srinivas . . . . (850) 410-6164
palanki@eng.fsu.edu
Peters, Michael H. (Chair) . . . . (850) 410-6149
peters@eng.fsu.edu

Associate Professors

Chella, Ravindran . . . . (850) 410-6170
chella@eng.fsu.edu
Gibbs, Steve . . . . (850) 644-3393
gibbs@magnet.fsu.edu
Kalu, Eric . . . . (850) 410-6148
ekalu@eng.fsu.edu
Telotte, John C. . . . . (850) 410-6168
telotte@eng.fsu.edu
Vinals, Jorge . . . . (850) 644-7028
vinals@eng.fsu.edu

Assistant Professors

Chen, Kevin . . . . (850) 410-6684
kevinchen@eng.fsu.edu
Chin, Wei-Chun . . . . (850) 410-6578
wchin@eng.fsu.edu
Kwon, Soonjo . . . . (850) 410-6411
skwon@eng.fsu.edu
Ma, Teng . . . . (850) 410-6149
teng@eng.fsu.edu
Schrieber, Loren (Adjunct) . . . . (850) 410-6684
schreiber@eng.fsu.edu
Wesson, G. Dale . . . . (850) 410-6170
wesson@eng.fsu.edu

Other Faculty

Finney, Wright C. . . . . (850) 410-6309
finney@eng.fsu.edu

Accredited by: ABET

Degrees granted 2003-2004:

B.S.: 33 M.S.: 2

Graduate advisor: Srinivas Palanki

Undergraduate advisor: Wright C. Finney

Student organization: AIChE
Advisor: Wright C. Finney

Department reports to: Ching-Jen Chen, Dean, (850) 410-6439

Placement service: Todd Kramer, (850)410-6388

University of South Florida

Chemical Engineering Department
4202 East Fowler Avenue
ENB 118
Tampa, FL 33620-5350

University phone .... (813) 974-2011
Department phone .... (813) 974-3997
Department FAX .... (813) 974-3651

Professors

Busot, J. Carlos (Emeritus) .... (813) 974-2141
busot@eng.usf.edu
Campbell, Scott W. .... (813) 974-3970
campbell@eng.usf.edu
Gilbert, Richard .... (813) 974-2139
gilbert@eng.usf.edu
Joseph, Babu (Chair) .... (813) 974-0692
joseph@eng.usf.edu
Lee, William E. III .... (813) 974-2136
lee@eng.usf.edu
Llewellyn, J. Anthony .... (813) 974-1780
tony@eng.usf.edu
Smith, Carlos A. (Associate Dean) .... (813) 974-3780
csmith@eng.usf.edu
Sunol, Aydin K. .... (813) 974-3566
sunol@eng.usf.edu

Associate Professors

Bhethanabotla, Venkat R. .... (813) 974-2116
bethana@eng.usf.edu
Jaroszeski, Mark .... (813) 974-4662
mjarosze@eng.usf.edu

Assistant Professors

Alcantar, Norma .... (813) 974-3997
alcantar@eng.usf.edu
VanAuker, Michael .... (813) 974-3186
vanauker@eng.usf.edu

Instructors

Biver, Carl J. .... (813) 944-2112
cbiver@eng.usf.edu

Accredited by: ABET

Degrees granted 2003-2004:
B.S.: 35 M.S.: 13 Ph.D.: 6

Graduate advisor: Dr. Wolan

Undergraduate advisor: Dr. Campbell

Student organization: AIChE/OCE/BMES
Advisor: Dr. Bhethanabotla

Department reports to: Dr. Louis Martin-Vega, Dean of Engineering

Placement service: Drema K. Howard (813) 974-2171, Director

Georgia

Georgia Institute of Technology

School of Chemical & Biomolecular Engineering
Atlanta, GA 30332-0100
Deliveries: 311 Ferst Drive

University phone .... (404) 894-2000
Department phone .... (404) 894-2865
Department FAX .... (404) 894-2866

Professors

Allen, Mark G. (Joint) .... (404) 894-9419
mark.allen@ece.gatech.edu
Banerjee, Sujit .... (404) 894-9709
sujit.banerjee@che.gatech.edu
Baum, Gary (Adjunct) .... (404) 894-7830
gary.baum@ipst.gatech.edu
Bidstrup Allen, Sue Ann (Assoc. Chair) (404) 894-2872
sue.bidstrup@che.gatech.edu
Bommarius, Andreas .... (404) 385-1334
andreas.bommarius@che.gatech.edu
Eckert, Charles A. .... (404) 894-7070
charles.eckert@che.gatech.edu
Empie, Howard L. .... (404) 894-9704
howard.empie@che.gatech.edu
Ernst, William R. (Emeritus) .... (404) 894-2889
william.ernst@che.gatech.edu
Frederick, William .... (404) 894-2082
jim.frederick@ipst.gatech.edu
Fuller, Thomas (Joint) .... (770) 528-7075
tom.fuller@chbe.gatech.edu
Hess, Dennis W. .... (404) 894-5922
dennis.hess@che.gatech.edu
Hsieh, Jeffery .... (404) 894-3556
jeffery.hsieh@che.gatech.edu
Kohl, Paul .... (404) 894-2893
paul.kohl@che.gatech.edu
Koros, William J. .... (404) 385-2684
william.koros@che.gatech.edu
Lee, Jay .... (404) 385-2148
jay.lee@che.gatech.edu
Liotta, Charles L. (Joint) .... (404) 894-4048
charles.liotta@carnegie.gatech.edu
Muzzy, John D. .... (404) 894-2882
john.muzzy@che.gatech.edu
Nerem, Robert M. (Joint) .... (404) 894-2768
robert.nerem@ibb.gatech.edu
Rousseau, Ronald W. (Chair) .... (404) 894-2867
ronald.rousseau@che.gatech.edu
Sambanis, Athanassios .... (404) 894-2869
athanassios.sambanis@che.gatech.edu
Samuels, Robert J. (Emeritus) .... (404) 894-2885
robert.samuels@che.gatech.edu
Schork, F. Joseph (Associate Chair) ... (404) 894-3274
joseph.schork@che.gatech.edu
Skelland, A. H. Peter (Emeritus) ...... (404) 894-2889
anthony.skelland@che.gatech.edu
Sommerfeld, Jude T. (Emeritus) ...... (404) 894-2873
jude.sommerfeld@che.gatech.edu
Stancell, Arnold (Emeritus) .... (404) 894-0316
arnold.stancell@che.gatech.edu
Teja, Amyn (Associate Chair) .... (404) 894-3098
amyn.teja@che.gatech.edu
White, Mark G. .... (404) 894-2822
mark.white@che.gatech.edu
Wick, Timothy M. .... (404) 894-8795
timothy.wick@che.gatech.edu
Winnick, Jack (Emeritus) .... (404) 894-2839
jack.winnick@che.gatech.edu
Yoganathan, Ajit P. (Joint) .... (404) 894-7063
ajit.yoganathan@bme.gatech.edu

Associate Professors

Agrawal, Pradeep K. .... (404) 894-2826
pradeep.agrawal@che.gatech.edu
Chaikof, Elliot L. (Adjunct) .... (404) 727-8413
elliot.chaikof@che.gatech.edu
Chen, Rachel .... (404) 894-1838
rachel.chen@chbe.gatech.edu
Deng, Yulin .... (404) 894-5759
yulin.deng@che.gatech.edu
Forney, Larry J. .... (404) 894-2825
larry.forney@che.gatech.edu
Henderson, Clifford .... (404) 385-0525
cliff.henderson@che.gatech.edu
Prausnitz, Mark R. .... (404) 894-5135
mark.prausnitz@che.gatech.edu
Realff, Matthew J. .... (404) 894-1834
matthew.realff@che.gatech.edu
Roberts, Ronnie S. (Emeritus) .... (404) 894-2889
ronnie.roberts@che.gatech.edu
Tedder, Dan W. .... (404) 894-2856
daniel.tedder@che.gatech.edu

Assistant Professors

Breedveld, L. Victor .... (404) 894-5134
victor.breedveld@che.gatech.edu
Gallivan, Martha .... (404) 894-2878
martha.gallivan@che.gatech.edu
Jones, Christopher .... (404) 385-1683
chris.jones@che.gatech.edu
Nair, Sankar .... (404) 894-4826
sankar.nair@che.gatech.edu
Nenes, Athanasios (Joint) .... (404) 894-9225
thanos.nenes@che.gatech.edu

Other Faculty

Bayer, Charlene W. (Adjunct) .... (404) 894-3825
charlene.bayer@gtri.gatech.edu

Accredited by: ABET

Degrees granted 2003-2004:

B.S.: 99 M.S.: 24 Ph.D.: 15

Graduate advisor: Amyn S. Teja

Undergraduate advisor: F. Joseph Schork

Student organization: AIChE
Advisor: Pradeep K. Agrawal

Department reports to: Don P. Giddens, Dean of Engineering

Placement service: Ralph Mobley, Director

Idaho

University of Idaho

Chemical Engineering Department
PO Box 441021
6th and Urquhart
Moscow, ID 83844-1021

University phone .... (208) 885-6111
Department phone .... (208) 885-6793
Department FAX .... (208) 885-7462

Professors

Edwards, Louis L. Jr. .... (208) 885-6793
jkidd@uidaho.edu
Jackson, Melbourne L. (Emeritus) ..... (208) 885-7575
Korus, Roger A. .... (208) 885-6005
rkorus@uidaho.edu
Park, Jin Y. .... (208) 885-6970
jinpark@uidaho.edu
Scheldorf, Jay J. (Emeritus) .... (208) 885-7282
von Braun, Margrit C. .... (208) 885-7838
vonbraun@uidaho.edu

Associate Professors

Admassu, Wudneh (Chair) .... (208) 885-8918
wadmassu@uidaho.edu
Drown, David C. .... (208) 885-7848
ddrown@uidaho.edu

Assistant Professors

Aston, Eric .... (208) 885-6953
aston@uidaho.edu
Thomas, Aaron .... (208) 885-7652
Utgikar, Vivek .... (208) 282-7720

Accredited by ABET

Graduate advisor: Eric Aston/Aaron Thomas

Undergraduate advisor: Wudneh Admassu

Student organization: AIChE
Advisor: David C. Drown

Department reports to: David Thompson, Dean of Engineering

Placement service: Daniel A. Blanco, Career Services Center

Illinois

University of Illinois at Chicago

Chemical Engineering Department
810 S. Clinton M/C 110
Chicago, IL 60607

University phone .... (312) 996-7000
Department phone .... (312) 996-3424
Department FAX .... (312) 996-0808

Professors

Diwekar, Urmila .... (312) 355-3449
urmila@uic.edu
Kiefer, John H. (Emeritus) .... (312) 996-5711
Kiefer@uic.edu
Mansoori, G. Ali .... (312) 996-5592
mansoori@uic.edu
Murad, Sohail .... (312) 996-5593
Murad@uic.edu
Saxena, Satish C. (Emeritus) .... (312) 996-2341
saxena@uic.edu
Takoudis, Christos .... (312) 996-3424
Takoudis@uic.edu
Turian, Raffi .... (312) 996-8734
Turian@uic.edu
Uslenghi, P.L.E. (Interim Head) .... (312) 413-3777
uslenghi@uic.edu

Associate Professors

Linninger, Andreas .... (312) 996-2581
Linninge@uic.edu
Nitsche, Ludwig .... (312) 996-3469
LCN@uic.edu
Regalbuto, John .... (312) 996-0288
JRR@uic.edu
Szepe, Stephen (Emeritus) .... (312) 996-2342
SSzepe@uic.edu
Wedgewood, Lewis .... (312) 996-5228
Wedge@uic.edu

Accredited by: ABET

Student organization: AIChE
Advisor: G. A. Mansoori

Department reports to: Lawrence A. Kennedy (312) 996-2400, Dean, College of Engineering

Placement service: Andres Garza (312) 996-2300, Director

University of Illinois at Urbana-Champaign

Department of Chemical and Biomolecular Engineering
114 Roger Adams Lab, Box C-3
600 South Mathews Avenue
Urbana, IL 61801-3602

University phone . . . . (217) 333-1000
Department phone . . . . (217) 333-3640
Department FAX . . . . (217) 333-5052

Professors

Alkire, Richard C. (217) 333-0063
r-alkire@uiuc.edu
Braatz, Richard D. . . . . (217) 333-5073
braatz@uiuc.edu
Hanratty, Thomas J. (Emeritus) . . . . . . (217) 333-1318
thanratt@uiuc.edu
Higdon, Jonathan J. L. . . . . (217) 333-1479
jhigdon@uiuc.edu
Leckband, Deborah E. (Head) . . . . (217) 244-9214
leckband@uiuc.edu
Masel, Richard I. . . . . (217) 244-2819
masel@scs.uiuc.edu
McHugh, Anthony J. (Emeritus)
Sahinidis, Nikolaos V. . . . . (217) 244-1304
nikos@uiuc.edu
Schowalter, William R. (Emeritus)
schow@uiuc.edu
Seebauer, Edmund G. . . . . (217) 333-4402
eseebaue@uiuc.edu
Westwater, James W. (Emeritus)
Zukoski, Charles F. . . . . (217) 333-0034
czukoski@uiuc.edu

Associate Professors

Hammack, William S. . . . . (217) 244-4146
hammack@netbox.com

Assistant Professors

Kenis, Paul J. A. . . . . (217) 265-0523
kenis@uiuc.edu
Pack, Daniel W. . . . . (217) 244-2816
dpack@uiuc.edu
Rao, Christopher V.
Strano, Michael S. (217) 333-3634
strano@uiuc.edu
Zhao, Huimin . . . . (217) 333-2631
zhao5@uiuc.edu

Instructors

Miletic, Marina . . . . (217) 244-6730
marina@uiuc.edu

Accredited by: ABET

Degrees granted 2003-2004:

B.S.: 61 M.S.: 22 Ph.D.: 12

Undergraduate advisor: N. V. Sahinidis

Student organization: AIChE
Advisor: M. Miletic

Department reports to: Jesse G. Delia, Dean, College of Liberal Arts and Sciences

Placement service: Debe Deeb Williams, (217) 333-1050, 105 Noyes Lab

Illinois Institute of Technology

Department of Chemical and Environmental Engineering
10 W. 33rd St., 127 Perlstein Hall
Chicago, IL 60616-3793
Deliveries: 112 Perlstein Hall

University phone .... (312) 567-3000
Department phone .... (312) 567-3040
Department FAX .... (312) 567-8874

Professors

Arastoopour, Hamid (Chair) .... (312) 567-3038
arastoopour@iit.edu
Beissinger, Richard (Research)
Bernstein, Barry .... (312) 567-3166
bernsteinb@iit.edu
Cinar, Ali .... (312) 567-3042
cinar@iit.edu
Gidaspow, Dimitri .... (312) 567-3045
gidaspow@iit.edu
Linden, Henry (Research) .... (312) 567-3095
linden@iit.edu
Moschandreas, Demetrios .... (312) 567-3532
djm@iit.edu
Myerson, Allan .... (312) 567-3101
myerson@iit.edu
Nikolov, Alex .... (312) 567-5980
nikolov@iit.edu
Noll, Kenneth .... (312) 567-3536
noll@iit.edu
Parulekar, Satish J. (Associate Chair) .. (312) 567-3044
parulekar@iit.edu
Selman, J. Robert (Emeritus) .... (312) 567-3037
selman@iit.edu
Swanson, Bernard (Emeritus)
Venerus, David C. .... (312) 567-5177
venerus@iit.edu
Wasan, Darsh T. .... (312) 567-3001
wasan@iit.edu

Associate Professors

Abbasian, Javad .... (312) 567-3047
abbasian@iit.edu
Al-Hallaj, Said (Research) .... (312) 567-5118
alhallaj@iit.edu
Anderson, Paul .... (312) 567-3531
andersonp@iit.edu
Pagilla, Krishna .... (312) 567-5717
pagilla@iit.edu
Prakash, Jai .... (312) 567-3639
prakash@iit.edu
Schieber, Jay D. .... (312) 567-3046
schieber@iit.edu
Smotkin, Eugene S. .... (312) 567-3453
smotkin@iit.edu
Teymour, Fouad A. (Associate Chair) .. (312) 567-8947
teymour@iit.edu

Assistant Professors

Chmielewski, Donald .... (312) 567-3537
chmielewski@iit.edu
Khalili, Nasrin R. .... (312) 567-3534
khalili@iit.edu
Perez-Luna, Victor .... (312) 567-3963
perezluna@iit.edu

Other Faculty

Aderangi, Nader (Senior Lecturer) ..... (312) 567-8874
aderangi@iit.edu

Accredited by: ABET

Graduate advisor: Dr. Satish Parulekar (312) 567-3044

Undergraduate advisor: Dr. Fouad Teymour (312) 567-8947

Student organization: AIChE
Advisor: Dr. Don Chmielewski (312) 567-3537

Department reports to: Allan Myerson (312) 567-3163, Dean of Armour College

Placement service: Director (312) 567-6800

Northwestern University

Department of Chemical and Biological Engineering
2145 Sheridan Road, Room E136
Evanston, IL 60208-3120

University phone . . . . . (847) 491-3741
Department phone . . . . . (847) 491-7398
Department FAX . . . . . (847) 491-3728

Professors

Bankoff, S. George (Emeritus) . . . . . (847) 491-5267
gbankoff@northwestern.edu
Broadbelt, Linda . . . . . (847) 491-5351
broadbelt@northwestern.edu
Burghardt, Wesley R. (Associate Chair) (847) 467-1401
w-burghardt@northwestern.edu
Butt, John B. (Emeritus)
Carr, Stephen H. (Associate Dean) . . . . (847) 491-7379
s-carr@northwestern.edu
Cohen, William C. (Emeritus)
bill-cohen@northwestern.edu
Crist, Buckley Jr. . . . . . (847) 491-3279
b-crist@northwestern.edu
Dranoff, Joshua S. . . . . . (847) 491-5252
j-dranoff@northwestern.edu
Goldstick, Thomas K. (Emeritus) . . . . . (847) 491-5518
t-goldstick@northwestern.edu
Kung, Harold H. . . . . . (847) 491-7492
hkung@northwestern.edu
Miller, William M. (Chair) . . . . . (847) 491-4828
wmmiller@northwestern.edu
Mockros, Lyle F. (Emeritus) . . . . . (847) 491-3172
lmockros@northwestern.edu
Ottino, Julio M. . . . . . (847) 491-3558
ottino@chem-eng.northwestern.edu
Papoutsakis, E. Terry . . . . . (847) 491-7455
e-paps@northwestern.edu
Stevens, William F. (Emeritus)
Torkelson, John M. . . . . . (847) 491-7449
j-torkelson@northwestern.edu

Associate Professors

Amaral, Luis A. N. . . . . . (847) 491-1750
amaral@northwestern.edu
Barron, Annelise . . . . . (847) 491-2778
a-barron@northwestern.edu
Brown, George M. (Emeritus)
Snurr, Randall . . . . . (847) 467-2977
snurr@northwestern.edu

Assistant Professors

Grzybowski, Bartosz . . . . . (847) 491-3024
grzybor@northwestern.edu
Hatzimanikatis, Vassily . . . . . (847) 491-5357
vassily@northwestern.edu
Shea, Lonnie D. . . . . . (847) 491-7043
l-shea@northwestern.edu
Swartz, Melody A. (Adjunct) . . . . . (847) 467-6668
m-swartz2@northwestern.edu

Other Faculty

Graessley, William W. (Adjunct) . . . . . (847) 491-7398
Haug, Warren (Adjunct) . . . . . (847) 467-5712
w-haug@northwestern.edu
Kung, Mayfair . . . . . (847) 491-5085
m-kung@northwestern.edu

Accredited by: ABET

Degrees granted 2003-2004:

B.S.: 23 M.S.: 8 Ph.D.: 9

Student organization: AIChE
Advisor: Vassily Hatzimanikatis

Department reports to: John R. Birge, Dean, MEAS

Placement service: William J. Banis, Director

Indiana

University of Notre Dame

Department of Chemical & Biomolecular Engineering
Notre Dame, IN 46556-5637
Deliveries: 182 Fitzpatrick Hall

University phone . . . . . . (574) 631-5000
Department phone . . . . . . (574) 631-5580
Department FAX . . . . . . (574) 631-8366

Professors

Brennecke, Joan F. . . . . . . (574) 631-5847
jfb@nd.edu
Chang, Hsueh-Chia . . . . . . (574) 631-5697
chang2@nd.edu
Kantor, Jeffrey C. . . . . . . (574) 631-8052
Kantor.1@nd.edu
Leighton, David T. Jr. . . . . . . (574) 631-6698
dtl@nd.edu
McCready, Mark J. (Chair) . . . . . . (574) 631-7146
mjm@nd.edu
McGinn, Paul J. . . . . . . (574) 631-6151
mcginn.1@nd.edu
Miller, Albert E. . . . . . . (574) 631-8307
miller.3@nd.edu
Schmitz, Roger A. . . . . . . (574) 631-7798
schmitz.1@nd.edu
Stadtherr, Mark A. . . . . . . (574) 631-9318
markst@nd.edu
Strieder, William C. . . . . . . (574) 631-5648
strieder.1@nd.edu
Wolf, Eduardo E. . . . . . . (574) 631-5897
wolf.1@nd.edu

Associate Professors

Hill, Davide A. . . . . . . (574) 631-8487
hill.1@nd.edu
Maginn, Edward J. . . . . . . (574) 631-5687
ed@nd.edu

Assistant Professors

Ostafin, Agnes E. . . . . . . (574) 631-3798
aostafin@nd.edu
Palmer, Andre F. . . . . . . (574) 631-4776
apalmer@nd.edu

Other Faculty

Aki, Sudhir . . . . (574) 631-5848
Aki.1@nd.edu
Demekhin, Evgeny . . . . (574) 631-3661
edemekhi@darwin.helios.nd.edu
Kamat, Prashant (Joint) . . . . (574) 631-5411
pkamat@nd.edu
Mukasyan, Alexander . . . . (574) 631-9825
amoukasi@nd.edu
Saddawi, Salma . . . . (574) 631-3324
Saddawi.1@nd.edu

Accredited by: ABET

Degrees granted 2003-2004:

B.S.: 39 M.S.: 2 Ph.D.: 8

Graduate advisor: Mark A. Stadtherr

Student organization: AIChE
Advisor: Agnes E. Ostafin

Department reports to: Frank Incropera, McCloskey Dean of Engineering

Placement service: Tricia Ford (219) 631-3315, 248 Flanner Hall

Purdue University

School of Chemical Engineering
Forney Hall of Chemical Engineering
480 Stadium Mall Drive, 104
West Lafayette, IN 47907-2100
Deliveries: FRNY Hall of Chemical Engineering

University phone (765) 494-4600
Department phone (765) 494-4050
Department FAX (765) 494-0805

Professors

Albright, Lyle F. (Emeritus) (765) 494-4087 albright@ecn.purdue.edu
Andres, Ronald P. (Emeritus) (765) 494-4047 ronald@ecn.purdue.edu
Blau, Gary E. (Visiting) (765) 494-9472 blau@ecn.purdue.edu
Caruthers, James M. (765) 494-6625 caruther@ecn.purdue.edu
Chao, K. C. (Emeritus) (765) 494-4050 kwang@ecn.purdue.edu
Delgass, W Nicholas (765) 494-4059 delgass@ecn.purdue.edu
Emery, Alden H. (Emeritus) (765) 494-4050 emery@ecn.purdue.edu
Franses, Elias I. (765) 494-4078 franses@ecn.purdue.edu
Greenkorn, Robert A. (Emeritus) (765) 494-4051 greenkor@ecn.purdue.edu
Hannemann, Robert (Visiting) (765) 494-4079 hanneman@ecn.purdue.edu
Houze, R. Neal (765) 494-4076 houze@ecn.purdue.edu
Kim, Sangtae (765) 494-5692 kim55@ecn.purdue.edu
Pekny, Joseph F. (765) 494-7901 pekny@ecn.purdue.edu
Pipes, Byron (765) 494-4050
Ramkrishna, D. (Associate Head) (765) 494-4066 ramkrish@ecn.purdue.edu
Reklaitis, Gintaras V. (765) 494-9662 reklaiti@ecn.purdue.edu
Squires, Robert G. (Emeritus) (765) 494-4093 squires@ecn.purdue.edu
Tsao, George T. (765) 494-4068 tsaogt@ecn.purdue.edu
Venkatasubramanian, V. (765) 494-0734 venkat@ecn.purdue.edu
Wang, Linda Nien-Hwa (765) 494-4081 wangn@ecn.purdue.edu
Wankat, Phillip C. (765) 494-0814 wankat@ecn.purdue.edu

Associate Professors

Beaudoin, Stephen (765) 494-7944 sbeaudoi@ecn.purdue.edu
Corti, David S. (765) 496-6064 dscorti@ecn.purdue.edu
Harris, Michael T. (765) 494-0963 mtharris@ecn.purdue.edu
Lee, Gil U. (765) 494-0492 gl@ecn.purdue.edu
Ribeiro, Fabio (765) 494-7799 fabio@ecn.purdue.edu

Assistant Professors

Baertsch, Chelsey D. (765) 496-7826 baertsch@ecn.purdue.edu
Hillhouse, Hugh W. (765) 496-6056 hugh@ecn.purdue.edu
Morgan, John A. (765) 494-4088 jamorgan@ecn.purdue.edu
Thomson, Kendall T. (765) 496-6706 thomsonk@ecn.purdue.edu
Won, You-Yeon (765) 494-4077 yywon@ecn.purdue.edu

Other Faculty

Okos, M. (765) 494-1211 okos@ecn.purdue.edu

Accredited by: ABET

Degrees granted 2003-2004:

B.S.: 105 M.S.: 5 Ph.D.: 10

Graduate advisor: Osman Basaran

Undergraduate advisor: Jan Siebenthal

Student organization: AIChE
Advisor: Stephen P. Beaudoin

Department reports to: Linda P.B. Katehi, John A. Edwardson Dean of Engineering

Placement service: Center for Career Opportunities, Placement Service, STEWART Center

Rose-Hulman Institute of Technology

Chemical Engineering Department
5500 Wabash Ave.
Terre Haute, IN 47803-3999

University phone . . . . (812) 877-1511
Department phone . . . . (812) 877-8430
Department FAX . . . . (812) 877-8992

Professors

Abegg, Carl F. (Emeritus) . . . . . . (812) 877-8430
Carl.F.Abegg@rose-hulman.edu
Artigue, Ronald S. . . . . (812) 877-8369
Ronald.S.Artigue@rose-hulman.edu
Bowden, Warren W. (Emeritus) . . . . (812) 877-8430
Trebol1125@aol.com
Carlson, Alfred . . . . (812) 877-8423
Alfred.Carlson@rose-hulman.edu
Caskey, Jerry A. (Emeritus) . . . . . (812) 877-8430
Jerry.A.Caskey@rose-hulman.edu
Hariri, M. Hossein (Head) . . . . . . (812) 877-8381
M.H.Hariri@rose-hulman.edu
Leipziger, Stuart (Emeritus) . . . . . (812) 877-8430
Stuart.Leipziger@rose-hulman.edu
Moore, Noel E. (Emeritus) . . . . . . (812) 877-8430
Noel.E.Moore@rose-hulman.edu

Associate Professors

Anklam, Mark R. . . . . (812) 877-8098
Mark.R.Anklam@rose-hulman.edu
Coronell, Daniel G. . . . . (812) 877-8419
Daniel.G.Coronell@rose-hulman.edu
Serbezov, Atanas . . . . (812) 877-8097
Atanas.Serbezov@rose-hulman.edu

Assistant Professors

Miller, David C. . . . . (812) 877-8506
David.Miller@rose-hulman.edu
Sauer, Sharon G. . . . . (812) 877-8527
Sharon.G.Sauer@rose-hulman.edu

Accredited by: ABET

Degrees granted 2003-2004:

B.S.: 51 M.S.: 5

Graduate advisor: David Miller

Student organization: AIChE
Advisor: Atanas Serbezov

Department reports to: Art Western,Dean of Faculty

Placement service: Jan Ford (812) 877-8338, Associate Director

Tri-State University

Department of Chemical Engineering
1 University Ave.
Angola, IN 46703-0307

University phone . . . . (260) 665-4100
Department phone . . . . (260) 665-4815
Department FAX . . . . (260) 665-4814

Professors

Tucker, W. Henry (Emeritus)

Associate Professors

Finley, David (Dean) . . . . (260) 665-4224
finleyd@tristate.edu
Salim, Majid . . . . (260) 665-4223
salimm@tristate.edu
Wagner, John E. (Chair) . . . . (260) 665-4226
wagnerj@tristate.edu

Assistant Professors

Hersel, Allen . . . . (260) 665-4252
hersela@tristate.edu

Other Faculty

Holman, Chad . . . . (260) 665-4225
cholman@tristate.edu

Accredited by: ABET

Degrees granted 2003-2004:

B.S.: 10

Undergraduate advisor: Dr. Majid Salim/Dr. John Wagner/Dr. Allen Hersel

Student organization: AIChE
Advisor: Dr. Majid Salim

Department reports to: Dr. John Wagner (260) 665-4226, Chair, Chemical Engineering Dept.

Placement service: Ms. Debbie Roemke (219) 665-4123, Asst. Director of Placement

Iowa

University of Iowa

Chemical and Biochemical Engineering Department
College of Engineering
Iowa City, IA 52242-1527
Deliveries: 4133 Seamans Center

University phone .... (319) 335-3500
Department phone .... (319) 335-1400
Department FAX .... (319) 335-1415

Professors

Beddow, John K. (Emeritus) .... (319) 335-1400
Carmichael, Gregory R. .... (319) 335-1399
gcarmich@icaen.uiowa.edu
Murhammer, David W. .... (319) 335-1228
murham@icaen.uiowa.edu
Rethwisch, David G. .... (319) 335-1413
drethwis@icaen.uiowa.edu
Rodgers, Victor G. J. .... (319) 335-1401
vrodgers@icaen.uiowa.edu
Scranton, Alec .... (319) 335-1400
alec-scranton@uiowa.edu
Wiencek, John (Chair) .... (319) 353-2377
wiencek@icaen.uiowa.edu

Associate Professors

Peeples, Tonya .... (319) 335-2251
peeples@icaen.uiowa.edu

Assistant Professors

Guymon, C. Allan .... (319) 335-5015
allan-guymon@uiowa.edu
Jessop, Julie .... (319) 335-0681
julie-jessop@uiowa.edu

Instructors

Aurand, Gary .... (319) 384-0970
gary-aurand@uiowa.edu
Butler, Audrey .... (319) 384-0538
audrey-butler@uiowa.edu

Accredited by: ABET

Degrees granted 2003-2004:

B.S.: 19 M.S.: 6 Ph.D.: 5

Graduate advisor: Allan Guymon

Student organization: AIChE
Advisor: David W. Murhammer

Department reports to: P. Barry Butler, (319) 335-5766, Dean of Engineering

Placement service: Engineering Career Services (319) 335-5774

Iowa State University of Science and Technology

Chemical Engineering Department
2114 Sweeney Hall
Ames, IA 50011-2230

University phone .... (515) 294-4111
Department phone .... (515) 294-7642
Department FAX .... (515) 294-2689

Professors

Brown, Robert C. .... (515) 294-3759
rcbrown@iastate.edu
Burnet, George (Emeritus) .... (515) 294-8670
gxb@iastate.edu
Doraiswamy, L. K. (Emeritus) .... (515) 294-4117
dorai@iastate.edu
Fox, Rodney O. .... (515) 294-9104
rofox@iastate.edu
Glatz, Charles E. (Chair) .... (515) 294-8472
cglatz@iastate.edu
Hebert, Kurt R. .... (515) 294-6763
krhebert@iastate.edu
Hill, James C. .... (515) 29-4959
jchill@iastate.edu
Jolls, Kenneth R. .... (515) 294-5222
jolls@iastate.edu
Porter, Marc D. (Joint) .... (515) 294-6433
porter@ameslab.gov
Reilly, Peter J. .... (515) 294-5968
reilly@iastate.edu
Schrader, Glenn L. .... (515) 294-0519
schrader@iastate.edu
Seagrave, Richard C. (Emeritus) ...... (515) 294-0518
seagrave@iastate.edu
Shanks, Jackie V. .... (515) 294-4828
jshanks@iastate.edu
Ulrichson, Dean L. (Emeritus) .... (515) 294-8670
dlulrich@iastate.edu
Wheelock, Thomas D. (Emeritus) ..... (515) 294-5226
wheel@iastate.edu
Youngquist, Gordon R. (Emeritus) .... (515) 294-8670
gyoungqu@iastate.edu

Associate Professors

Hillier, Andrew .... (515) 294-3678
hillier@iastate.edu
Mallapragada, Surya K. .... (515) 294-7407
suryakm@iastate.edu
Narasimhan, Balaji .... (515) 294-8019
nbalaji@iastate.edu
Rollins, Derrick K. .... (515) 294-5516
drollins@iastate.edu
Shanks, Brent H. .... (515) 294-1895
bshanks@iastate.edu
Vigil, R. Dennis .... (515) 294-6438
vigil@iastate.edu

Assistant Professors

Gonzalez, Ramon .... (515) 294-1516
ramong@iastate.edu
Lamm, Monica .... (515) 294-6533
mhlamm@iastate.edu

Instructors

Loveland, Stephanie .... (515) 294-3024
prairie@iastate.edu

Other Faculty

Hanneman, Larry F. (Director) .... (515) 294-0253
lfhannem@iastate.edu

Accredited by: ABET

Degrees granted 2003-2004:

B.S.: 73 M.S.: 4 Ph.D.: 3

Graduate advisor: Balaji Narasimhan

Undergraduate advisor: R. Dennis Vigil & Brenda Kutz

Student organization: AIChE
Advisor: Kenneth R. Jolls

Department reports to: James L. Melsa, (515) 294-5933; Dean, College of Engineering

Placement service: Larry F. Hanneman, (515) 294-0253; Director, Engineering Career Services

Kansas

University of Kansas

Department of Chemical and Petroleum Engineering
Learned Hall, Room 4132
1530 W. 15th St.
Lawrence, KS 66045-7609

University phone .... (785) 864-2700
Department phone .... (785) 864-4965
Department FAX .... (785) 864-4967

Professors

Bishop, Kenneth A. (Emeritus) .... (785) 864-2918
kbishop@ku.edu
Camarda, Kyle V. .... (785) 864-2908
camarda@ku.edu
Gehrke, Stevin H. .... (785) 864-4956
shgehrke@ku.edu
Green, Don W. .... (785) 864-2911
dgreen@ku.edu
Locke, Carl E. Jr. .... (785) 864-2929
lok@ku.edu
Maloney, James O. (Emeritus) .... (785) 864-4942
Mesler, Russell B. (Emeritus) .... (785) 843-0863
Preston, Floyd W. (Emeritus) .... (785) 843-6212
petlep@ku.edu
Rosson, Harold F. (Emeritus) .... (785) 331-0747
Subramaniam, Bala (Director) .... (785) 330-4490
bsubramaniam@ku.edu
Swift, George W. (Emeritus) .... (785) 842-2751
cbswift@ku.edu
Vossoughi, Shapour .... (785) 864-2902
shapour@ku.edu
Walas, Stanley M. (Emeritus) .... (785) 843-0547
Weatherley, Laurence R. (Chair) ...... (785) 864-3553
Willhite, G. Paul .... (785) 864-2906
willhite@ku.edu

Associate Professors

Howat, Colin S. .... (785) 218-3718
cshowat@ku.edu
Liang, Jenn-Tai .... (785) 864-2669
jtliang@ku.edu
Nguyen, Trung Van .... (785) 864-3938
cptvn@ku.edu
Nordheden, Karen J. .... (785) 864-8820
nordhed@ku.edu
Ostermann, Russell D. (Associate Chair) (785) 864-2907
ostermann@ku.edu
Southard, Marylee Z. .... (785) 864-3868
marylee@ku.edu

Assistant Professors

Berkland, Cory J. .... (785) 864-4949
berkland@ku.edu
Detamore, Michael S. .... (785) 864-4943
detamore@ku.edu
Guzman, Javier .... (785) 864-4920
Laurence, Jennifer S. .... (785) 864-4949
laurencj@ku.edu
Scurto, Aaron M. .... (785) 864-4947
ascurto@ku.edu
Williams, Susan M. .... (785) 864-2919
smwilliams@ku.edu

Instructors

Howat, Julie .... (785) 218-3260
jnhowat@ku.edu

Accredited by: ABET

Degrees granted 2003-2004:

B.S.: 30 M.S.: 13 Ph.D.: 2

Graduate advisor: Trung V. Nguyen

Undergraduate advisor: All faculty

Student organization: AIChE
Advisor: Colin S. Howat

Department reports to: Stuart R. Bell, Dean, School of Engineering, (785) 864-3881

Placement service: Cathy Schwabauer, Coordinator, (785) 864-3891

Kansas State University

Department of Chemical Engineering
105 Durland
Manhattan, KS 66506-5102

University phone . . . . (785) 532-6011
Department phone . . . . (785) 532-5584
Department FAX . . . . (785) 532-7372

Professors

Akins, Richard G. (Emeritus) . . . . (785) 532-5584
akins@ksu.edu
Edgar, James H. . . . . (785) 532-4320
edgarjh@ksu.edu
Erickson, Larry E. . . . . (785) 532-4313
lerick@ksu.edu
Fan, L. T. . . . . (785) 532-4327
fan@cheme.ksu.edu
Glasgow, Larry A. . . . . (785) 532-4314
glasgow@cheme.ksu.edu
King, Terry S. (Dean) . . . . (785) 532-5590
tsking@ksu.edu
Kyle, Benjamin G. (Emeritus) . . . . (785) 532-5584
chedept@cheme.ksu.edu
Matthews, John C. (Emeritus) . . . . (785) 532-5584
chedept@cheme.ksu.edu
Rezac, Mary (Head) . . . . (785) 532-5584
rezac@ksu.edu
Schlup, John R. . . . . (785) 532-4319
jrsch@ksu.edu
Walawender, Walter P. . . . . (785) 532-4318
walawen@cheme.ksu.edu

Associate Professors

Hohn, Keith L. . . . . (785) 532-4315
hohn@ksu.edu
Pfromm, Peter . . . . (785) 532-4312
pfromm@ksu.edu

Accredited by: ABET

Degrees granted 2003-2004:

B.S.: 17 M.S.: 2 Ph.D.: 6

Graduate advisor: James H. Edgar

Undergraduate advisor: Walter P. Walawender

Student organization: AIChE
Advisor: Walter P. Walawender

Department reports to: Terry S. King, Dean of Engineering

Placement service: Marcia Schuley, Associate Director
(785) 532-1685

Kentucky

University of Kentucky

Department of Chemical and Materials Engineering
177 Anderson Hall
Lexington, KY 40506-0046

University phone .... (859) 257-9000
Department phone .... (859) 257-8028
Department FAX .... (859) 323-1929

Professors

Anderson, Kimberly .... (859) 257-2300x232
kanderson@engr.uky.edu
Bhattacharyya, Dibakar .... (859) 257-2794
db@engr.uky.edu
Gillis, Peter P. (Emeritus) .... (859) 257-8884
Grulke, Eric A. .... (859) 257-2300x289
egrulke@engr.uky.edu
Hamrin, Charles E. Jr. (Emeritus) .... (859) 257-4959
hamrin@engr.uky.edu
Keane, Mark .... (859) 257-5857
makeane@engr.uky.edu
Kermode, Richard I. .... (859) 257-2823
Morris, James G. (Emeritus) .... (859) 257-8090
Okazaki, Kenji .... (859) 257-1307
Penn, Lynn .... (859) 257-7897
penn@engr.uky.edu
Ray, Asit K. .... (859) 257-7990
akray@engr.uky.edu
Reucroft, Philip .... (859) 257-8723
reuc@engr.uky.edu
Schrodt, J. Thomas (Emeritus) .... (859) 257-3969
che207@ukcc.uky.edu
Tsang, Tate T. H. (Chair) .... (859) 257-2825
tsang@engr.uky.edu

Associate Professors

Geertsema, Arie .... (859) 257-0305
ari@caer.uky.edu
Kalika, Douglass .... (859) 257-5507
kalika@engr.uky.edu
Knutson, Barbara L. .... (859) 257-5715
bknutson@engr.uky.edu
Smart, Jimmy .... (270) 534-3119
jsmart@engr.uky.edu

Assistant Professors

Balk, John ....(859)257-4582
balk@engr.uky.edu
Dunbar, Paul .... (270) 534-3105
pdunbar@engr.uky.edu
Hilt, J. Zachary ....(859)257-9844
hilt@engr.uky.edu
Hinds, Bruce .... (859) 257-5507
bhind0@engr.uky.edu
Lee-DeSaumels, Rhonda .... (270) 534-3122
rlee@engr.uky.edu
Rankin, Stephen .... (859) 257-9799
srankin@engr.uky.edu
Silverstein, David .... (270) 534-3132
silverdl@engr.uky.edu
Yang, Fuqian .... (859) 257-2994
fyang0@engr.uky.edu
Zhai, Tongguang .... (859) 257-4958
tzhai@engr.uky.edu

Other Faculty

Huffman, Gerald .... (859) 257-4029
cffls@pop.uky.edu
Huggins, Frank .... (859) 257-4045
frank@funky.cffls.uky.edu
Shah, Naresh .... (859) 257-5119
naresh@pop.uky.edu

Accredited by: ABET

Degrees granted 2003-2004:

B.S.: 28 M.S.: 4 Ph.D.: 4

Graduate advisor: Dr. Barbara Knutson

Undergraduate advisor: Dr. Kim Anderson

Student organization: AIChE
Advisor: Dr. D. Bhattacharyya

Department reports to: Thomas W. Lester, Dean of Engineering

University of Louisville Speed Scientific School

Chemical Engineering Department
Ernst Hall 106
Louisville, KY 40292
Deliveries: 2301 South Third Street

University phone . . . . (502) 852-5555
Department phone . . . . (502) 852-6347
Department FAX . . . . (502) 852-6355

Professors

Collins, Dermot J. (502) 852-6737
dermot@louisville.edu
Deshpande, Pradeep B. . . . . (502) 852-0885
pbdesh01@gwise.louisville.edu
Fleischman, Marvin (Emeritus) . . . . (502) 852-6347
m0flei01@gwise.louisville.edu
Gerhard, Earl R. (Emeritus)
Harper, Dean O. (Emeritus) . . . . (502) 852-6347
doharp01@gwise.louisville.edu
Laukhuf, W. L. S. (Associate Chair) . . . (502) 852-6350
scrap@louisville.edu
Plank, Charles A. (Emeritus)
Ralston, Patricia A. S. . . . . (502) 852-0479
parals01@gwise.louisville.edu
Spencer, Hugh T. (Emeritus)
Starr, Thomas L. (Chair) . . . . (502) 852-6347
tom.starr@louisville.edu
Watters, James C. . . . . (502) 852-6349
jim.watters@louisville.edu

Associate Professors

Kang, Kyung A. . . . . (502) 852-2094
kyung.kang@louisville.edu
Sunkara, Mahendra K. . . . . (502) 852-1558
mksunk01@gwise.louisville.edu

Accredited by: ABET

Degrees granted 2003-2004:

B.S.: 18 M.S.: 18 Ph.D.: 4

Graduate advisor: Pradeep Deshpande

Undergraduate advisor: James C. Watters

Student organization: AIChE
Advisor: James C. Watters

Department reports to: Mickey R. Wilhelm, Dean (502) 852-6281

Placement service: Penny Hoerter (502) 852-0394

Louisiana

Louisiana State University

Chemical Engineering Department Baton Rouge, LA 70803-7303

University phone .... (225) 578-5011
Department phone .... (225) 578-1426
Department FAX .... (225) 578-1476

Professors

Corripio, Armando B. .... (225) 578-3061
corripio@che.lsu.edu
Dooley, Kerry M. .... (225) 578-3063
dooley@che.lsu.edu
Griffin, Gregory L. .... (225) 578-3064
griffin@che.lsu.edu
Groves, Frank R. (Emeritus) .... (225) 578-3060
groves@che.lsu.edu
Harrison, Douglas P. .... (225) 578-3066
harrison@che.lsu.edu
Hjortso, Martin .... (225) 578-3058
hjortso@che.lsu.edu
Knopf, F. Carl (Chair) .... (225) 578-1426
knopf@che.lsu.edu
McCoy, Benjamin J. .... (225) 578-1377
bjmccoy@lsu.edu
McLaughlin, Edward (Emeritus) ...... (225) 578-3546
emcl@eng.lsu.edu
Pike, Ralph W. .... (225) 578-3428
pike@che.lsu.edu
Sterling, Arthur (Emeritus) .... (225) 578-3057
sterling@che.lsu.edu
Thibodeaux, Louis J. .... (225) 578-3055
thibod@che.lsu.edu
Valsaraj, Kalliat T. .... (225) 578-6522
valsaraj@che.lsu.edu

Associate Professors

Podlaha, Elizabeth J. .... (225) 578-3056
knopf@che.lsu.edu
Spivey, James J. .... (225) 578-3690
jjspivey@lsu.edu
Thompson, Karsten E. .... (225) 578-3069
karsten@che.lsu.edu
Wetzel, David M. .... (225) 578-3071
dwetzel@che.lsu.edu
Wornat, Mary J. .... (225) 578-7509
mjwornat@lsu.edu

Instructors

Cygan, Margaret .... (225) 578-3062
cygan@che.lsu.edu
Mowrey, Daniel B. .... (225) 578-1321
dbmowrey@lsu.edu
Toups, Harry 225-578-3068
hjtoups@lsu.edu

Accredited by: ABET

Degrees granted 2003-2004:

B.S.: 62 M.S.: 9 Ph.D.: 4

Graduate advisor: K.T. Valsaraj

Undergraduate advisor: Gregory L. Griffin

Student organization: AIChE
Advisor: Karsten E. Thompson

Department reports to: Pius Egbelu (225) 578-5731, Dean of Engineering

Placement service: Mary Feduccia (225) 578-2162

Louisiana Tech University

Chemical Engineering Department
P. O. Box 10348 T.S.
Ruston, LA 71272
Deliveries: 600 W. Arizona St., Bogard Hall, Room 222

University phone .... (318) 257-0211
Department phone .... (318) 257-2483
Department FAX .... (318) 257-2562

Professors

Thompson, Ronald H. .... (318) 257-2603
thompson@engr.latech.edu

Associate Professors

Elmore, Bill B. (Chair) .... (318) 257-2902
belmore@engr.latech.edu

Assistant Professors

Mainardi, Daniela .... (318) 257-4714
mainardi@latech.edu
Palmer, Jim .... (318) 257-5126
jpalmer@coes.latech.edu
Wesson, Laura .... (318) 257-2483
lwesson@coes.latech.edu

Instructors

Behbahani, Ahmed (Visiting) .... (318) 257-3075
sbehbaha@coes.latech.edu

Accredited by: ABET

Graduate advisor: James Palmer

Undergraduate advisor: Bill B. Elmore

Student organization: AIChE
Advisor: James Palmer/Bill B. Elmore

Department reports to: Les Guice (318) 257-4647, Dean of Engineering and Science

Placement service: Cheryl B. Myers (318) 257-4336, Dean of Student Services
Louisiana Tech University, Ruston, LA 71272

McNeese State University

Engineering Department
P.O. Box 91735
Lake Charles, LA 70609-1735
Deliveries: 143 Drew Hall

University phone . . . . . . (337) 475-5000
Department phone . . . . . . (337) 475-5874
Department FAX . . . . . . (337) 475-5286

Professors

Barkat, Omar . . . . . . (337) 475-5817
obarkat@mail.mcneese.edu
Dautenhahn, Pamela C. . . . . . . (337) 475-5859
pdauten@mail.mcneese.edu
Griffith, D. John . . . . . . (337) 475-5858
griffith@mail.mcneese.edu
Karkalits, O. C. (Dean) . . . . . . (337) 475-5876
ckarkal@mail.mcneese.edu

Associate Professors

Robinson, Richard L. (Coordinator) . . . (337) 475-5864
rrobin@mail.mcneese.edu

Accredited by: ABET (general engineering curriculum)

Student organization: AIChE
Advisor: Dr. Pamela Dautenhahn

Department reports to: O. C. Karkalits, Dean, College of Engineering and Technology

Placement service: Kathy Bond

University of Louisiana at Lafayette

Chemical Engineering Department
P.O. Box 44130
Lafayette, LA 70504-4130
Deliveries: Madison Hall, Room 217, 131 Rex Street

University phone . . . . . . (337) 482-1000
Department phone . . . . . . (337) 482-6562
Department FAX . . . . . . (337) 482-1220

Professors

Fang, Cheng S. . . . . . . (337) 482-5350
Farshad, Fred F. . . . . . . (337) 482-5862
Garber, James D. (Head) . . . . . . (337) 482-6151
garber@louisiana.edu
Misra, Devesh . . . . . . (337) 482-6430
Ponter, Anthony B. (Dean) . . . . . . (337) 482-6685
Reinhardt, James R. . . . . . . (337) 482-5351

Associate Professors

Hill, Archibald G. . . . . . . (337) 482-5761

Assistant Professors

Newman, Bernard . . . . . . (337) 482-6564

Accredited by: ABET

Degrees granted 2003-2004:

B.S.: 13 M.S.: 20

Graduate advisor: Devesh Misra

Student organization: AIChE
Advisor: J.R. Reinhardt

Department reports to: Anthony B. Ponter (337) 482-6685, Dean of Engineering

Placement service: Mrs. Evelyn Rogers (337) 262-5300

Tulane University

Department of Chemical and Biomolecular Engineering
New Orleans, LA 70118
Deliveries: 300 Lindy Boggs Building

University phone .... (504) 862-8000
Department phone .... (504) 865-5772
Department FAX .... (504) 865-6744

Professors

Bailey, Raymond V. (Emeritus) .... (504) 865-5772
De Kee, Daniel .... (504) 865-5620
ddekee@tulane.edu
Gonzalez, Richard .... (504) 865-5741
richard.gonzalez@tulane.edu
John, Vijay T. (Chair) .... (504) 865-5883
vj@tulane.edu
Law, Victor J. .... (504) 865-5773
law@tulane.edu
Mitchell, Brian S. .... (504) 862-8257
brian@tulane.edu
O'Connor, Kim C. .... (504) 865-5740
koc@tulane.edu
Papadopoulos, Kyriakos D. .... (504) 865-5826
pops@tulane.edu

Associate Professors

Sullivan, Samuel L. Jr. (Emeritus) ..... (504) 865-5772

Assistant Professors

Ashbaugh, Henry S. .... (504) 862-8258
hanka@tulane.edu
Godbey, W T. .... (504) 865-5872
godbey@tulane.edu
Lu, Yunfeng .... (504) 865-5827
ylu@tulane.edu

Accredited by: ABET

Degrees granted 2003-2004:

B.S.: 18 Ph.D.: 4

Graduate advisor: Kyriakos Papadopoulos

Undergraduate advisor: Kyriakos Papadopoulos

Student organization: AIChE
Advisor: Henry Ashbaugh

Department reports to: Nicholas J. Altiero (504) 865-5764, Dean of Engineering

Placement service: Jack Grubbs, Assoc. Dean for External Programs, (504) 865-5742

Maine

University of Maine

Chemical and Biological Engineering Department
5737 Jenness Hall
Orono, ME 04469-5737
Deliveries: 5737 Jenness Hall

Department phone .... (207) 581-2276
Department FAX .... (207) 581-2323

Professors

Bousfield, Douglas W. .... (207) 581-2300
bousfld@maine.edu
Genco, Joseph M. .... (207) 581-2284
genco@maine.edu
Hassler, John .... (207) 581-2285
hassler@maine.edu
Mumme, Kenneth (Emeritus)
mumme@maine.edu
Pendse, Hemant (Chair) .... (207) 581-2290
pendse@maine.edu
Ruthven, Douglas M. .... (207) 581-2283
druthven@umche.maine.edu
Thompson, Edward V. (Emeritus)
ethompson@umche.maine.edu
van Heiningen, Adriaan .... (207) 581-2278
avanheiningen@umche.maine.edu

Associate Professors

Co, Albert .... (207) 581-2282
albertco@maine.edu
Donahue, Darrell .... (207) 581-2728
darrell_donahue@umit.maine.edu
Hwalek, John J. .... (207) 581-2302
hwalek@maine.edu

Assistant Professors

DeSisto, William .... (207) 581-2291
wdesisto@umche.maine.edu
Millard, Paul .... (207) 581-2277
paul.millard@maine.edu
Neivandt, David .... (207) 581-2288
dneivandt@umche.maine.edu
Wheeler, Clayton .... (207) 581-2280
cwheeler@umche.maine.edu

Other Faculty

Hill, Marquita .... (207) 581-2301
mhill@umche.maine.edu

Accredited by: ABET

Student organization: AIChE
Advisor: John J. Hwalek

Department reports to: Larryl Matthews, Dean

Placement service: Patricia Counihan (207) 581-1359

Maryland

Johns Hopkins University

Department of Chemical and Biomolecular Engineering
3400 N. Charles Street
221 Maryland Hall
Baltimore, MD 21218

University phone .... (410) 516-8000
Department phone .... (410) 516-7170
Department FAX .... (410) 516-5510

Professors

Betenbaugh, Michael J. (Chair) .... (410) 516-5461
beten@jhu.edu
Donohue, Marc D. (Associate Dean) ... (410) 516-7761
mdd@jhu.edu
Katz, Joseph L. .... (410) 516-8484
JLK@jhu.edu
Stebe, Kathleen J. .... (410) 516-7769
kjs@jhu.edu
Wirtz, Denis .... (410) 516-7006
wirtz@jhu.edu

Associate Professors

Gracias, David .... (410) 516-5284
dgracias@jhu.edu
Hanes, Justin .... (410) 516-3484
hanes@jhu.edu
Konstantopoulos, Konstantinos .... (410) 516-6290
konst_k@jhu.edu

Assistant Professors

Gray, Jeffrey .... (410) 516-5313
jgray@jhu.edu
Harden, James .... (410) 516-0170
harden@jhu.edu
Ostermeier, Marc .... (410) 516-7144
oster@jhu.edu

Other Faculty

Aranovich, Gregory .... (410) 516-7079
aranovich@jhu.edu
Gaddy, Glen .... (410) 516-8747
ggaddy@jhu.edu
Hoh, Jan .... (410) 614-3795
Karweit, Michael .... (410) 516-8415
mjk@jhu.edu
Krambeck, Fredrick (Visiting) .... (410) 516-
fjkrambeck@jhu.edu
Leong, Kam W. .... (410) 614-5644
Searson, Peter .... (410) 516-8774
Tseng, Yiider .... (410) 516-5287
yiider@jhu.edu

Accredited by: ABET

Degrees granted 2003-2004:

B.S.: 14 M.S.: 5 Ph.D.: 6

Graduate advisor: Kathleen J. Stebe

Undergraduate advisor: Joseph L. Katz

Student organization: AIChE
Advisor: Jeffrey Gray

Department reports to: Nicholas Jones, Dean of Engineering

Placement service: Patricia Matteo, Director, Career Planning and Development

University of Maryland, College Park

Chemical Engineering Department
College Park, MD 20742-2111
Deliveries: Chemical & Nuclear Engineering Building, Room 2113

University phone .... (301) 405-1000
Department phone .... (301) 405-1935
Department FAX .... (301) 405-0523

Professors

Anisimov, Mikhail A. .... (301) 405-8049
anisimov@eng.umd.edu
Barbari, Timothy A. .... (301) 405-2983
barbari@eng.umd.edu
Bentley, William E. .... (301) 405-1906
bentley@eng.umd.edu
Calabrese, Richard V. .... (301) 405-1908
rvc@eng.umd.edu
Choi, Kyu Yong .... (301) 405-1907
choi@eng.umd.edu
Gentry, James W. (Emeritus) .... (301) 405-1915
gentry@eng.umd.edu
Gomezplata, Albert (Emeritus)
Greer, Sandra C. .... (301) 405-1895
sgreer@umd.edu
McAvoy, Thomas J. (Emeritus) .... (301) 405-1939
mcavoy@eng.umd.edu
Regan, Thomas M. (Emeritus) .... (301) 405-1936
tregan@eng.umd.edu
Sengers, Jan V. (Emeritus) .... (301) 405-4805
js45@umail.umd.edu
Smith, Theodore G. (Emeritus) .... (301) 405-1918
tgsmith@eng.umd.edu
Weigand, William A. .... (301) 405-1916
weigand@eng.umd.edu

Associate Professors

Adomaitis, Raymond A. (Chair) ...... (301) 405-2969
adomaiti@isr.umd.edu
Ehrman, Sheryl H. .... (301) 405-1917
sehrman@eng.umd.edu
Kofinas, Peter .... (301) 405-7335
kofinas@eng.umd.edu
Wang, Nam Sun .... (301) 405-1910
nsw@eng.umd.edu
Zafiriou, Evanghelos .... (301) 405-6625
zafiriou@isr.umd.edu

Assistant Professors

Dimitrakopoulos, Panagiotis .... (301) 405-8166
dimitrak@eng.umd.edu
Fisher, John P. .... (301) 405-7475
jpfisher@eng.umd.edu
Klapa, Maria I. .... (301) 405-1320
mklapa@eng.umd.edu
Pulliam Holoman, Tracey R. .... (301) 405-1909
tholoman@eng.umd.edu
Raghavan, Srinivasa R. .... (301) 405-8164
sraghava@eng.umd.edu

Other Faculty

DiMarzio, Edmund A. (Adjunct)
Ranade, Madhav B. (Adjunct)
Wesson, Rosemarie D. (Adjunct)
Yang, Arthur J.-M. (Adjunct)

Accredited by: ABET

Degrees granted 2003-2004:

B.S.: 40 M.S.: 5 Ph.D.: 5

Graduate advisor: Sheryl H. Ehrman

Undergraduate advisor: Nam S. Wang

Student organization: AIChE
Advisor: Raymond A. Adomaitis

Department reports to: Nariman Farvardin (301) 405-3868, Dean of Engineering

Placement service: Heidi Sauber (301) 405-3863, Director of Engineering Career Services

University of Maryland, Baltimore County

Department of Chemical and Biochemical Engineering
1000 Hilltop Circle
ECS 101
Baltimore, MD 21250
Deliveries: UMBC/CBE

University phone . . . . (410) 455-1000
Department phone . . . . (410) 455-3400
Department FAX . . . . (410) 455-1049

Professors

Bayles, Taryn . . . . (410) 455-3428
tbayles@umbc.edu
Frey, Douglas D. . . . . (410) 455-3400
dfrey1@umbc.edu
Moreira, Antonio R. . . . . (410) 455-6576
moreira@umbc.edu
Payne, Gregory F. . . . . (410) 455-3413
payne@research.umbc.edu
Rao, Govind (Chair) . . . . (410) 455-3415
grao@umbc.edu

Associate Professors

Good, Theresa . . . . (410) 455-3405
tgood@umbc.edu
Marten, Mark . . . . (410) 455-3439
marten@umbc.edu
Ross, Julia M. . . . . (410) 455-3414
jross@umbc.edu

Assistant Professors

Castellanos, Mariajose . . . . (410) 455-8151
Owens, Camelia (Visiting) . . . . (410) 455-8152
owens@umbc.edu

Accredited by: ABET

Degrees granted 2003-2004:

B.S.: 25 M.S.: 8 Ph.D.: 5

Graduate advisor: Theresa Good

Undergraduate advisor: Mark Marten

Student organization: AIChE
Advisor: Gregory Payne

Department reports to: Carmi Shlomo, Dean of Engineering (410) 455-3270

Placement service: Betty J. Glascoe (410) 455-2069, Director

Massachusetts

University of Massachusetts Amherst

159 Goessmann Lab
Chemical Engineering Department
Amherst, MA 01003-93030
Deliveries: 686 North Pleasant Street

University phone .... (413) 545-0111
Department phone .... (413) 545-2507
Department FAX .... (413) 545-1647

Professors

Douglas, James M. (Emeritus) .... (413) 545-2252
Henson, Michael A. .... (413) 545-3481
henson@ecs.umass.edu
Laurence, Robert L. (Emeritus) .... (413) 545-0470
rlaurence@ecs.umass.edu
Malone, Michael F. .... (413) 545-2359
mmalone@ecs.umass.edu
Maroudas, Dimitrios .... (413) 545-3617
maroudas@ecs.umass.edu
Monson, Peter A. .... (413) 545-0661
monson@ecs.umass.edu
Watkins, James J. .... (413) 545-2569
watkins@ecs.umass.edu
Westmoreland, Phillip R. (Head) .... (413) 545-1750
westm@ecs.umass.edu
Winter, Horst H. .... (413) 545-0922
winter@ecs.umass.edu

Associate Professors

Conner, William C. Jr. .... (413) 545-0316
wconner@ecs.umass.edu

Assistant Professors

Bhatia, Surita R. .... (413) 545-0096
sbhatia@ecs.umass.edu
Davis, Jeffrey M. .... (413) 545-2916
jmdavis@ecs.umass.edu
Forbes, Neil S. .... (413) 577-0132
forbes@ecs.umass.edu
Roberts, Susan C. .... (413) 545-1660
sroberts@ecs.umass.edu

Other Faculty

Auerbach, Scott M. (Adjunct) .... (413) 545-1240
auerbach@chem.umass.edu
Doherty, Michael F. (Adjunct) .... (413) 545-2507
mfd@engineering.ucsb.edu
Hoagland, David A. (Adjunct) .... (413) 577-1513
Santore, Maria M. (Adjunct) .... (413) 577-1417
santore@mail.pse.umass.edu

Accredited by: ABET

Degrees granted 2003-2004:

B.S.: 32 M.S.: 5 Ph.D.: 6

Graduate advisor: James J. Watkins

Undergraduate advisor: Michael A. Henson

Student organization: AIChE
Advisor: Neil S Forbes

Department reports to: Michael F. Malone (413) 545-0300, Dean of Engineering

Placement service: (413) 545-2224, Director of Mather Career Center

University of Massachusetts-Lowell

Department of Chemical and Nuclear Engineering
1 University Avenue
Lowell, MA 01854
Deliveries: One University Avenue

University phone .... (978) 934-4000
Department phone .... (978) 934-3171
Department FAX .... (978) 934-3047

Professors

Bonner, Francis J. .... (978) 934-3154
Francis_Bonner@uml.edu
Brown, Gilbert Jay .... (978) 934-3166
Gilbert_Brown@uml.edu
Donatelli, Alfred Anthony (Chair) ..... (978) 934-3156
Alfred_Donatelli@uml.edu
Higgins, Charles J. (Emeritus) .... (978) 934-3171
Phelps, James Parkhurst (Emeritus) ... (978) 934-3171
Sama, Dominick Anthony (Emeritus) .. (978) 934-3161
Dominick_Sama@uml.edu
Sheff, James Robert .... (978) 934-3169
James_Sheff@uml.edu
Vasilos, Thomas .... (978) 934-3162
Thomas_Vasilos@uml.edu
Walkinshaw, John W. .... (978) 934-3159
John_Walkinshaw@uml.edu
White, John Robert .... (978) 934-3165
John_White@uml.edu

Associate Professors

Flood, H. William (Emeritus) .... (978) 934-3171
Sung, Changmo .... (978) 934-3540
Changmo_Sung@uml.edu

Assistant Professors

Lawton, Carl W. .... (978) 934-3158
Carl_Lawton@uml.edu

Accredited by: ABET

Graduate advisor: Thomas Vasilos

Student organization: AIChE, ANS, TAPPI, OXE, Alpha Nu Sigma
Advisor: John W. Walkinshaw(AIChE), Gilbert J. Brown (ANS, Alpha Nu Sigma), John W. Walkinshaw (TAPPI), Alfred A. Donatelli (Omega Chi Epsilon)

Department reports to: Krishna Vedula (978) 934-2576, Dean of Engineering

Placement service: Patricia Yates (978) 934-2355, Director of Career Services

Massachusetts Institute of Technology

Department of Chemical Engineering
77 Massachusetts Avenue, 66-350
Cambridge, MA 02139-4307
Deliveries: 25 Ames Street, 66-350, Cambridge, MA 02142

University phone .... (617) 253-1000
Department phone .... (617) 253-4561
Department FAX .... (617) 258-8992

Professors

Armstrong, Robert C. (Head) .... (617) 253-4581
rca@mit.edu
Beér, János M. (Emeritus) .... (617) 253-6661
jmbeer@mit.edu
Blankschtein, E. Daniel .... (617) 253-4594
dblank@mit.edu
Brenner, Howard .... (617) 253-6687
hbrenner@mit.edu
Brown, Robert A. .... (617) 253-5726
rab@mit.edu
Cohen, Robert E. .... (617) 253-3777
recohen@mit.edu
Colton, Clark K. .... (617) 253-4585
ckcolton@mit.edu
Cooney, Charles L. .... (617) 253-5066
ccooney@mit.edu
Deen, William A. .... (617) 253-4535
wmdeen@mit.edu
Gast, Alice P. .... (617) 253-1403
gast@mit.edu
Gleason, Karen K. .... (617) 253-3108
kkg@mit.edu
Hatton, T. Alan .... (617) 253-4588
tahatton@mit.edu
Howard, Jack B. (Emeritus) .... (617) 253-4574
jbhoward@mit.edu
Jensen, Klavs F. .... (617) 253-4589
kfjensen@mit.edu
Karel, Marcus (Emeritus) .... (617) 253-6744
karel@mit.edu
Langer, Robert S. .... (617) 253-3107
rlanger@mit.edu
Lauffenburger, Douglas A. .... (617) 252-1629
lauffen@mit.edu
McRae, Gregory J. .... (617) 253-6564
mcrae@mit.edu
Merrill, Edward W. (Emeritus) .... (617) 253-4593
emerrill@mit.edu
Reid, Robert C. (Emeritus) .... (617) 253-4571
bobcreid@aol.com
Rutledge, Gregory C. .... (617) 253-4561
rutledge@mit.edu
Sarofim, Adel F. (Emeritus) .... (617) 253-4561
sarofim@mit.edu
Satterfield, Charles N. (Emeritus) ..... (617) 253-4584
cnsatter@mit.edu
Sawin, Herbert H. .... (617) 253-4570
hhsawin@mit.edu
Smith, Kenneth A. .... (617) 253-1973
kas@mit.edu
Stephanopoulos, George .... (617) 253-3904
geosteph@mit.edu
Stephanopoulos, Gregory .... (617) 253-4583
gregstep@mit.edu
Tester, Jefferson W. .... (617) 253-3401
testerel@mit.edu
Wang, Daniel I. C. .... (617) 253-2126
dicwang@mit.edu
Wittrup, K. Dane .... (617) 253-4578
wittrup@mit.edu
Ying, Jackie Y. .... (617) 253-2899
jyying@mit.edu

Associate Professors

Barton, Paul I. .... (617) 253-6526
pib@mit.edu
Green, William H. .... (617) 253-4580
whgreen@mit.edu
Hammond, Paula T. .... (617) 258-7577
hammond@mit.edu
Trout, Bernhardt L. .... (617) 258-5021
trout@mit.edu
Virk, Preetinder S. .... (617) 253-3177
psvirk@mit.edu

Assistant Professors

Beers, Kenneth J. .... (617) 258-8986
kbeers@mit.edu
Doyle, Patrick S. .... (617) 253-4534
pdoyle@mit.edu
Prather, Kristala J. .... (617) 253-1950

Other Faculty

Dalzell, William H. .... (617) 253-5273
wdalzell@mit.edu
Friedly, John C. .... (617) 253-6600
friedly@mit.edu
Johnston, Barry S. .... (617) 253-6600
bsjohnst@mit.edu
Mohr, C. Michael .... (617) 253-2015
cmmohr@mit.edu

Accredited by: ABET

Degrees granted 2003-2004:
B.S.: 41 M.S.: 36 Ph.D.: 28

Graduate advisor: Daniel Blankschtein

Undergraduate advisor: C. Michael Mohr

Student organization: AIChE
Advisor: C. Michael Mohr

Department reports to: Thomas L. Magnanti, Dean of Engineering, (617) 253-6604

Placement service: Elizabeth Reed (617) 253-4733, Director, Career Planning and Placement

Northeastern University

Chemical Engineering Department - 342SN
360 Huntington Ave.
Boston, MA 02115
Deliveries: 342 Snell Engineering Center

Department phone .... (617) 373-2989
Department FAX .... (617) 373-2209

Professors

Sacco, Albert Jr. .... (617) 373-7912
asacco@coe.neu.edu
Willey, Ronald J. .... (617) 373-3962
r.willey@neu.eud
Williams, John A. (Emeritus)
Wise, Donald L. (Emeritus)

Associate Professors

Barabino, Gilda A. .... (617) 373-3900
g.barabino@neu.edu
Buonopane, Ralph A. (Emeritus) ...... (617) 373-2996
r.buonopane@neu.edu
Goodwin, Bernard M. (Emeritus) ..... (617) 373-2990
b.goodwin@neu.edu
Stewart, Richard R. (Emeritus) .... (617) 373-2990
Thorgerson, Eric (Department Officer) . (617) 373-2991
e.thorgerson@neu.edu

Assistant Professors

Burkey, Daniel .... (617) 373-2989
Carrier, Rebecca .... (617) 373-2989
Lee, Carolyn W. T. .... (617) 373-3634
clee@coe.neu.edu
Zeimer, Katherine S. .... (617) 373-2990
k.zeimer@neu.edu

Instructors

Satvat, Behrooz (Joint) .... (617) 373-3461
b.satvat@neu.edu

Other Faculty

Gutoff, Edgar B. (Adjunct) .... (617) 373-2989
ebgutof@coe.neu.edu
Manning, Michael (Adjunct) .... (617) 373-2991
mmanning@coe.neu.edu

Accredited by: ABET

Degrees granted 2003-2004:

B.S.: 34 M.S.: 5 Ph.D.: 2

Graduate advisor: Albert Sacco, Jr.

Undergraduate advisor: M. Bates

Student organization: AIChE
Advisor: K. S. Ziemer

Department reports to: Allen Soyster (617) 373-2685,
Dean of Engineering

Placement service: Behrooz Satvat (617) 373-3461

Tufts University

Chemical & Biological Engineering Department
4 Colby Street
Medford, MA 02155
Deliveries: Room 148

University phone .... (617) 628-5000
Department phone .... (617) 627-3900
Department FAX .... (617) 627-3991

Professors

Botsaris, Gregory B. (Emeritus) .... (617) 627-3445
gregory.botsaris@tufts.edu
Flytzani-Stephanopoulos, Maria .... (617) 627-3048
mflytzan@tufts.edu
Georgakis,Christos (Chair) .... (617) 627-2573
Christos.Georgakis@tufts.edu
Juda, Walter (Adjunct) .... (617) 627-2786
walterjuda@aol.com
Kaplan, David .... (617) 627-3251
david.kaplan@tufts.edu
Sung, Nak-Ho .... (617) 627-3447
nak.sung@tufts.edu
VanWormer, Kenneth A. .... (617) 627-2598
kenneth.vanwormer@tufts.edu

Associate Professors

Kelley, Brian (Adjunct) .... (617) 627-3900
bkelley@wyeth.com
Meldon, Jerry .... (617) 627-3570
jerry.meldon@tufts.edu
Ryder, Daniel .... (617) 627-3446
daniel.ryder@tufts.edu

Assistant Professors

Lee, Kyongbum .... (617) 627-4323
kyongbum.lee@tufts.edu
Pfeifer, Blaine 617-627-3900
blaine.pfeifer@tufts.edu

Instructors

Gyure, Dale C. (Adjunct) .... (617) 627-3900
105365.2054@compuserve.com

Other Faculty

Edwards, Aurelie .... (617) 627-3731
aurelie.edwards@tufts.edu
Saltsburg, Howard .... (617) 627-5658
hsalts01@tufts.edu
Valluzzi, Regina .... (617) 627-2580
regina.valluzzi@tufts.edu
Volloch, Vladimir .... (617) 627-3186
vladimir.volloch@tufts.edu
Vunjak-Novakovic, Gordana (Adjunct) . (617) 627-3900
gordana@MIT.edu
Winkler, Stefan (Adjunct) .... (617) 627-3900
stefan.winkler@alumni.tufts.edu

Accredited by: ABET

Degrees granted 2003-2004:
B.S.: 33 M.S.: 9 Ph.D.: 1

Graduate advisor: Prof. Nak Ho Sung

Student organization: AIChE
Advisor: Kenneth Van Wormer

Department reports to: Linda Abriola (617) 627-3237, Dean of Engineering

Placement service: Edward Benford (617) 627-3299

Worcester Polytechnic Institute

Chemical Engineering Department
100 Institute Road
Worcester, MA 01609-2280
Deliveries: Goddard Hall, Room 127

University phone .... (508) 831-5000
Department phone .... (508) 831-5250
Department FAX .... (508) 831-5853

Professors

Datta, Ravindra (Head) .... (508) 831-6036
rdatta@wpi.edu
Dixon, Anthony G. .... (508) 831-5350
agdixon@wpi.edu
Ma, Yi Hua .... (508) 831-5398
yhma@wpi.edu
Moser, William (Emeritus) .... (508) 831-5250
Thompson, Robert W. .... (508) 831-5525
rwt@wpi.edu
Wagner, Robert (Emeritus) .... (508) 831-5250
Weiss, Alvin (Emeritus) .... (508) 831-5250

Associate Professors

Clark, William M. .... (508) 831-5259
wmclark@wpi.edu
DiBiasio, David .... (508) 831-5372
dibiasio@wpi.edu

Assistant Professors

Camesano, Terri A. .... (508) 831-5380
terric@wpi.edu
Kazantzis, Nikolaos K. .... (508) 831-5666
nikolas@wpi.edu
Wilcox, Jennifer L. .... (508) 831-5493
jwilcox@wpi.edu

Accredited by: ABET

Degrees granted 2003-2004:

M.S.: 10 Ph.D.: 29

Student organization: AIChE
Advisor: Professor William Clark

Department reports to: John F. Carney, Provost

Placement service: Yvonne V. Harrison (508) 831-5260, Director

Michigan

University of Michigan

The Herbert H. Dow Building
2300 Hayward St., Room 3074
Ann Arbor, MI 48109-2136

University phone .... (734) 764-1817
Department phone .... (734) 764-2383
Department FAX .... (734) 763-0459

Professors

Briggs, Dale E. (Emeritus) .... (734) 763-1331
briggs@umich.edu
Burns, Mark A. .... (734) 764-4315
mab@umich.edu
Carnahan, Brice (Emeritus) .... (734) 764-3366
carnahan@umich.edu
Curl, Rane (Emeritus) .... (734) 764-3489
ranecurl@umich.edu
Donahue, Francis (Emeritus) .... (734) 764-4304
fdonahue@umich.edu
Fogler, H. Scott .... (734) 763-1361
sfogler@umich.edu
Gulari, Erdogan .... (734) 763-5941
gulari@umich.edu
Kadlec, Robert H. (Emeritus) .... (734) 764-3362
rhkad@umich.edu
Larson, Ronald G. (Chair) .... (734) 936-0772
rlarson@umich.edu
Linderman, Jennifer J. .... (734) 763-5253
linderma@umich.edu
Powers, John E. (Emeritus) .... (801) 259-5060
Savage, Phillip E. .... (734) 764-3386
psavage@umich.edu
Schwank, Johannes W. .... (734) 764-3374
schwank@umich.edu
Tek, M. Rasin (Emeritus) .... (808) 325-7618
Thompson, Levi T. .... (734) 936-2015
ltt@umich.edu
Wang, Henry Y. .... (734) 763-5659
hywang@umich.edu
Weber, Walter J. Jr. .... (734) 763-1464
wjwjr@umich.edu
Wilkes, James O. (Emeritus) .... (734) 764-3378
wilkes@umich.edu
Yang, Ralph T. .... (734) 936-0771
yang@umich.edu
Young, Edwin H. (Emeritus) .... (734) 764-4316
young@umich.edu
Ziff, Robert M. .... (734) 764-5498
rziff@umich.edu

Associate Professors

Glotzer, Sharon C. .... (734) 615-6296
sglotzer@umich.edu
Kotov, Nicholas .... (734) 763-8768
kotov@umich.edu
Solomon, Michael J. .... (734) 764-3119
mjsolo@umich.edu

Assistant Professors

Kim, Jingsang (Joint) .... (734) 936-4681
jingsang@umich.edu
Lahann, Joerg .... (734) 763-7543
lahann@umich.edu
Linic, Suljo .... (734) 764-2383
Mayer, Michael (Joint) .... (734) 763-4609
mimayer@umich.edu
Woolf, Peter .... (734) 764-2383

Instructors

Montgomery, Susan M. .... (734) 936-1890
smontgom@umich.edu

Accredited by ABET

Degrees granted 2003-2004:

B.S.: 124 M.S.: 15 Ph.D.: 3

Undergraduate advisor: Susan Montgomery

Student organization: AIChE
Advisor: Sharon Glotzer

Department reports to: Stephen W. Director, Dean College of Engineering

Placement service: Cynthia Redwine, Director (734) 647-7168, 230 Chrysler Center, 2121 Bonisteel Blvd. N. Campus

Michigan State University

Chemical Engineering & Materials Science Department
East Lansing, MI 48824-1226
Deliveries: 2527 Engineering Building

University phone . . . . (517) 355-1855
Department phone . . . . (517) 355-5135
Department FAX . . . . (517) 432-1105

Professors

Berglund, Kris A. . . . . (517) 353-4565
berglund@cem.msu.edu
Case, Eldon . . . . (517) 353-6715
casee@egr.msu.edu
Dale, Bruce E. . . . . (517) 353-6777
bdale@egr.msu.edu
Drzal, Lawrence T. (Director) . . . . (517) 353-7759
drzal@egr.msu.edu
Grummon, David . . . . (517) 353-4688
grummon@egr.msu.edu
Hawley, Martin C. (Chair) . . . . (517) 355-5135
hawley@egr.msu.edu
Jayaraman, Krishnamurthy . . . . (517) 355-5138
jayarama@egr.msu.edu
Mackay, Michael . . . . (517) 432-4495
mackay@egr.msu.edu
Miller, Dennis J. . . . . (517) 353-3928
millerd@egr.msu.edu
Narayan, Ramani . . . . (517) 432-0775
narayan@msu.edu
Petty, Charles A. . . . . (517) 353-5486
petty@egr.msu.edu
Subramanian, K.N. . . . . (517) 353-5397
subraman@egr.msu.edu
Worden, R. Mark . . . . (517) 353-9015
worden@egr.msu.edu

Associate Professors

Baumann, Melissa . . . . (517) 432-1243
mbaumann@egr.msu.edu
Bieler, Thomas . . . . (517) 353-9767
beiler@egr.msu.edu
Briedis, Daina M. . . . . (517) 353-3861
briedis@egr.msu.edu
Chan, Christina . . . . (517) 432-4530
krischan@egr.msu.edu
Crimp, Martin . . . . (517) 355-0294
crimp@egr.msu.edu
Lee, Andre . . . . (517) 355-5112
leea@egr.msu.edu
Lira, Carl T. . . . . (517) 355-9731
lira@egr.msu.edu
Lucas, James . . . . (517) 432-2883
lucas@egr.msu.edu
Ofoli, Robert Y. . . . . (517) 432-1575
ofoli@egr.msu.edu

Assistant Professors

Lee, Ilsoon . . . . (517) 355-5125
Walton, S. Patrick . . . . (517) 432-8733
spwalton@egr.msu.edu

Other Faculty

Bender, Timothy . . . . (517) 355-3444
bendert@egr.msu.edu

Accredited by: ABET

Degrees granted 2003-2004:

B.S.: 52 M.S.: 5 Ph.D.: 4

Undergraduate advisor: Cynthia Sarver

Student organization: AIChE
Advisor: Robert Y. Ofoli

Department reports to: Janie M. Fouke, Dean of Engineering

Placement service: Jim Novak, Director

Michigan Technological University

Department of Chemical Engineering
1400 Townsend Drive
Houghton, MI 49931-1295
Deliveries: Adminstration Building Room G31

University phone .... (906) 487-1885
Department phone .... (906) 487-3132
Department FAX .... (906) 487-3213

Professors

Barna, Bruce A. .... (906) 487-2569
bbarna@mtu.edu

Crowl, Daniel A. .... (906) 487-3221
crowl@mtu.edu

Kawatra, S. Komar .... (906) 487-2064
skkawatr@mtu.edu

Mullins, Michael E. (Chair) .... (906) 487-3132
memullin@mtu.edu

Associate Professors

Caneba, Gerard T. .... (906) 487-2051
caneba@mtu.edu

Co, Tomas B. .... (906) 487-2144
tbco@mtu.edu

Kim, Nam K. .... (906) 487-2298
kimnk@mtu.edu

King, Julia A. .... (906) 487-3106
jaking@mtu.edu

Morrison, Faith A. .... (906) 487-2050
fmorriso@mtu.edu

Nesbitt, Carl .... (906) 487-2796
cnesbitt@mtu.edu

Sandell, John .... (906) 487-2557
jfsandel@mtu.edu

Shonnard, David R. .... (906) 487-3468
drshonna@mtu.edu

Assistant Professors

Holles, Joseph H. .... (906) 487-1956
jhholles@mtu.edu

Keith, Jason M. .... (906) 487-2106
jmkeith@mtu.edu

Instructors

Clancey, M. Sean .... (906) 487-3338
msclance@mtu.edu

Other Faculty

Cornilsen, Bahne C.
bccornil@mtu.edu

Fisher, Edward R. (Emeritus)

Kolkka, Robert W.
rwkolkka@mtu.edu

Pintar, Anton J. (Emeritus)

Williams, Edwin T. (Emeritus)

Accredited by: ABET

Degrees granted 2003-2004:

B.S.: 64 M.S.: 7 Ph.D.: 2

Undergraduate advisor: Faith A. Morrison

Student organization: AIChE
Advisor: Jason M. Keith

Department reports to: Dean Engr. (906) 487-2005

Placement service: Univ. Career Ctr. (906) 487-2313

Wayne State University

Department of Chemical Engineering and Materials Science
Detroit, MI 48202
Deliveries: Room 1100, 5050 Anthony Wayne Dr.

University phone .... (313) 577-2424
Department phone .... (313) 577-3800
Department FAX .... (313) 577-3810

Professors

Gulari, Esin .... (313) 577-3767
egulari@che.eng.wayne.edu
Huang, Yinlun .... (313) 577-3771
yhuang@che.eng.wayne.edu
Kummler, Ralph H. (Dean) .... (313) 577-3861
rkummler@che.eng.wayne.edu
Louvar, Joeseph .... (313) 577-9358
jlouvar@che.eng.wayne.edu
Manke, Charles (Chair) .... (313) 577-3849
cmanke@che.eng.wayne.edu
Ng, K. Y. Simon .... (313) 577-3805
sng@che.eng.wayne.edu
Putatunda, Susil K. .... (313) 577-3808
sputa@che.eng.wayne.edu
Rothe, Erhard W. .... (313) 577-3865
erothe@che.eng.wayne.edu

Associate Professors

Mao, Guangzhao .... (313) 577-3804
gzmao@che.eng.wayne.edu
Matthew, Howard .... (313) 577-5238
hmatthew@che.eng.wayne.edu
McMicking, James H. (Emeritus) ...... (313) 577-3802
jmcmicking@che.eng.wayne.edu
Salley, Steven O. .... (313) 577-3755
ssalley@che.eng.wayne.edu
Shreve, Gina S. .... (313) 577-3874
gshreve@che.eng.wayne.edu
Van Tassel, Paul .... (313) 577-0899
vantasse@che.eng.wayne.edu

Assistant Professors

da Rocha, Sandro .... (313) 577-3800
Kannan, Rangaramanujam .... (313) 577-3879
rkannan@che.eng.wayne.edu
Potoff, Jeffrey J. .... (313) 577-9537
jpotoff@che.eng.wayne.edu

Accredited by: ABET

Graduate advisor: Yinlun Huang

Undergraduate advisor: Andrea Eisenberg

Student organization: AIChE
Advisor: Paul Van Tassel

Placement service: John Crusoe (313) 577-3390

Minnesota

University of Minnesota at Duluth

Chemical Engineering Department
176 Engineering Building
1303 Ordean Court
Duluth MN 55812-3025

University phone .... (218) 726-8000
Department phone .... (218) 726-7126
Department FAX .... (218) 726-6907

Professors

Davis, Richard .... (218) 726-6162
rdavis@d.umn.edu
Hasan, A. Rashid (Head) .... (218) 726-7127
rhasan@d.umn.edu

Associate Professors

Lodge, Keith B. .... (218) 726-6164
klodge@d.umn.edu
Sternberg, Steven P.K. .... (218) 726-6165
ssternbe@d.umn.edu

Assistant Professors

Rother, Michael .... (218) 726-6154
mrother@d.umn.edu
Rutkowski, Gregory .... (218) 726-7828
gear1016@yahoo.com

Instructors

DeWitte, Renee .... (218) 726-6578
rdewitt1@d.umn.edu
Khoroosi, Hossain .... (218) 726-6320
hkhoroos@d.umn.edu

Accredited by: ABET

Degrees granted 2003-2004:

B.S.: 26

Graduate advisor: A. Rashid Hasan

Undergraduate advisor: A. Rashid Hasan, Richard Davis, Keith Lodge, Michael Rother, Gregory Rutkowski, Steven Sternberg, Renee DeWitte

Student organization: AIChE
Advisor: Steven Sternberg

Department reports to: James Riehl (218) 726-6397, Dean, College of Science and Engineering

Placement service: Julie Westlund (218) 726-7985

University of Minnesota at Minneapolis St. Paul

Department of Chemical Engineering and Materials Science
151 Amundson Hall
421 Washington Avenue SE
Minneapolis, MN 55455

University phone .... (612) 625-5000
Department phone .... (612) 625-1313
Department FAX .... (612) 626-7246

Professors

Aris, Rutherford (Emeritus) .... (612) 625-1313
Bates, Frank S. (Head) .... (612) 625-4356
bates@cems.umn.edu
Caretta, Raul .... (612) 625-8066
caretta@umn.edu
Carr, Robert W. Jr. (Emeritus) .... (612) 625-2551
carrx002@umn.edu
Carter, C. Barry .... (612) 625-8805
carter@cems.umn.edu
Chelikowsky, James R. .... (612) 625-4837
jrc@msi.umn.edu
Cussler, Edward L. .... (612) 625-1596
cussler@cems.umn.edu
Dahler, John S. (Emeritus) .... (612) 625-1884
dahler@chemsun.chem.umn.edu
Daoutidis, Prodromos .... (612) 625-8818
daoutidi@cems.umn.edu
Davis, H. Ted (Dean) .... (612) 624-2006
davis@itdean.umn.edu
Derby, Jeffrey J. .... (612) 625-8881
derby@umn.edu
Francis, Lorraine F. .... (612) 625-0559
lfrancis@umn.edu
Fredrickson, Arnold G. (Emeritus) ..... (612) 625-6588
fredr001@umn.edu
Geankoplis Christie J. .... (612) 625-1586
geank001@umn.edu
Gerberich, William W. .... (612) 625-8548
wgerb@umn.edu
Hu, Wei-Shou .... (612) 625-0546
wshu@cems.umn.edu
Keller, Kenneth H. .... (612) 626-9547
khkeller@umn.edu
Lodge, Timothy P. .... (612) 625-0877
lodge@umn.edu
Macosko, Christopher W. .... (612) 625-0092
macosko@umn.edu
McCormick, Alon V. .... (612) 625-1822
mccormic@cems.umn.edu
Oriani, Richard A. (Emeritus) .... (612) 625-5862
orian001@umn.edu
Palmstrom, Christopher J. .... (612) 625-7558
palms001@umn.edu
Schmidt, Lanny D. .... (612) 625-9391
schmi001@umn.edu
Scriven, L. E. .... (612) 625-1058
pjensen@cems.umn.edu
Seidel, Robert W. .... (612) 624-8003
rws@umn.edu
Shores, David A. .... (612) 625-0014
dshores@umn.edu
Smyrl, William H. .... (612) 625-0717
smyrl001@umn.edu
Srienc, Friedrich .... (612) 624-9776
fried@biosci.cbs.umn.edu
Tranquillo, Robert T. .... (612) 625-6868
tranquillo@cems.umn.edu
Tsapatsis, Michael .... (612) 626-0920
tsapatsis@cems.umn.edu
Ward, Michael D. .... (612) 625-3062
wardx004@umn.edu

Associate Professors

Frisbie, C. Daniel .... (612) 625-0779
frisbie@cems.umn.edu
Morse, David C. .... (612) 625-0167
morse@cems.umn.edu
Norris, David .... (612) 625-2043
dnorris@cems.umn.edu
Sivertsen, John M. (Emeritus) .... (612) 625-1313
Wentzcovitch, Renata .... (612) 625-6345
wentzcov@cems.umn.edu

Assistant Professors

Kaznessis, Yiannis .... (612) 624-4197
yiannis@cems.umn.edu
Kokkoli, Efrosini .... (612) 626-1185
kokkoli@cems.umn.edu
Kumar, Satish .... (612) 625-2558
kumar@cems.umn.edu
Leighton, Christopher .... (612) 624-4018
leighton@umn.edu
Maynard, Jennifer .... (612) 626-0635
jmaynard@umn.edu
McClurg, Richard B. .... (612) 624-9056
mcclurg@cems.umn.edu

Accredited by ABET

Degrees granted 2003-2004:

B.S.: 78 M.S.: 12 Ph.D.: 25

Graduate advisor: David Norris

Undergraduate advisor: Alon McCormick

Student organization: AIChE
Advisor: David Norris

Department reports to: H. Ted Davis, Dean, Institute of Technology

Placement service: Sharon Kurtt

Mississippi

University of Mississippi

Department of Chemical Engineering
134 Anderson Hall
University, MS 38677-1840

University phone .... (662) 915-7211
Department phone .... (662) 915-7023
Department FAX .... (662) 915-7023

Professors

Anderson, Frank (Emeritus) .... (662) 915-7023
Aven, Russell E. (Emeritus) .... (662) 915-7023
Chen, Wei-Yin .... (662) 915-5651
cmchengs@olemiss.edu
Sadana, Ajit .... (662) 915-5349
cmsadana@olemiss.edu
Stasaik, Raymond (Adjunct) .... (662) 915-7023
Sukanek, Peter C. (Chair) .... (662) 915-7023
cmpcs@olemiss.edu

Associate Professors

O'Haver, John Howard .... (662) 915-5347
johaver@olemiss.edu
Williford, Clint W. Jr. .... (662) 915-5348
drwill@olemiss.edu

Assistant Professors

Scovazzo, Paul .... (662) 915-5354
scovazzo@olemiss.edu

Accredited by: ABET

Degrees granted 2003-2004:

B.S.: 7 M.S.: 4

Undergraduate advisor: Peter C. Sukanek

Student organization: AIChE
Advisor: Clint W. Williford, Jr.

Department reports to: Dean of Engineering

Placement service: Davud Gaston (662) 915-7174, Director of Career Center

Mississippi State University

Dave C. Swalm School of Chemical Engineering
Box 9595
Mississippi State, MS 39762-9595
Deliveries: 330 Swalm - President's Circle

University phone .... (662) 325-2131
Department phone .... (662) 325-2480
Department FAX .... (662) 325-2482

Professors

Cornell, David (Emeritus)
George, Clifford E. .... (662) 325-7205
george@che.msstate.edu
Hall, William B. (Emeritus)
Hill, Donald O. (Emeritus)
Koelling, Harold A. (Emeritus)
Kuo, Chiang-Hai (Emeritus)
Rogers, Rudy E. .... (662) 325-5106
rogers@che.msstate.edu
Schulz, Kirk H. (Director) .... (662) 325-2480
schulz@che.msstate.edu
Sparrow, Charles A. (Emeritus) .... (662) 325-7323
sparrow@che.msstate.edu

Associate Professors

Bricka, R. Mark .... (662) 325-1615
bricka@che.msstate.edu

Assistant Professors

Hernandez, Rafael .... (662) 325-0790
rhernandez@che.msstate.edu
Hill, Priscilla J. .... (662) 325-8249
phill@che.msstate.edu

Other Faculty

French, W. Todd .... (662) 325-4308
tfrench@che.msstate.edu

Accredited by: ABET

Degrees granted 2003-2004:

B.S.: 61 M.S.: 9

Undergraduate advisor: Irvin A. Jefcoat

Student organization: AIChE
Advisor: Irvin A. Jefcoat

Department reports to: Dean of the James Worth Bagley College of Engineering

Placement service: Gloria H. Reeves, Director of Placement, P. O. Drawer P

Missouri

University of Missouri-Columbia

Chemical Engineering Department
Columbia, MO 65211-2200
Deliveries: W2033 Engr. Bldg. East
Department phone .... (573) 882-3563
Department FAX .... (573) 884-4940

Professors

Bajpai, Rakesh K. .... (573) 882-3708
bajpair@missouri.edu
Lee, Sunggyu (Chair) .... (573) 882-4281
leesu@missouri.edu
Loyalka, Sudarshan .... (573) 882-3568
LoyalkaS@missouri.edu
Luebbers, Ralph H. (Emeritus) .... (573) 882-3563
Luecke, Richard H. (Emeritus) .... (573) 882-3691
luecker@missouri.edu
Marc de Chazal, L. E. (Emeritus) ..... (573) 882-3563
Marrero, Thomas R. .... (573) 882-3802
marrerot@missouri.edu
Storvick, Truman S. (Emeritus) .... (573) 882-3215
storvickt@missouri.edu
Viswanath, Dabir S. (Emeritus) .... (573) 884-0707
viswanthd@missouri.edu
Yasuda, H. K. (Emeritus) .... (573) 882-9602
yasudah@missouri.edu

Associate Professors

Chan, Paul C. H. .... (573) 882-7684
chanp@missouri.edu
Jacoby, William A. .... (573) 882-3563
jacobyw@missouri.edu
Krentsel, Evguenii B. .... (573) 882-1520
KrentselE@missouri.edu
Lombardo, Stephen J. .... (573) 882-3563
lombardos@missouri.edu
Retzloff, David G. (Associate Chair) ... (573) 882-4036
retzloffd@missouri.edu
Suppes, Galen J. .... (573) 884-0562
suppesg@missouri.edu

Assistant Professors

Darcy, Patricia A. .... (573) 882-8244
darcyp@missouri.edu
Doskocil, Eric .... (573) 884-2273
doskocile@missouri.edu
Yu, Qingsong .... (573) 882-8076
yuq@missouri.edu

Accredited by: ABET

Degrees granted 2003-2004:

B.S.: 36 M.S.: 7 Ph.D.: 1

Graduate advisor: David G. Retzloff

Undergraduate advisor: Paul C.H. Chan

Student organization: AIChE
Advisor: Paul C.H. Chan

Department reports to: James Thompson, Dean

Placement service: Matt Riske, Director

University of Missouri-Rolla

Chemical & Biological Engineering Department
Rolla, MO 65409-1230
Deliveries: 143 Schrenk Hall, 11th and State Str.
Department phone .... (573) 341-4416
Department FAX .... (573) 341-4377

Professors

Adams, Craig (Adjunct) .... (573) 341-4041
adams@umr.edu
Azbel, David S. (Emeritus)
Book, Neil L. .... (573) 341-4422
nbook@umr.edu
Crosser, Orrin K. (Emeritus) .... (573) 341-4459
okc@umr.edu
Liapis, Athanasios I. .... (573) 341-4414
ail@umr.edu
Ludlow, Douglas K. .... (573) 341-6477
dludlow@umr.edu
Manley, David B. (Emeritus) .... (573) 341-4428
dbm@umr.edu
Mollenkamp, Robert A. (Adjunct) ..... (573) 341-6709
robertm@umr.edu
Morosoff, Nicholas (Emeritus) .... (573) 341-4903
ncmoros@umr.edu
Neogi, Partho .... (573) 341-4460
neogi@umr.edu
Patterson, Gary K. (Emeritus) .... (573) 341-6941
garyp@umr.edu
Raper, Judy A. (Chair) .... (573) 341-7518
raperj@umr.edu
Reed, X B Jr. (Emeritus) .... (573) 341-4423
xbreed@umr.edu
Rosen, Stephen L. (Emeritus) .... (573) 341-4443
slr@umr.edu
Strunk, Mailand R. (Emeritus)
Waggoner, Raymond C. (Emeritus) .... (573) 341-4425
wagglr@umr.edu

Associate Professors

Forciniti, Daniel .... (573) 341-4427
forcinit@umr.edu
Sitton, Oliver C. .... (573) 341-4426
ocs@umr.edu

Assistant Professors

Henthorn, David .... (573) 341-4415
henthord@umr.edu
Henthorn, Kimberly .... (573) 341-4415
henthork@umr.edu
Wang, Jee-Ching .... (573) 341-6705
jcwang@umr.edu
Xing, Yangchuan .... (573) 341-6772
xingy@umr.edu

Accredited by: ABET

Degrees granted 2003-2004:

B.S.: 34 M.S.: 10

Graduate advisor: Judy A. Raper

Undergraduate advisor: Douglas K. Ludlow

Student organization: AIChE
Advisor: Robert Mollenkamp

Department reports to: Robert Mitchell, Dean, School of Engineering

Placement service: Lea Ann Morton, Director

Washington University

Chemical Engineering Department
Campus Box 1198
St. Louis, MO 63130-4899
Deliveries: One Brookings Drive, Urbauer Hall 208

University phone .... (314) 935-5000
Department phone .... (314) 935-6082
Department FAX .... (314) 935-7211

Professors

Biswas, Pratim .... (314) 935-5482
pratim.biswas@seas.wustl.edu
Dudukovic, Milorad (Chair) .... (314) 935-6021
dudu@poly1.che.wustl.edu
Kardos, John L. .... (314) 935-6062
kardos@che.wustl.edu
Khomami, Bamin .... (314) 935-6065
bam@poly1.che.wustl.edu
Ramachandran, P. A. .... (314) 935-6531
rama@che.wustl.edu

Associate Professors

Al-Dahhan, Muthanna .... (314) 935-7187
muthanna@che.wustl.edu
Gleaves, John T. .... (314) 935-4159
klatu_00@poly1.che.wustl.edu
Sureshkumar, R. .... (314) 935-4988
suresh@poly1.che.wustl.edu
Turner, Jay R. .... (314) 935-5480
jrturner@seas.wustl.edu

Assistant Professors

Angenent, Lars .... (314) 935-5663
angenent@seas.wustl.edu

Other Faculty

McKelvey, James M. .... (314) 935-4836
mckelvey@poly1.che.wustl.edu
Motard, Rodolphe L. .... (314) 935-6072
motard@che.wustl.edu
Thies, Curt (Emeritus) .... (314) 935-6082
thiesman@aol.com
Wolf, Clarence J. .... (314) 935-6052
cjw@mecf.wustl.edu
Yablonsky, Gregory .... (314) 935-4367
gy@che.wustl.edu

Accredited by: ABET

Degrees granted 2003-2004:

B.S.: 22 M.S.: 10 Ph.D.: 1

Student organization: AIChE
Advisor: Dr. Lars Angenent

Department reports to: Christopher I. Byrnes, Dean, School of Engineering and Applied Science

Placement service: Amanda Matheu (314) 935-4459

Montana

Montana State University

Chemical and Biological Engineering Department
Montana State University
P.O. Box 173920
306 Cobleigh Hall
Bozeman, MT 59717-3920
Deliveries: 306 Cobleigh Hall

University phone .... (406) 994-0211
Department phone .... (406) 994-2221
Department FAX .... (406) 994-5308

Professors

Cokelet, Giles (Visiting) .... (406) 994-7048
giles_c@coe.montana.edu
Larsen, Ronald W. (Head) .... (406) 994-2221
ronl@coe.montana.edu
Mandell, John F. .... (406) 994-4543
johnm@coe.montana.edu
McCandless, Frank P. (Emeritus) ...... (406) 994-5928
philm@coe.montana.edu
Nickelson, Robert L. (Emeritus)
Scarrah, Warren P. (Emeritus)
Stewart, Phillip .... (406) 994-2890
phil_s@erc.montana .edu

Associate Professors

Deibert, Max C. .... (406) 994-5990
maxd@coe.montana.edu
Shaffer, Daniel L. .... (406) 994-5922
dans@coe.montana.edu

Assistant Professors

Codd, Sarah (Research) .... (406) 994-1944
scodd@coe.montana.edu
Duffy, James E. .... (406) 994-5926
james_d@coe.montana.edu
King, Susan A. (Research) .... (406) 994-7474
susank@coe.montana.edu
Seymour, Joseph .... (406) 994-6853
jseymour@coe.montana.edu

Accredited by: ABET

Degrees granted 2003-2004:

B.S.: 30 M.S.: 4

Graduate advisor: Ronald W. Larsen

Undergraduate advisor: Ronald W. Larsen

Student organization: AIChE
Advisor: Jim Duffy

Department reports to: Robert Marley, Dean of Engineering

Placement service: Ron Larsen, Shelley I. Thomas

Nebraska

University of Nebraska

Department of Chemical Engineering
Lincoln, NE 68588-0643
Deliveries: Othmer 207A

Department phone .... (402) 472-2750
Department FAX .... (402) 472-6989

Professors

Clements, L. Davis (Emeritus) .... (402) 472-2750
Eakman, James M. (Emeritus) .... (402) 472-2752
jeakman1@unl.edu
Gilbert, Richard E. (Emeritus) .... (402) 472-2750
Hendrix, James L. .... (402) 472-0697
jhendrix1@unl.edu
Meagher, Michael M. (Director) .... (402) 472-2342
mmeagher1@unl.edu
Tao, Luh C. (Emeritus) .... (402) 472-2750
Timm, Delmar C. (Chair) .... (402) 472-3232
dtimm1@unl.edu
Viljoen, Hendrik J. .... (402) 472-9318
hviljoen1@unl.edu
Weber, James H. (Emeritus) .... (402) 472-2750

Associate Professors

Brand, Jennifer I. .... (402) 472-9320
jbrand1@unl.edu
Larsen, Gustavo .... (402) 472-9805
glarsen1@unl.edu
Lauderback, Lee L. .... (402) 472-5725
llauderback1@unl.edu

Assistant Professors

Noureddini, Hossein .... (402) 472-2751
hnouredd@unl.edu
Subramanian, Anu .... (402) 472-3463
ASUBRAMANIAN2@unlnotes.unl.edu

Instructors

Nelson, Donald (Adjunct) .... (402) 472-6029
dnelson3@unl.edu

Accredited by: ABET

Graduate advisor: G. Larsen

Undergraduate advisor: J. L. Hendrix

Student organization: AIChE
Advisor: Hossein Noureddini

Department reports to: David Allen, Dean, College of Engineering, and Technology

Placement service: Larry Routh, Nebraska Union 230

Nevada

University of Nevada, Reno

Chemical Engineering Department
Mailstop 170
Reno, NV 89557-0136
Deliveries: Room 310, LME Building

University phone .... (775) 784-1110
Department phone .... (775) 784-4307
Department FAX .... (775) 784-4764

Professors

Baglin, Frank G. (Adjunct) .... (775) 784-6651
baglin@chem.unr.edu

Farmer, Richard (Adjunct) .... (775) 784-4307
rfarmer@unr.edu

Roche, Edward C. (Adjunct) .... (775) 784-4307
roche@unr.edu

Whiting, Wallace B. (Chair) .... (775) 784-4307
wwhiting@unr.edu

Associate Professors

Coronella, Charles J. .... (775) 784-4253
coronell@unr.edu

Fuchs, Alan .... (775) 327-2227
afuchs@unr.edu

Zabaloy, Marcelo (Adjunct) .... (775) 784-4307
zabaloy@unr.edu

Assistant Professors

Gecol, Hatice .... (775) 784-1116
gecol@unr.edu

Vasquez, Victor R. .... (775) 784-6060
vvasquez@unr.edu

Accredited by: ABET

Degrees granted 2003-2004:

B.S.: 12 M.S.: 5 Ph.D.: 1

Student organization: AIChE
Advisor: C. J. Coronella

Department reports to: Ted Batchman, Dean (775) 784-6925

Placement service: David A. Hansen (775) 784-4898

New Hampshire

Dartmouth College

Biotechnology and Biochemical
Engineering Program
Thayer School of Engineering
8000 Cummings Hall
Hanover, NH 03755-8000

University phone . . . . (603) 646-1110
Department phone . . . . (603) 646-2230
Department FAX . . . . (603) 646-2277

Professors

Baker, Ian . . . . (603) 646-2184
Ian.Baker@Dartmouth.EDU
Collier, John . . . . (603) 646-2355
John.P.Collier@Dartmouth.EDU
Converse, Alvin O. (Emeritus) . . . . (603) 646-3060
Alvin.O.Converse@Dartmouth.EDU
Cushman-Roisin, Benoit . . . . (603) 646-3248
Benoit.R.Roisin@Dartmouth.EDU
Kennedy, Francis . . . . (603) 646-2094
Francis.E.Kennedy@Dartmouth.EDU
Lynd, Lee R. . . . . (603) 646-2231
Lee.Lynd@Dartmouth.EDU
Petrenko, Victor . . . . (603) 646-3526
Victor.F.Petrenko@Dartmouth.EDU
Queneau, Paul (Emeritus) . . . . (603) 646-2208
Richter, Horst . . . . (603) 646-2701
Horst.J.Richter@Dartmouth.EDU
Schulson, Erland . . . . (603) 646-2888
Erland.M.Schulson@Dartmouth.EDU
Sonnerup, Bengt (Emeritus) . . . . (603) 646-2883
Wallis, Graham B. (Emeritus) . . . . (603) 646-2789
Wyman, Charles E. . . . . (603) 646-3193
Charles.E.Wyman@Dartmouth.EDU

Associate Professors

Frost, Harold . . . . (603) 646-3444
Harold.J.Frost@Dartmouth.EDU
Gerngross, Tillman . . . . (603) 646-3161
Tillman.Gerngross@Dartmouth.EDU
Gibson, Urusla . . . . (603) 646-3243
Ursula.J.Gibson@Dartmouth.EDU

Assistant Professors

Proehl, Jeffrey A. (603) 646-3803
Jeffrey.A.Proehl@Dartmouth.EDU

Accredited by: ABET

Department reports to: Lewis Duncan (603) 646-2238, Dean of Engineering

Placement service: Kristin Roth (603) 646-2375, Director of Career Services

University of New Hampshire

Chemical Engineering Department
Kingsbury Hall, Rm 258
33 College Road
Durham, NH 03824

University phone . . . . (603) 862-1234
Department phone . . . . (603) 862-3654
Department FAX . . . . (603) 862-3747

Professors

Barkey, Dale P. . . . . (603) 862-1918
dpb@cisunix.unh.edu
Carr, Russell T. . . . . (603) 862-1429
rtc@cisunix.unh.edu
Fan, Stephen (Chair) . . . . (603) 862-3656
sstf@cisunix.unh.edu
Farag, Ihab H. . . . . (603) 862-2313
Ihab.Farag@unh.edu
Mathur, Virendra K. . . . . (603) 862-1917
vkm@cisunix.unh.edu
Ulrich, Gael D. (Emeritus)
gdu@cisunix.unh.edu
Vasudevan, Palligarnai T. . . . . (603) 862-2298
vasu@unh.edu

Assistant Professors

Gupta, Nivedita . . . . (603) 862-3655
N.Gupta@unh.edu

Accredited by: ABET

Degrees granted 2003-2004:

B.S.: 14 M.S.: 1 Ph.D.: 1

Graduate advisor: Stephen S.T. Fan

Undergraduate advisor: Stephen S.T. Fan

Student organization: AIChE
Advisor: P. T. Vasudevan

Department reports to: Arthur Greenberg, Dean, College of Engineering and Physical Sciences

Placement service: Bethany Cooper, Univ. Advis./Career Ctr, Hood House 102

New Jersey

New Jersey Institute of Technology

Otto H. York Department of Chemical Engineering
University Heights
Newark, NJ 07102
Deliveries: 150 Tiernan Hall, 161 Warren St.

University phone .... (973) 596-3000
Department phone .... (973) 596-3568
Department FAX .... (973) 596-8436

Professors

Armenante, Piero .... (973) 596-3548
armenante@adm.njit.edu
Baltzis, Basil (Chair) .... (973) 596-3619
baltzis@adm.njit.edu
Barat, Robert .... (973) 596-5605
barat@adm.njit.edu
Greenstein, Teddy .... (973) 596-5706
greenstein@adm.njit.edu
Hanesian, Deran .... (973) 596-3597
hanesian@adm.njit.edu
Huang, Ching-Rong (Emeritus)
Kimmel, Howard .... (973) 596-3574
kimmel@adm.njit.edu
Knox, Dana .... (973) 596-3599
knoxd@adm.njit.edu
Lewandowski, Gordon .... (973) 596-3573
lewandows@adm.njit.edu
Perna, Angelo .... (973) 596-3616
perna@adm.njit.edu
Pfeffer, Robert .... (973) 642-7496
pfeffer@adm.njit.edu
Sirkar, Kamalesh .... (973) 596-8447
sirkar@adm.njit.edu
Tomkins, Reginald (Associate Chair) .. (973) 596-5656
tomkins@adm.njit.edu
Xanthos, Marino .... (973) 642-4762
xanthos@admin.njit.edu
York, Otto (Honorary)

Associate Professors

Loney, Norman .... (973) 596-6598
loney@adm.njit.edu

Assistant Professors

Bart, Ernest .... (973) 596-2998
bart@adm.njit.edu
Hahn, Dittmar .... (973) 353-5235
dhahn@andromeda.rutgers.edu
Howley, Maureen .... (973) 596-3585
Maureen.a.howley@njit.edu
Huang, Michael .... (973) 596-5613
mhuang@adm.njit.edu
Simon, Laurent .... (973) 596-5263
laurent.simon@njit.edu
Wu, Jing .... (973) 642-7064
wu@adm.njit.edu

Instructors

Kisutcza, Joseph .... (973) 642-7610
kisutcza@adm.njit.edu

Other Faculty

Gogos, Costas .... (973) 642-7365
costas@polymers-ppi.org
Hyun, Kun Sup .... (973) 596-3267
hyun@adm.njit.edu
Todd, David .... (973) 596-5602
david@polymers-ppi.org

Accredited by: ABET

Degrees granted 2003-2004:

B.S.: 20 M.S.: 20 Ph.D.: 3

Graduate advisor: Reginald Tomkins

Undergraduate advisor: Reginald Tomkins

Student organization: AIChE
Advisor: Angelo Perna

Department reports to: T.R. Blake (973) 596-5534, Dean of Engineering

Placement service: Greg Mass, Director of Placement (973) 596-5745

Princeton University

Chemical Engineering Department
Olden Street
Princeton, NJ 08544-5263
Deliveries: Prospect Street

University phone . . . . (609) 258-3000
Department phone . . . . (609) 258-4581
Department FAX . . . . (609) 258-0211

Professors

Aksay, Ilhan A. . . . . (609) 258-4393
iaksay@princeton.edu
Benziger, Jay B. . . . . (609) 258-5416
benziger@princeton.edu
Debenedetti, Pablo G. . . . . (609) 258-5480
pdebene@princeton.edu
Floudas, Christodoulos A. . . . . (609) 258-4595
floudas@titan.princeton.edu
Gillham, John K. (Emeritus) . . . . (609) 258-4581
gillham@princeton.edu
Graessley, William W. (Emeritus) . . . . . (609) 258-4581
Jackson, Roy (Emeritus) . . . . (609) 258-4581
rjackson@always-online.com
Johnson, Ernest F. (Emeritus) . . . . (609) 258-4581
Kevrekidis, Yannis G. . . . . (609) 258-2818
yannis@arnold.princeton.edu
Kostin, Morton D. . . . . (609) 258-4586
kostin@princeton.edu
Panagiotopoulos, Athanassios Z. . . . . . . (609) 258-4591
azp@princeton.edu
Prud'homme, Robert K. . . . . (609) 258-4577
prudhomm@princeton.edu
Register, Richard A. . . . . (609) 258-4691
register@princeton.edu
Russel, William B. . . . . (609) 258-4590
wbrussel@princeton.edu
Saville, Dudley A. . . . . (609) 258-4585
dsaville@princeton.edu
Schowalter, William (Emeritus) . . . . (609) 258-3553
schowalt@princeton.edu
Sundaresan, Sankaran . . . . (609) 258-4583
sundar@princeton.edu
Troian, Sandra M. . . . . (609) 258-4574
stroian@princeton.edu
Vanderlick, T. Kyle (Chair) . . . . (609) 258-4891
vandertk@princeton.edu
Wei, James . . . . (609) 258-5618
jameswei@princeton.edu

Assistant Professors

Carbeck, Jeffrey D. . . . . (609) 258-1331
jcarbeck@princeton.edu
Shvartsman, Stanislav Y. . . . . (609) 258-4694
stas@princeton.edu
Wood, David W. (609) 258-5721
dwood@princeton.edu

Other Faculty

Mills, Robert G. (Emeritus) . . . . (609) 258-4573

Accredited by: ABET

Degrees granted 2003-2004:
B.S.: 9 Ph.D.: 12

Graduate advisor: Athanassios Z. Panagiotopoulos

Undergraduate advisor: Jay B. Benziger

Student organization: AIChE
Advisor: Jay B. Benziger

Department reports to: Maria Klawe, Dean, School of Engineering and Applied Science

Placement service: Minnie H. Reed, Director

Rowan University

Department of Chemical Engineering
201 Mullica Hill Road
Glassboro, NJ 08028-1701
Deliveries: 312 Henry M. Rowan Hall

University phone . . . . (856) 256-4000
Department phone . . . . (856) 256-5310
Department FAX . . . . (856) 256-5242

Professors

Dorland, Dianne (Dean) . . . . (856) 256-5300
dorland@rowan.edu
Hesketh, Robert P. (Chair) . . . . (856) 256-5313
hesketh@rowan.edu
Newell, James A. . . . . (856) 256-5316
newell@rowan.edu
Slater, C. Stewart . . . . (856) 256-5312
slater@rowan.edu

Associate Professors

Dahm, Kevin D. . . . . (856) 256-5318
dahm@rowan.edu
Farrell, Stephanie . . . . (856) 256-5315
farrell@rowan.edu
Gephardt, Zenaida Otero . . . . (856) 256-5314
gephardtzo@rowan.edu
Savelski, Mariano J. . . . . (856) 256-5317
savelski@rowan.edu

Assistant Professors

Lefebvre, Brian G. (856) 256-5338
lefebvre@rowan.edu

Other Faculty

Hoffman, Robert (Adjunct) . . . . (856) 256-5310
Natoli, John (Adjunct) . . . . (856) 256-5310
VanKirk, Jesse (Adjunct) . . . . (856) 256-5310

Accredited by: ABET

Degrees granted 2003-2004:

B.S.: 18 M.S.: 6

Graduate advisor: Mariano Savelski

Undergraduate advisor: Robert P. Hesketh

Student organization: AIChE
Advisor: Kevin D. Dahm

Department reports to: Dianne Dorland (856) 256-5300, Dean of Engineering

Placement service: Melanie Basantis, Director of Engineering Outreach (856) 256-5307

Rutgers, The State University of New Jersey

Department of Chemical and Biochemical Engineering
School of Engineering
98 Brett Road
Piscataway, NJ 08854-8058

University phone . . . . (732) 932-1766
Department phone . . . . (732) 445-2228
Department FAX . . . . (732) 445-2581

Professors

Chiew, Yee C. . . . . (732) 445-0315
ychiew@sol.rutgers.edu
Constantinides, Alkis . . . . (732) 445-3678
constant@sol.rutgers.edu
Couchman, Peter (Emeritus) . . . . (732) 445-4944
Davidson, Burton Z. . . . . (732) 445-2203
burtond@sol.rutgers.edu
Dittman, F. W. (Emeritus) . . . . (732) 445-5612
Hara, Masanori . . . . (732) 445-3817
mhara@rci.rutgers.edu
Klein, Michael T. (Dean) . . . . (732) 445-2214
mtklein@jove.rutgers.edu
Muzzio, Fernando J. . . . . (732) 445-3357
muzzio@sol.rutgers.edu
Pedersen, Henrik (Chair) . . . . (732) 445-2568
hpederse@sol.rutgers.edu
Sauer, John A. (Emeritus) . . . . (732) 445-5612
Scheinbeim, Jerry . . . . (732) 445-3669
jis@rci.rutgers.edu
Wang, Shaw S. . . . . (732) 445-3360
shaww@sol.rutgers.edu

Associate Professors

Buettner, Helen M. . . . . (732) 445-2231
buettner@sol.rutgers.edu
Glasser, Benjamin J. . . . . (732) 445-4243
bglasser@sol.rutgers.edu
Ierapetritou, Marianthi G. . . . . (732) 445-2971
marianth@sol.rutgers.edu
Khinast, Johannes . . . . (732) 445-2970
khinast@sol.rutgers.edu
Moghe, Prabhas . . . . (732) 445-4951
moghe@sol.rutgers.edu

Assistant Professors

Roth, Charles . . . . (732) 445-4109
cmroth@rci.rutgers.edu
Tomassone, M. Silvina . . . . (732) 445-2972
silvina@sol.rutgers.edu

Accredited by: ABET

Degrees granted 2003-2004:

B.S.: 38 M.S.: 14 Ph.D.: 7

Graduate advisor: Yee Chiew

Undergraduate advisor: Marianthi Ierapetritou

Student organization: AIChE
Advisor: Henrik Pedersen

Department reports to: Michael T. Klein, Dean, School of Engineering

Placement service: Ellen Azevedo (732) 445-6127 x19

Stevens Institute of Technology

Department of Chemical, Biomedical and Materials Engineering
Mclean Building
Castle Point on Hudson
Hoboken, NJ 07030

University phone .... (201) 216-5000
Department phone .... (201) 216-5546
Department FAX .... (201) 216-8306

Professors

Besser, Ronald .... (201) 216-5257
rbesser@stevens.edu
Blanks, Robert F. .... (201) 216-5588
rblanks@stevens.edu
DeLancey, George B. .... (201) 216-5060
gdelance@stevens.edu
Du, Henry .... (201) 216-5262
hdu@stevens.edu
Kalyon, Dilhan .... (201) 216-8225
dkalyon@stevens.edu
Kovenklioglu, Suphan .... (201) 216-5519
skoven@stevens.edu
Lee, Woo Y. (Director) .... (201) 216-8307
wlee@stevens.edu
Libera, Matthew R. .... (201) 216-5259
mlibera@stevens.edu
Ritter, Arthur .... (201) 216-8290
aritter@steven.edu
Rothberg, Gerald .... (201) 216-5269
grothber@stevens.edu

Associate Professors

Lawal, Adeniyi .... (201) 216-8241
alawal@stevens.edu

Accredited by: ABET

Degrees granted 2003-2004:

B.S.: 19 M.S.: 10 Ph.D.: 2

Graduate advisor: Suphan Kovenklioglu

Undergraduate advisor: George DeLancey

Student organization: AIChE
Advisor: Robert Blanks

Department reports to: George Korfiatis (201) 216-5348

Placement service: Donna Beyer (201)216-8943

New Mexico

New Mexico Institute of Mining and Technology

Department of Chemical Engineering
801 Leroy Place
Socorro, NM 87801

University phone 800-428-8324
Department phone .... (505) 835-5412
Department FAX .... (505) 835-5210

Professors

Lee, Robert L (Adjunct) .... (505) 835-5408
lee@prrc.nmt.edu
McCoy, John Adjunct) .... (505) 835-5379
mccoy@nmt.edu

Associate Professors

Bretz, Robert E. .... (505) 835-5436
bretz@nmt.edu
Cal, Mark (Adjunct) .... (505) 835-5059
mcal@nmt.edu
Dunston, Doug (Adjunct) .... (505) 835-5200
ddunston@nmt.edu
Jeon, Hyun-Sik (Sam) .... (505) 835-5761
jeon@nmt.edu
Weinkauf, Donald H. (Chair) .... (505) 835-5689
weinkauf@nmt.edu

Assistant Professors

Dong, Junhang .... (505) 835-5293
jhdong@nmt.edu

Instructors

Marshall, John .... (505) 835-5926
jmarshal@nmt.edu

Accredited by ABET

Degrees granted 2003-2004:

B.S.: 14

Undergraduate advisor: D. H. Weinkauf

Student organization: AIChE
Advisor: J. H. Dong

Department reports to: VP of Academic Affairs

Placement service: D. H. Weinkauf

University of New Mexico

Chemical and Nuclear Engineering Department
MSC01 1120
1 University of New Mexico
Albuquerque, NM 87131-0001
Deliveries: 209 Farris Engineering Center

University phone . . . . . (505) 277-0111
Department phone . . . . . (505) 277-5431
Department FAX . . . . . (505) 277-5433

Professors

Brinker, C. Jeffrey . . . . . (505) 272-7627
jbrinker@unm.edu
Cecchi, Joseph L. (Dean) . . . . . (505) 277-5522
cecchi@unm.edu
Datye, Abhaya (Director) . . . . . (505) 277-0477
datye@unm.edu
El-Genk, Mohamed S. (Director) . . . . . (505) 277-5442
mgenk@unm.edu
Fulghum, Julia E. (Chair) . . . . . (505) 277-8670
jfulghum@unm.edu
Kauffman, David (Emeritus) . . . . . (505) 277-5431
kauffman@unm.edu
Lopez, Gabriel P. . . . . . (505) 277-4939
gplopez@unm.edu
Nuttall, H. Eric . . . . . (505) 277-6112
nuttall@unm.edu
Prinja, Anil K. (Associate Chair) . . . . . (505) 277-4600
prinja@unm.edu
Roderick, Norman F. . . . . . (505) 277-2209
roderick@unm.edu
Wilkins, Ebtisam S. . . . . . (505) 277-5906
wilkins@unm.edu

Associate Professors

Anderson, Harold M. . . . . . (505) 277-5661
anderson@unm.edu
Cooper, Gary W. . . . . . (505) 277-2557
garywc@unm.edu
Mead, Richard W. . . . . . (505) 277-3221
rmead@unm.edu
Ward, Timothy L. . . . . . (505) 277-2067
tlward@unm.edu

Assistant Professors

Atanassov, Plamen . . . . . (505) 277-2640
plamen@unm.edu
Han, Sang . . . . . (505) 277-3118
meister@unm.edu
Ueki, Taro . . . . . (505) 277-7964
tueki@unm.edu

Instructors

Busch, Robert D. . . . . . (505) 277-8027
busch@unm.edu

Other Faculty

Kroenke, William . . . . . (505) 277-6824
yonder@unm.edu
Loehman, Ronald E. . . . . . (505) 272-7601
loehman@unm.edu
Phillips, Jonathan . . . . . (505) 277-8026
jphillips@lanl.gov
Saber, Hamed . . . . . (505) 277-0338
hsaber@unm.edu
Tournier, Jean-Michel . . . . . (505) 277-7961
tournier@unm.edu
Weaver, Harry . . . . . (505) 277-8446
hweaver@unm.edu
Yang, Yi . . . . . (505) 272-7634
yiyang@unm.edu

Accredited by: ABET

Degrees granted 2003-2004:

B.S.: 25 M.S.: 18 Ph.D.: 6

Graduate advisor: Jeff Brinker

Undergraduate advisor: Abhaya Datye

Student organization: AIChE, ANS
Advisor: Richard W. Mead

Department reports to: Joseph L. Cecchi, Dean of Engineering

Placement service: Eleanor Sanchez

New Mexico State University

Chemical Engineering Department
Post Office Box 30001, MSC 3805
Las Cruces, NM 88003-8001
Deliveries: Jett Hall, Room 259

University phone . . . . (505) 646-0111
Department phone . . . . (505) 646-1214
Department FAX . . . . (505) 646-7706

Professors

Bhada, Rohinton K. (Ron) (Emeritus) . (505) 646-7697
rbhada@nmsu.edu
Johnson, Charles L. (Head) . . . . (505) 646-8637
cjohnson@nmsu.edu
Long, Richard L. Jr. (Associate Chair) . (505) 646-2503
rilong@nmsu.edu
Munson-McGee, Stuart . . . . (505) 646-6439
stumcgee@nmsu.edu
Patton, John T. (Emeritus) . . . . (505) 646-1516
Roubicek, Rudolf (Emeritus) . . . . (505) 646-4146
Thode, Edward F. (Emeritus) . . . . (505) 646-1214
Wilson, Bruce (Emeritus) . . . . (505) 646-1214

Associate Professors

Andersen, Paul . . . . (505) 646-8153
pka@nmsu.edu
Mitchell, Martha C. . . . . (505) 646-1214
martmitc@nmsu.edu
Rockstraw, David A. . . . . (505) 646-7705
drockstr@nmsu.edu

Assistant Professors

Creed, Joe L. . . . . (505) 646-2913
jcreed@nmsu.edu
Deng, Shuguang 646-4346
sdeng@nmsu.edu

Other Faculty

Del Valle, Francisco . . . . (505) 646-4204
fdelvall@nmsu.edu

Accredited by: ABET

Degrees granted 2003-2004:

B.S.: 14 M.S.: 6 Ph.D.: 2

Graduate advisor: Martha Mitchell

Undergraduate advisor: Paul Andersen

Student organization: AIChE
Advisor: Richard L. Long, Jr. (505) 646-2503

Department reports to: Steve Castillo, (505) 646-2914, Dean of Engineering

Placement service: Steve Salway (505) 646-1631, Box 3509

New York

The City College of The City University of New York

Chemical Engineering Department
Convent Avenue at 140th St.
Room 325
New York, NY 10031

Department phone .... (212) 650-7232
Department FAX .... (212) 650-6660

Professors

Acrivos, Andreas (Emeritus) .... (212) 650-8159
acrivos@scisun.sci.ccny.cuny.edu
Denn, Morton M. (Director) .... (212) 650-7444
denn@levdec.engr.ccny.cuny.edu
Graff, Robert A. (Chair) .... (212) 650-8136
graff@chemail.engr.ccny.cuny.edu
Isaacs, Leslie L. .... (212) 650-7146
chelli@chemail.engr.ccny.cuny.edu
Kolodney, Morris (Emeritus)
List, Harvey L. (Emeritus)
Maldarelli, Charles .... (212) 650-8160
charles@chemail.engr.ccny.cuny.edu
Pfeffer, Robert (Emeritus)
Rinard, Irven H. .... (212) 650-7135
rinard@chemail.engr.ccny.cuny.edu
Rumschitzki, David .... (212) 650-5430
david@ees1s0.engr.ccny.cuny.edu
Shinnar, Reuel .... (212) 650-6679
shinnar@chemail.engr.ccny.cuny.edu
Steiner, Carol .... (212) 650-7230
steiner@chemail.engr.ccny.cuny.edu
Tardos, Gabriel I. .... (212) 650-6665
tardos@chemail.engr.ccny.cuny.edu
Weinstein, Herbert .... (212) 650-8279
hweinst@chemail.engr.ccny.cuny.edu

Associate Professors

Couzis, Alexander .... (212) 650-6701
acouzis@chemail.engr.ccny.cuny.edu

Assistant Professors

Gilchrist, M. Lane .... (212) 650-6664
lane@chemail.engr.ccny.cuny.edu
Lee, Jae W. .... (212) 650-6688
lee@chemail.engr.ccny.cuny.edu

Accredited by: ABET

Graduate advisor: David Rumschitzki

Undergraduate advisor: Herbert Weinstein

Student organization: AIChE
Advisor: Herbert Weinstein

Department reports to: Mohammad Karim, Dean

Placement service: Sophia Demetriou (212) 650-6507

Clarkson University

Department of Chemical Engineering
Box 5705
Potsdam, NY 13699-5705
Deliveries: 8 Clarkson Ave.

University phone .... (315) 268-6400
Department phone .... (315) 268-6650
Department FAX .... (315) 268-6654

Professors

Babu, S. V. .... (315) 268-3999
babu@clarkson.edu
Baltus, Ruth E. .... (315) 268-2368
baltus@clarkson.edu
Campbell, Gregory A. .... (315) 268-6577
gac@clarkson.edu
Chin, Der-Tau (Emeritus) .... (315) 268-7930
chin@clarkson.edu
Cole, Robert (Emeritus)
Hopke, Philip K. .... (315) 268-3861
hopkepk@clarkson.edu
McLaughlin, John B. .... (315) 268-6663
jmclau@clarkson.edu
Rasmussen, Don H. .... (315) 268-3820
rasmu@clarkson.edu
Subramanian, R. Shankar .... (315) 268-6648
shankar@clarkson.edu
Taylor, Ross (Chair) .... (315) 268-6650
taylor@clarkson.edu
Wilcox, William R. .... (315) 268-7672
wilcox@clarkson.edu

Associate Professors

Harris, Sandra L. .... (315) 268-2284
slharris@clarkson.edu
Jachuck, Roshan J.J. .... (315) 268 6325
rjachuck@clarkson.edu
McCluskey, Richard J. (Associate Chair) (315) 268-2303
bq02@clarkson.edu
Rengaswamy, Raghunathan .... (315) 268-4423
raghu@clarkson.edu
Suni, Ian I. .... (315) 268-4471
isuni@clarkson.edu

Accredited by: ABET

Degrees granted 2003-2004:

B.S.: 28 M.S.: 9 Ph.D.: 1

Undergraduate advisor: Richard J. McCluskey

Student organization: AIChE
Advisor: I.I. Suni

Department reports to: S.S. Venkata, Dean of Engineering

Placement service: Kathryn Briggs Johnson

Columbia University

Department of Chemical Engineering
801 S. W. Mudd, Mail Code 4721
500 W. 120th St.
New York, NY 10027

University phone . . . . . (212) 854-1754
Department phone . . . . . (212) 854-4453
Department FAX . . . . . (212) 854-3054

Professors

Durning, C. J. . . . . . (212) 854-8161
cjd2@columbia.edu
Flynn, George W. . . . . . (212) 854-4162
flynn@chem.columbia.edu
Gryte, Carl C. . . . . . (212) 854-2470
ccg2@columbia.edu
Ju, Jingyue . . . . . (212) 854-2487
dj222@columbia.edu
Koberstein, J. (Chair) . . . . . (212) 854-3120
jk1191@columbia.edu
Leonard, Edward F. . . . . . (212) 854-4448
leonard@columbia.edu
O'Shaughnessy, B. . . . . . (212) 854-3203
bo8@columbia.edu
Spencer, Jordan L. (Emeritus) . . . . . (212) 854-4471
spencer@columbia.edu
Turro, Nicholas J. . . . . . (212) 854-2175
turro@chem.columbia.edu
West, A. C . . . . . (212) 854-4452
acw17@columbia.edu

Assistant Professors

Banta, Scott . . . . . (214) 854-4453
sab2373@columbia.edu
Calabrese Barton, Scott . . . . . (212) 854-4655
scb2001@columbia.edu
Levicky, Rasti . . . . . (212) 854-2869
rl268@columbia.edu
Shapley, Nina . . . . . (212) 854-1095
ncs2101@columbia.edu

Accredited by: ABET

Degrees granted 2003-2004:

B.S.: 20 M.S.: 10 Ph.D.: 7

Graduate advisor: Alan C. West

Undergraduate advisor: Christopher Durning

Student organization: AIChE

Department reports to: Zvi Galil (212) 854-2993, Dean

Cooper Union

Albert Nerken School of Engineering
Chemical Engineering Department
51 Astor Place
New York, NY 10003

University phone . . . . . (212) 353-4100
Department phone . . . . . (212) 353-4370
Department FAX . . . . . (212) 353-4341

Professors

Ahmed, Zikri . . . . . (212) 353-4379
ahmed@cooper.edu
Brazinsky, Irv (Chair) . . . . . (212) 353-4373
brazin@cooper.edu
Cheng, Shang-I (Emeritus)

Associate Professors

Okorator, Charles . . . . . (212) 353-4371
okoraf@cooper.edu
Sidebotham, George . . . . . (212) 353-4308
sidebo@cooper.edu
Stock, Richard . . . . . (212) 353-4317
stock@cooper.edu

Assistant Professors

Lam-Anderson, Marca . . . . . (212) 353-4393
mjlam@cooper.edu

Accredited by: ABET

Student organization: AIChE
Advisor: Zikri Ahmed

Department reports to: E. Baum, Dean, School of Engineering

Placement service: Melissa Benca

Cornell University

School of Chemical and Biomolecular Engineering
Cornell University
Ithaca, NY 14853-5201
Deliveries: 120 Olin Hall

University phone .... (607) 255-2000
Department phone .... (607) 255-8656
Department FAX .... (607) 255-9166

Professors

Clancy, Paulette (Director) .... (607) 255-4430
pc@cheme.cornell.edu
Cohen, Claude .... (607) 255-7292
cc@cheme.cornell.edu
Finn, Robert K. (Emeritus) .... (607) 255-8656
Gubbins, Keith E. (Emeritus)
Harriott, Peter (Emeritus) .... (607) 255-3529
ph@cheme.cornell.edu
Koch, Donald L. .... (607) 255-3484
don@cheme.cornell.edu
Olbricht, William L. .... (607) 255-4362
wlo@cheme.cornell.edu
Rodriguez, Ferdinand (Emeritus) ...... (607) 255-8656
fr@cheme.cornell.edu
Shuler, Michael L. .... (607) 255-7577
mls@cheme.cornell.edu
Smith, Julian C. (Emeritus) .... (607) 255-8656
jcs29@cornell.edu
Steen, Paul H. .... (607) 255-4749
steen@cheme.cornell.edu
Streett, William B. (Emeritus) .... (607) 255-8656
Thorpe, R. G. (Emeritus) .... (607) 255-7461
rgt3@cornell.edu
VonBerg, Robert L. (Emeritus) .... (607) 255-8656
Wiegandt, H. F. (Emeritus) .... (607) 255-8656

Associate Professors

Anton, A. Brad .... (607) 255-3629
aba@cheme.cornell.edu
Archer, Lynden A. .... (607) 254-8825
archer@cheme.cornell.edu
Duncan, T. Michael .... (607) 255-8715
tmd@cheme.cornell.edu
Engstrom, James R. .... (607) 255-9934
jre@cheme.cornell.edu
Lee, Kelvin H. .... (607) 255-4215
khlee@cheme.cornell.edu

Assistant Professors

DeLisa, Matthew P. .... (607) 254-8560
Escobedo, Fernando .... (607) 255-8243
escobedo@cheme.cornell.edu
Joo, Yong Lak .... (607) 255-8591
Putnam, David A. .... (607) 255-4352
dap43@cornell.edu
Stroock, Abraham D. .... (607) 255-4276
ads10@conrell.edu

Other Faculty

Baxter, Gregory (Adjunct)
Carter, John N. (Adjunct)
Center, Alfred M. .... (607) 255-3422
amc@cheme.cornell.edu
Hatzimanikatis, Vassily (Adjunct)
Vaeth, Kathleen (Adjunct)

Accredited by: ABET

Degrees granted 2003-2004:

B.S.: 45 M.S.: 1 Ph.D.: 7

Graduate advisor: L. A. Archer

Undergraduate advisor: T. M. Duncan

Student organization: AIChE
Advisor: J. R. Engstrom

Department reports to: W. Kent Fuchs, Dean of Engineering

Manhattan College

Chemical Engineering Department
Riverdale, NY 10471-4099
Deliveries: 3825 Corlear Ave., Riverdale, NY 10463

University phone .... (718) 862-8000
Department phone .... (718) 862-7185
Department FAX .... (718) 862-7819

Professors

Burris, Conrad T. (Emeritus)
Heist, Richard (Dean) .... (718) 862-7307
richard.heist@manhattan.edu
Hollein, Helen C. (Emeritus) .... (718) 862-7281
helen.hollein@manhattan.edu
Reynolds, Joseph .... (718) 862-7187
jreynold@manhattan.edu
Theodore, Louis .... (718) 862-7188
chmldept@manhattan.edu
Zenz, Frederick A. (Emeritus)

Associate Professors

Assaf-Anid, Nada (Chair) .... (718) 862-7420
nada.assaf-anid@manhattan.edu
Famularo, Jack (Emeritus)
jackf507@rcn.com
Marnell, Paul .... (718) 862-7194
paul.marnell@manhattan.edu

Assistant Professors

Castaldi, Marco (Adjunct) .... (718) 862-7185
mc2352@columbia.edu
Devine, Kevin (Adjunct) .... (718) 862-7185
kdevine@kraft.com
Feintuch, Howard (Adjunct)
Howard_Feintuch@fwc.com
Flynn, Ann Marie .... (718) 862-7286
annmarie.flynn@manhattan.edu

Instructors

Lucas, Robert (Adjunct) .... (718) 862-7185
robert.lucas@manhattan.edu

Accredited by: ABET

Degrees granted 2003-2004:

B.S.: 16 M.S.: 4

Graduate advisor: Dr. Nada Assaf-Anid

Undergraduate advisor: Richard Schneider

Student organization: AIChE
Advisor: Dr. Joseph Reynolds

Department reports to: Dr. Richard Heist, Dean of Engineering

Placement service: Marjorie Apel, Career Services

Polytechnic University

Dept. of Chemical and Biological Sciences and Engineering
Six Metrotech Center
Brooklyn, NY 11201

Department phone .... (718) 260-3470
Department FAX .... (718) 260-3125

Professors

Arnold, Stephen .... (718) 260-3085
arnold@photon.poly.edu
Garetz, Bruce .... (718) 260-3278
bgaretz@duke.poly.edu
Georgakis, Christos (Head) .... (718) 260-3579
c.georgakis@poly.edu
Green, Mark .... (718) 260-3177
mgreen@duke.poly.edu
Gross, Richard .... (718) 260-3024
rgross@duke.poly.edu
Levon, Kalle .... (718) 260-3339
klevon@duke.poly.edu
Mijovic, Jovan S. .... (718) 260-3097
jmijovic@duke.poly.edu
Motzkin, Shirley (Emeritus) .... (718) 260-3662
smotzkin@photon.poly.edu
Ulman, Abraham .... (718) 260-3119
aulman@duke.poly.edu
Wolf, Edward .... (718) 260-3629
ewolf@duke.poly.edu
Zlatanova, Jordanka .... (718) 260-3176
jzlatano@duke.poly.edu

Associate Professors

Cowman, Mary .... (718) 260-3054
mcowman@duke.poly.edu
Pinto, Jose M. .... (718) 260-3470
jpinto@poly.edu
Stiel, Leonard I. (Emeritus) .... (718) 260-3638
lstiel@duke.poly.edu
Teraoka, Iwao .... (716) 260-3466
teraoka@duke.poly.edu
Tooney, Nancy (Emeritus) .... (718) 260-3121
ntooney@duke.poly.edu
Ziegler, Edward N. .... (718) 260-3276
eziegler@duke.poly.edu
Zurawsky, Walter P. .... (718) 260-3725
zurawsky@duke.poly.edu

Accredited by: ABET

Graduate advisor: Jovan Mijovic

Undergraduate advisor: Edward Ziegler

Student organization: AIChE
Advisor: Walter Zurawsky

Department reports to: Budd Griffis, Vice President & Dean of Engineering

Rensselaer Polytechnic Institute

Chemical & Biological Engineering Department
110-8th Street
Troy, NY 12180-3590
Deliveries: Ricketts Building

University phone . . . . . (518) 276-6000
Department phone . . . . . (518) 276-6377
Department FAX . . . . . (518) 276-4030

Professors

Abbott, Michael M. (Emeritus) . . . . . (518) 276-6764
abbotm2@rpi.edu
Altwicker, Elmar R. (Emeritus) . . . . . (518) 276-6927
altwie@rpi.edu
Belfort, Georges . . . . . (518) 276-6948
belfog@rpi.edu
Bequette, B. Wayne . . . . . (518) 276-6683
bequette@rpi.edu
Bungay, Henry R. III (Emeritus) . . . . . (518) 276-6799
bungah@rpi.edu
Cale, Timothy S. . . . . . (518) 276-6059
calet@rpi.edu
Cramer, Steven M. . . . . . (518) 276-6198
crames@rpi.edu
Dordick, Jonathan S. . . . . . (518) 276-2899
dordick@rpi.edu
Fontijn, Arthur . . . . . (518) 276-6508
fontia@rpi.edu
Gill, William N. . . . . . (518) 276-2880
gillw@rpi.edu
Kumar, Sanat . . . . . (528) 276-3032
kumar@rpi.edu
Littman, Howard (Emeritus) . . . . . (518) 276-6039
littmh@rpi.edu
Nauman, E. Bruce . . . . . (518) 276-6726
nauman@rpi.edu
Plawsky, Joel L. . . . . . (518) 276-6049
plawsky@rpi.edu
Van Ness, Hendrick C. (Emeritus) . . . . . (518) 276-6727
vanneh@rpi.edu
Wayner, Peter C. . . . . . (518) 276-6199
wayner@rpi.edu

Associate Professors

Garde, Shekhar . . . . . (518) 276-6048
gardes@rpi.edu

Assistant Professors

Kane, Ravi . . . . . (518) 276-2356
kaner@rpi.edu
Martin, Lealon . . . . . (518) 276-3327
lealon@rpi.edu
Sharfstein, Susan . . . . . (518) 276-2166
sharfs@rpi.edu

Accredited by: ABET

Degrees granted 2003-2004:

B.S.: 51 M.S.: 3 Ph.D.: 10

Graduate advisor: B. Wayne Bequette

Undergraduate advisor: B. Wayne Bequette

Student organization: AIChE
Advisor: B. Wayne Bequette

Department reports to: Lester Gerhardt, Acting Dean of Engineering

Placement service: Thomas Tarantelli (518) 276-6234

University of Rochester

Department of Chemical Engineering
Rochester, NY 14627-0166
Deliveries: 206 Gavett Hall

University phone . . . . (716) 275-2121
Department phone . . . . (716) 275-4041
Department FAX . . . . (716) 273-1348

Professors

Chen, Shaw-Horng (Chair) . . . . (585) 275-0909
shch@lle.rochester.edu
Chimowitz, Eldred H. (Associate Chair) (585) 275-8497
chim@che.rochester.edu
Eisenberg, Richard F. (Emeritus) . . . . . . (585) 275-4041
Ferron, John R. (Emeritus) . . . . (585) 275-4041
Greener, Jehuda (Adjunct) . . . . (585) 477-3723
jehuda.greener.kodak.com
Jacobs, Stephen D. (Joint) . . . . (585) 275-4837
sjac@lle.rochester.edu
Jorne, Jacob . . . . (585) 275-4584
jorne@che.rochester.edu
Rothberg, Lewis J. (Joint) . . . . (585) 273-4725
ljr@chem.chem.rochester.edu
Shapir, Yonathon (Joint) . . . . (585) 275-7291
ysha@pas.rochester.edu
Sotirchos, Stratis V. . . . . (585) 275-4626
svs2@che.rochester.edu
Wu, J. H. David . . . . (585) 275-8499
davidwu@che.rochester.edu

Associate Professors

Foster, David (Adjunct) . . . . (585) 275-4041
Harding, David R. (Joint) . . . . (585) 275-5850
dhar@lle.rochester.edu

Assistant Professors

Yang, Hong . . . . (585) 275-2110
hyang@che.rochester.edu
Yates, Matthew . . . . (585) 273-2335
myates@che.rochester.edu

Instructors

Olsen, Thor . . . . (585) 275-7885
thor@che.rochester.edu

Accredited by: ABET

Graduate advisor: J. H. David Wu

Student organization: AIChE
Advisor: Eldred H. Chimowitz

Department reports to: Kevin J. Parker, Dean, School of Engineering and Applied Sciences

Placement service: Burton Nadler

State University of New York at Buffalo

Chemical & Biological Engineering Department
Clifford C. Furnas Hall
Buffalo, NY 14260-4200
Deliveries: 303 Furnas Hall

University phone .... (716) 645-2000
Department phone .... (716) 645-2911
Department FAX .... (716) 645-3822

Professors

Alexandridis, Paschalis .... x2210 †
palexand@eng.buffalo.edu
Good, Robert J. (Emeritus) x2219
Hlavacek, Vladimir x2208
hlavacek@acsu.buffalo.edu
Kofke, David A. x2209
kofke@eng.buffalo.edu
Lund, Carl R. F. (Chair) x2211
lund@eng.buffalo.edu
Mountziaris, T. J. x2212
tjm@eng.buffalo.edu
Nicholson, Bruce J. (Adjunct) .... (716) 645-3344
bjn@acsu.buffalo.edu
Nitsche, Johannes M. x2213
nitsche@eng.buffalo.edu
Petrou, Athos (Adjunct) .... (716) 645-2987
petrou@acsu.buffalo.edu
Ruckenstein, Eli x2214
feaeliru@acsu.buffalo.edu
Ryan, Michael E. (Associate Dean) x1130
meryan@eng.buffalo.edu
van Oss, Carel J. (Adjunct) .... (716) 831-2900
cjvanoss@acsu.buffalo.edu
Weber, Thomas W. (Emeritus) x2219
twweber@eng.buffalo.edu
Weller, Sol W. (Emeritus) x2219

Associate Professors

Andreadis, Stelios T. x2204
sandread@eng.buffalo.edu
Mihalko, William M. (Adjunct) .... (716) 829-2982
mihalk@acsu.buffalo.edu
Neelamegham, Sriram x2220
neel@acsu.buffalo.edu
Swihart, Mark x2205
swihart@eng.buffalo.edu

Assistant Professors

Errington, Jeffrey R. x2222
jerring@eng.buffalo.edu
Koffas, Mattheos x2221
mkoffas@eng.buffalo.edu
Tzanakakis, Emmanouhl x2206
emtzan@buffalo.edu
Zhou, Yaoqi (Adjunct) .... (716) 829-2985
yqzhou@acsu.buffalo.edu

Instructors

Kofke, Tamara x2168
tgkofke@eng.buffalo.edu

Other Faculty

Tsianou, Marina x2257
mtsianou@eng.buffalo.edu

† Unless noted faculty numbers are extensions of (716) 645-2911.

Accredited by: ABET

Degrees granted 2003-2004:

B.S.: 34 M.S.: 12 Ph.D.: 10

Graduate advisor: M. T. Swihart

Undergraduate advisor: D. A. Kofke

Student organization: AIChE
Advisor: Mark T. Swihart

Department reports to: Mark H. Karwan, Dean of Engineering

Placement service: Daniel J. Ryan, Director

Syracuse University

Department of Chemical Engineering and Materials Science
373 Link Hall
Syracuse, NY 13244-1190

University phone . . . . (315) 443-1870
Department phone . . . . (315) 443-2558
Department FAX . . . . (315) 443-1243

Professors

Amato, Wayne (Adjunct) . . . . (315) 696-8651
dukeamato@a-znet.com
Martin, George C. (Chair) . . . . (315) 443-4467
gcmartin@syr.edu
Ramarao, Bamdaru (Adjunct) . . . . (315) 470-6513
bvramara@syr.edu
Rice, Philip A. (Emeritus) . . . . (315) 443-9444
parice@syr.edu
Sangani, Ashok . . . . (315) 443-4502
asangani@syr.edu
Santanam, Suresh (Adjunct) . . . . (315) 443-2311
ssantana@syr.edu
Schroder, Klaus (Emeritus) . . . . (315) 443-2846
kschrode@syr.edu
Schwarz, James A. . . . . (315) 443-4575
jschwarz@syr.edu
Stern, S. Alexander (Emeritus) . . . . (315) 443-2558
sasternou@aol.com
Tavlarides, Lawrence L. . . . . (315) 443-1883
lltavlar@syr.edu
Tien, Chi (Emeritus) . . . . (315) 443-4050
ctien@syr.edu

Associate Professors

Heydweiller, John C. . . . . (315) 443-4468
jcheydwe@syr.edu

Assistant Professors

Hasenwinkel, Julie (Honorary) . . . . (315) 443-3064
jmhasen@syr.edu
Luk, Yan Yeung (Honorary)

Accredited by: ABET

Degrees granted 2003-2004:

B.S.: 7 M.S.: 3 Ph.D.: 1

Student organization: AIChE

Department reports to: Eric Spina (315) 443-4341, Dean

Placement service: Karen Kenty, ECS Opportunities Center, (315) 443-2239

North Carolina

North Carolina Agricultural and Technical State University

Department of Mechanical and Chemical Engineering
618 McNair Hall
Greensboro, NC 27411

University phone . . . . (336) 334-7500
Department phone . . . . (336) 334-7564
Department FAX . . . . (336) 334-7417

Professors

Adewuyi, Yusuf G. . . . . x54 †
adewuyi@ncat.edu
Ilias, Shamsuddin 312
ilias@ncat.edu
Kabadi, Vinayak N. x46
kabadi@ncat.edu
King, Franklin G. 311
king@ncat.edu
Tatterson, Gary B. 326
gbt@ncat.edu
Uitenham, Leonard (Chair) 310
u10ham@ncat.edu

Associate Professors

Lou, Jianzhong 323
lou@ncat.edu
Roberts, Kenneth L. 321
kroberts@ncat.edu
Schimmel, Keith A. (Director) 327
schimmel@ncat.edu

† Faculty numbers are extensions of department phone

Accredited by ABET

Degrees granted 2003-2004:

B.S.: 18 M.S.: 13

Graduate advisor: Leonard Uitenham

Undergraduate advisor: Leonard Uitenham

Student organization: AIChE
Advisor: Kenneth Roberts

Department reports to: Joseph Monroe (336) 334-7589, Dean, College of Engineering

Placement service: Joyce Edwards (336) 334-7755, Director

North Carolina State University

Department of Chemical Engineering
Campus Box 7905
Raleigh, NC 27695-7905
Deliveries: 113 Riddick Labs, 2401 Stinson Dr.

University phone .... (919) 515-2011
Department phone .... (919) 515-2324
Department FAX .... (919) 515-3465

Professors

Beatty, Kenneth O. (Emeritus) .... (919) 515-6398
kobeatty@eos.ncsu.edu
Carbonell, Ruben G. .... (919) 515-5118
ruben@ncsu.edu
DeSimone, Joseph M. .... (919) 962-2166
desimone@unc.edu
Fedkiw, Peter S. (Associate Head) ..... (919) 515-3572
fedkiw@eos.ncsu.edu
Felder, Richard M. (Emeritus) .... (919) 515-2327
felder@eos.ncsu.edu
Gubbins, Keith E. .... (919) 513-2262
keg@ncsu.edu
Hall, Carol K. .... (919) 515-3571
hall@turbo.che.ncsu.edu
Hopfenberg, Harold B. .... (919) 515-5118
hbh@ncsu.edu
Kelly, Robert M. .... (919) 515-6396
rmkelly@eos.ncsu.edu
Khan, Saad A. .... (919) 515-4519
khan@eos.ncsu.edu
Kilpatrick, Peter K. (Head) .... (919) 515-7121
peter-k@eos.ncsu.edu
Lim, Phooi K. .... (919) 515-2328
lim@eos.ncsu.edu
Ollis, David F. .... (919) 515-2329
ollis@eos.ncsu.edu
Overcash, Michael R. .... (919) 515-2325
overcash@eos.ncsu.edu
Parsons, Gregory N. .... (919) 515-7553
gnp@ncsu.edu
Roberts, George W. .... (919) 515-7328
groberts@eos.ncsu.edu
Setzer, C. John (Emeritus) .... (919) 515-2520
setzer@eos.ncsu.edu
Spontak, Richard J. .... (919) 515-4200
rich_spontak@ncsu.edu

Associate Professors

Genzer, Jan .... (919) 515-2069
jgenzer@eos.ncsu.edu
Grant, Christine S. .... (919) 515-2317
grant@eos.ncsu.edu
Lamb, H. Henry .... (919) 515-6395
lamb@eos.ncsu.edu
Peretti, Steven W. .... (919) 515-6397
peretti@eos.ncsu.edu

Assistant Professors

Haugh, Jason .... (919) 513-3851
jmhaugh@eos.ncsu.edu
Van Zanten, John .... (919) 515-2324
jhvanzan@eos.ncsu.edu
Velev, Orlin .... (919) 513-4318
odvelev@unity.ncsu.edu

Instructors

Bullard, Lisa .... (919) 515-7455
lgbullar@eos.ncsu.edu

Accredited by: ABET

Degrees granted 2003-2004:

B.S.: 103 M.S.: 15 Ph.D.: 12

Graduate advisor: Saad Khan

Undergraduate advisor: Lisa Bullard

Student organization: AIChE
Advisor: John Van Zanten

Department reports to: Nino A. Masnari (919) 515-2311, Dean of Engineering

Placement service: Beverly J. Marchi (919) 515-2396

North Dakota

University of North Dakota

Department of Chemical Engineering
P. O. Box 7101
Grand Forks, ND 58202-7101
Deliveries: Harrington Hall 323

University phone .... (701) 777-2011
Department phone .... (701) 777-4244
Department FAX .... (701) 777-3773

Professors

Erjavec, John (Chair) .... (701) 777-3797
john_erjavec@mail.und.nodak.edu
Owens, Thomas C. (Emeritus) .... (701) 777-4245
tom_owens@mail.und.nodak.edu
Watson, John, L. (Dean) .... (701) 777-3411
john_watson@mail.und.nodak.edu

Associate Professors

Mann, Michael D. .... (701) 777-3852
mike_mann@mail.und.nodak.edu
Muggli, Darrin S. .... (701) 777-2337
darrin_muggli@und.nodak.edu
Seames, Wayne S. .... (701) 777-2958
wayne_seames@mail.und.nodak.edu

Assistant Professors

Kolodka, Edward B. .... (701) 777-3798
edward_kolodka@mail.und.nodak.edu

Accredited by: ABET

Degrees granted 2003-2004:

B.S.: 16 M.S.: 4

Graduate advisor: Erjavec, John

Undergraduate advisor: Erjavec, John

Student organization: AIChE
Advisor: Muggli, Darrin S.

Department reports to: John L. Watson, Dean, School of Engineering and Mines

Placement service: Mark Thompson, Director, Career Services

Ohio

University of Akron

Chemical Engineering Department
Whitby Hall 211
Akron, OH 44325-3906
Deliveries: 200 E. Buchtel Commons

University phone .... (330) 972-7111
Department phone .... (330) 972-7250
Department FAX .... (330) 972-5856

Professors

Atwood, Glenn A. (Emeritus) .... (330) 972-7250
Chase, George G. .... (330) 972-7943
gchase@uakron.edu
Cheung, H. Michael .... (330) 972-7282
cheung@uakron.edu
Chuang, Steven (Chair) .... (330) 972-6993
schuang@uakron.edu
Elliott, J. Richard .... (330) 972-7253
jelliott@uakron.edu
Greene, Howard (Emeritus) .... (330) 972-7250
Ju, Lu-Kwang .... (330) 972-7252
lukeju@uakron.edu
Roberts, Robert W. (Emeritus) .... (330) 972-7250

Associate Professors

Evans, Edward A. .... (330) 972-8292
evanse@uakron.edu
Lopina, Stephanie T. .... (330) 972-7255
lopina@uakron.edu
Qammar, Helen C. .... (330) 972-5917
hqammar@uakron.edu

Assistant Professors

Newby, Bi-min Zhang .... (330) 972-2510
bmznewby@uakron.edu
Wang, Ping .... (330) 972-2096
wangp@uakron.edu

Accredited by: ABET

Degrees granted 2003-2004:

B.S.: 36 M.S.: 7 Ph.D.: 5

Graduate advisor: Dr. J Richard Elliott

Undergraduate advisor: Dr. Helen Qammar

Student organization: AIChE
Advisor: George Chase

Department reports to: Dr. George K. Haritos, Dean of Engineering

Placement service: Dr. Kim Beyer, Interim Co-Director of Center for Career Management

Case Western Reserve University

Department of Chemical Engineering
10900 Euclid Avenue
Cleveland, OH 44106-7217
Deliveries: A.W. Smith Building, Room 116

University phone . . . . (216) 368-2000
Department phone . . . . (216) 368-4182
Department FAX . . . . (216) 368-3016

Professors

Anderson, John L. (Provost) . . . . (216) 368-4346
jla24@case.edu
Angus, John C. (Emeritus) . . . . (216) 368-4133
jca3@case.edu
Edwards, Robert V. . . . . (216) 368-4151
rve2@case.edu
Feke, Donald L. (Vice Provost) . . . . (216) 368-2750
dlf4@case.edu
Harris, Robert E. (Adjunct) . . . . (216) 368-1145
reh@case.edu
Landau, Uziel . . . . (216) 368-4132
uxl@case.edu
Liu, Chung-Chiun . . . . (216) 368-2935
cxl9@case.edu
Mann, J. Adin Jr. (216) 368-4122
jam12@case.edu
Pintauro, Peter N. (Chair) . . . . (216) 368-4150
pnp3@case.edu
Qutubuddin, Syed . . . . (216) 368-2764
sxq@case.edu
Savinell, Robert F. (Dean) . . . . (216) 368-4436
rfs2@case.edu
Wnek, Gary . . . . (216) 368-2782
gew5@case.edu
Zawodzinski, Thomas . . . . (216) 368-5547
taz5@case.edu

Assistant Professors

Baskaran, Harihara . . . . (216) 368-1029
hari@case.edu
Martin, Heidi B. . . . . (216) 368-3810
hbm@case.edu

Accredited by: ABET

Degrees granted 2003-2004:

B.S.: 31 M.S.: 8 Ph.D.: 1

Graduate advisor: Heidi Martin

Student organization: AIChE
Advisor: Peter N. Pintauro

Department reports to: Robert F. Savinell, Dean of Engineering, Case School of Engineering

Placement service: Eric L. Dicken

University of Cincinnati

Department of Chemical and Materials Engineering
Mail Location 12
Cincinnati, Ohio 45221-0012
Deliveries: 697 Rhodes Hall

University phone .... (513) 556-6000
Department phone .... (513) 556-3096
Department FAX .... (513) 556-2569

Professors

Delcamp, Robert M. (Emeritus) .... (513) 556-2761
Fried, Joel R. .... (513) 556-2767
Joel.Fried@uc.edu
Govind, Rakesh .... (513) 556-2666
Rakesh.Govind@UC.Edu
Greenberg, David B. (Emeritus) .... (513) 556-2741
dgreenbe@uceng.uc.edu
Hershey, Daniel .... (513) 556-2740
Daniel.Hershey@UC.edu
Hwang, Sun-Tak .... (513) 556-2791
Sun-Tak.Hwang@UC.Edu
Kao, Y. K. .... (513) 556-2762
ykao@alpha.che.uc.edu
Khang, Soon-Jai .... (513) 556-2789
skhang@alpha.che.uc.edu
Krantz, William .... (513) 556-4021
bkrantz@alpha.che.uc.edu
Lemlich, Robert (Emeritus) .... (513) 556-2761
Licht, William (Emeritus) .... (513) 556-2761
Lin, Jerry Y.S. .... (513) 556-2769
jlin@alpha.che.uc.edu
Phillips, Paul (Chair) .... (513) 556-1178
paul.phillips@uc.edu
Smirniotis Panagiotis (Director) ...... (513) 556-1474
Panagiotis.Smirniotis@uc.edu

Associate Professors

Cosgrove, Stanley (Emeritus) .... (513) 556-2761
Guliants, Vadim (Director) .... (513) 556-0203
vguliant@alpha.che.uc.edu

Assistant Professors

Co, Carlos .... (513) 556-2731
cco@alpha.che.uc.edu
Ho, Chia-Chi .... (513) 556-2438
cho@alpha.che.uc.edu

Accredited by: ABET

Degrees granted 2003-2004:

B.S. 38 M.S.: 8 Ph.D.: 5

Graduate advisor: P.Smirniotis

Undergraduate advisor: V. Guliants

Student organization: AIChE
Advisor: Carlos Co

Department reports to: Stephen Kowel, Dean

Placement service: Stephen Gilby

Cleveland State University

Chemical and Biomedical Engineering Department
2121 Euclid Avenue, Stilwell Hall 455
Cleveland, OH 44115
Deliveries: Stilwell Hall 455

University phone .... (216) 687-2000
Department phone .... (216) 687-2569
Department FAX .... (216) 687-9220

Professors

Coulman, George A. (Emeritus) .... (216) 687-3526
Gatica, Jorge E. .... (216) 523-7274
j.gatica@csuohio.edu
Ghorashi, Bahman .... (216) 687-2562
b.ghorashi@csuohio.edu
Godleski, Edward S. (Emeritus) .... (216) 687-2555
Shah, Dhananjai B. .... (216) 687-3569
d.shah@csuohio.edu
Talu, Orhan (Chair) .... (216) 687-3539
o.talu@csuohio.edu
Tewari, Surendra N. .... (216) 687-7342
s.tewari@csuohio.edu

Associate Professors

Belovich, Joanne M. .... (216) 687-3502
j.belovich@csuohio.edu
Lustig, Rolf .... (216) 687-3526
r.lustig@csuohio.edu

Assistant Professors

Chatzimavroudis, George P. .... (216) 687-5396
g.chatzimavroudis@csuohio.edu
Holland, Nolan B. .... (216) 687-2569
N.Holland@csuohio.edu
Ungarala, Sridhar .... (216) 687-9368
s.ungarala@csuohio.edu

Other Faculty

Rekhson, Simon .... (216) 687-5283
s.rekhson@csuohio.edu

Accredited by: ABET

Degrees granted 2003-2004:

B.S.: 8 M.S.: 12 Ph.D.: 6

Graduate advisor: Dr. Jorge E. Gatica

Undergraduate advisor: Jorge E. Gatica

Student organization: AIChE
Advisor: Sridhar Ungarala

Department reports to: Dr. Charles Alexander, Dean of Engineering

Placement service: Paul B. Klein (216) 687-2246, Director

University of Dayton

Chemical & Materials Engineering Department
KL 445
300 College Park
Dayton, OH 45469-0246
Deliveries: Dr. Tony E. Saliba

University phone .... (937) 229-4111
Department phone .... (937) 229-2627
Department FAX .... (937) 229-3433

Professors

Eylon, Daniel (Director) .... (937) 229-2679
Daniel.Eylon@notes.udayton.edu
Flach, Lawrance .... (937) 229-2627
Larry.Flach@notes.udayton.edu
Lee, C. William .... (937) 229-2627
Chih-Kuo.Lee@notes.udayton.edu
Myers, Kevin J. .... (937) 229-2627
Kevin.Myers@notes.udayton.edu
Saliba, Tony E. (Chair) .... (937) 229-2627
TSaliba@udayton.edu
Sandhu, Sarwan S. .... (937) 229-2627
Sarwan.Sandhu@notes.udayton.edu
Snide, James A. (Emeritus) .... (937) 229-2679
James.Snide@notes.udayton.edu

Associate Professors

Ciric, Amy .... (937) 229-2627
Amy.Ciric@notes.udayton.edu
Lu, Christopher C. (Emeritus) .... (937) 229-2627
Chris.Lu@notes.udayton.edu

Assistant Professors

Wilkens, Robert .... (937) 229-2627
Wilkens@notes.udayton.edu

Accredited by: ABET

Degrees granted 2003-2004:

B.S.: 41 M.S.: 7

Graduate advisor: Dr. Tony E. Saliba

Student organization: AIChE
Advisor: Robert J. Wilkens

Department reports to: Dr. Joseph Saliba, Dean of Engineering

Placement service: Gregory Hayes, Director

Ohio State University

Chemical & Biomolecular Engineering Department
125 Koffolt Labs
140 West 19th Avenue
Columbus, OH 43210-1180

University phone .... (614) 292-3980
Department phone .... (614) 292-3309
Department FAX .... (614) 292-3769

Professors

Brodkey, Robert (Emeritus) .... (614) 292-2609
brodkey@chbmeng.ohio-state.edu
Chalmers, Jeffrey .... (614) 292-2727
chalmers@chbmeng.ohio-state.edu
Cooper, Stuart (Chair) .... (614) 292-7907
coopers@chbmeng.ohio-state.edu
Fan, L. S. .... (614) 688-3262
fan@chbmeng.ohio-state.edu
Geankoplis, Christie J. (Emeritus)
Haering, Edwin R. (Emeritus)
Hershey, Harry C. (Emeritus) .... (614) 292-6591
hershey@che.eng.ohio-state.edu
Ho, W.S. Winston .... (614) 292-9970
ho@chbmeng.ohio-state.edu
Lee, James Ly .... (614) 292-2408
leelj@chbmeng.ohio-state.edu
Ozkan, Umit (Associate Dean) .... (614) 292-6623
ozkan@chbmeng.ohio-state.edu
Rathman, James F. .... (614) 292-3760
rathman@chbmeng.ohio-state.edu
Slider, H. C (Emeritus) .... (614) 292-2698
slider@che.eng.ohio-state.edu
Smith, Edwin E. (Emeritus) .... (614) 292-6033
smith@che.eng.ohio-state.edu
Sweeney, Thomas L. (Emeritus) .... (614) 292-7907
sweeney.3@osu.edu
Yang, Shang-Tian .... (614) 292-6611
yangst@chbmeng.ohio-state.edu
Zakin, Jacques L. (Emeritus) .... (614) 688-4113
zakin@chbmeng.ohio-state.edu

Associate Professors

Bakshi, Bhavik .... (614) 292-4904
bakshi@chbmeng.ohio-state.edu
Koelling, Kurt W. .... (614) 292-2256
koelling@chbmeng.ohio-state.edu
Lee, Stephen .... (614) 688-5447
lee@bme.ohio-state.edu
Svanks, Karlis (Emeritus)
Tomasko, David .... (614) 292-4249
tomasko@chbmeng.ohio-state.edu
Wyslouzil, Barbara .... (614) 688-3583
wyslouzil@chbmeng.ohio-state.edu

Assistant Professors

Kusaka, Isamu .... (614) 292-7907
kusaka@chbmeng.ohio-state.edu

Instructors

Corn, John (Adjunct) .... (614) 688-8254
corn@chbmeng.ohio-state.edu

Accredited by: ABET

Degrees granted 2003-2004:

B.S.: 73 M.S.: 14 Ph.D.: 8

Graduate advisor: Winston Ho

Undergraduate advisor: Mary Lamont

Student organization: AIChE
Advisor: Kurt Koelling

Department reports to: Bud Baeslack (614) 292-2836, Dean of Engineering

Placement service: Rosemary Hill (614) 292-6651

Ohio University

Department of Chemical Engineering
172 Stocker Center
Athens, OH 45701-2979

University phone . . . . (740) 593-1000
Department phone . . . . (740) 593-1492
Department FAX . . . . (740) 593-0873

Professors

Dinos, Nicholas (Emeritus) . . . . (740) 593-0033
dinos@ohio.edu
Nesic, Srdjan . . . . (740) 593-1498
nesic@ohio.edu
Prudich, Michael E. (Chair) . . . . (740) 593-1501
prudich@ohio.edu

Associate Professors

Chen, Wen-Jia . . . . (740) 593-1497
chenw@ohio.edu
Crist, Kevin C. . . . . (740) 593-4751
cristk@ohio.edu
Goetz, Douglas J. (740) 593-1494
goetzd@ohio.edu
Gu, Tingyue . . . . (740) 593-1499
gu@ohio.edu
Gulino, Daniel A. (740) 593-1495
gulino@ohio.edu
Ridgway, Darin . . . . (740) 593-1504
ridgway@ohio.edu
Sampson, Kendree J. (Associate Dean) . (740) 593-1503
sampson@ohio.edu
Young, Valerie L. (740) 593-1496
youngv@ohio.edu

Assistant Professors

Botte, Gerardine G. . . . . (740) 593-9670
botte@ohio.edu

Accredited by: ABET

Degrees granted 2003-2004:

B.S.: 27 M.S.: 2 Ph.D.: 1

Graduate advisor: Valerie L. Young

Undergraduate advisor: Michael E. Prudich

Student organization: AIChE
Advisor: Darin Ridgway

Department reports to: Dennis Irwin, Dean, Russ College of Engineering and Technology

Placement service: Thomas F. Korvas, Career Services

University of Toledo

Chemical and Environmental Engineering
Department
NI 3048, Mail Stop 305
2801 West Bancroft Street
Toledo, OH 43606-3390
Deliveries: 3048 Nitschke Hall

University phone . . . . (419) 530-4242
Department phone . . . . (419) 530-8080
Department FAX . . . . (419) 530-8086

Professors

Abraham, Martin A. (Dean) . . . . (419) 530-8092
martin.abraham@utoledo.edu
Bennett, Gary F. (Emeritus)
gbennett@eng.utoledo.edu
Coleman, Maria R. . . . . (419) 530-8091
macolema@eng.utoledo.edu
De Witt, Kenneth J. (Emeritus) . . . . (419) 530-8094
kdewitt@uoft02.utoledo.edu
Dismukes, John P. . . . . (419) 530-8065
john.dismukes@utoledo.edu
Jabarin, Saleh A. (Director) . . . . (419) 530-5005
sjabari@utnet.utoledo.edu
LeBlanc, Steven E. . . . . (419) 530-8264
steven.leblanc@utoledo.edu
Lipscomb, G. Glenn (Chair) . . . . (419) 530-8088
glenn.lipscomb@utoledo.edu
Nadarajah, Arunan (Associate Dean) . . (419) 530-8031
arunan.nadarajah@utoledo.edu
Poling, Bruce E. (Associate Dean) . . . . . (419) 530-8040
bpoling@uoft02.utoledo.edu
Varanasi, Sasidhar . . . . (419) 530-8093
svarana@ufot02.utoledo.edu

Associate Professors

Azad, Abdul-Majeed . . . . (419) 530-8103
aazad@eng.utoledo.edu
Cameron, Michael R. (Adjunct) . . . . (419) 530-5009
mcameron@eng.utoledo.edu
Schall, Constance A. . . . . (419) 530-8097
constance.schall@utoledo.edu
Xu, Xinglong . . . . (419) 530-8101
xxu@eng.utoledo.edu

Assistant Professors

Escobar, Isabel C. . . . . (419) 530-8267
isabel.escobar@utoledo.edu
Kim, Dong-Shik . . . . (419) 530-8084
dong.kim@utoledo.edu

Accredited by: ABET

Degrees granted 2003-2004:

B.S.: 28 M.S.: 7 Ph.D.: 3

Undergraduate advisor: Maria R. Coleman

Student organization: AIChE
Advisor: Dong-Shik Kim

Department reports to: Nagi Naganathan, Dean of Engineering

Placement service: Tamara S. Williams

Youngstown State University

Chemical Engineering Program
One University Plaza
Youngstown, OH 44555

University phone . . . . (330) 941-3000
Department phone . . . . (330) 941-3019
Department FAX . . . . (330) 941-3265

Professors

Hirtzel, Cynthia (Dean) . . . . (330) 941-3009
cshirtzel@ysu.edu
Lim, Soon-Sik . . . . (330) 941-3022
sslim@ysu.edu

Associate Professors

Garr, Jeanette . . . . (330) 941-1737
jgarr@ysu.edu

Assistant Professors

Price, Douglas (Director) . . . . (330) 941-3019
dmprice@ysu.edu

Accredited by ABET

Degrees granted 2003-2004:

B.S.: 12

Graduate advisor: Dr. Douglas M. Price

Undergraduate advisor: Dr. Douglas M. Price

Student organization: AIChE
Advisor: Dr. Soon-Sik Lim

Department reports to: Dr. Cynthia Hirtzel, Dean, College of Engineering & Technology (330) 941-3009

Placement service: Diane Hritz, Career Services (330) 742-3515

Oklahoma

University of Oklahoma

School of Chemical, Biological and Materials Engineering
100 E. Boyd, T-335
Sarkeys Energy Center
Norman, OK 73019-1004

University phone .... (405) 325-0311
Department phone .... (405) 325-5811
Department FAX .... (405) 325-5813

Professors

Bagajewicz, Miguel J. .... (405) 325-5458
bagajewicz@ou.edu
Block, Robert J. (Emeritus) .... (405) 325-5811
Crynes, Billy L. (Emeritus) .... (405) 325-3700
crynes@ou.edu
Daniels, Raymond D. (Emeritus) ...... (405) 325-5811
Ellington, Rex T. (Emeritus) .... (405) 325-5811
Harwell, Jeffrey H. (Associate Dean) ... (405) 325-4488
jharwell@ou.edu
Lee, Lloyd L. .... (405) 325-4374
lle@ou.edu
Lobban, Lance (Director) .... (405) 325-5814
llobban@ou.edu
Mallinson, Richard G. .... (405) 325-4378
mallinson@ou.edu
O'Rear, Edgar A. III .... (405) 325-4379
eorear@ou.edu
Resasco, Daniel E. .... (405) 325-4370
resasco@ou.edu
Scamehorn, John F. .... (405) 325-4382
scamehor@ou.edu
Shambaugh, Robert L. .... (405) 325-6070
shambaugh@ou.edu
Sliepcevich, C. M. (Emeritus) .... (405) 325-7263
Starling, Kenneth E. (Emeritus) .... (405) 325-5811

Associate Professors

Grady, Brian P. .... (405) 325-4369
bpgrady@ou.edu
Harrison, Roger G. .... (405) 325-4367
rharrison@ou.edu
Nollert, Matthias U. .... (405) 325-4366
nollert@ou.edu

Assistant Professors

McFetridge, Peter S. .... (405) 325-7193
mcfetridge@ou.edu
Papavassiliou, Dimitrios .... (405) 325-0574
dvpapava@ou.edu
Schmidtke, David W. .... (405) 325-7944
dschmidtke@ou.edu
Sikavitsas, Vassilios I. .... (405) 325-1511
vis@ou.edu

Accredited by: ABET

Degrees granted 2003-2004:
B.S.: 70 M.S.: 11 Ph.D.: 9

Graduate advisor: M. U. Nollert

Undergraduate advisor: V. I. Sikavitsas

Student organization: AIChE
Advisor: V. I. Sikavitsas

Department reports to: Dr. Skip Porter (405) 325-2621, Dean of Engineering

Placement service: Bette Scott (405) 325-1974

Oklahoma State University

School of Chemical Engineering - 423EN
Oklahoma State University
Stillwater, OK 74078-5021
Deliveries: 423 Engineering North

University phone . . . . (405) 744-5000
Department phone . . . . (405) 744-5280
Department FAX . . . . (405) 744-6338

Professors

Bell, Kenneth J. (Emeritus) . . . . (405) 744-5280
kbell@okstate.edu
Foutch, Gary L. . . . . (405) 744-5280
foutch@okstate.edu
Gasem, Khaled A. M. . . . . (405) 744-5280
gasem@okstate.edu
Johannes, Arland H. . . . . (405) 744-9118
aj@okstate.edu
Lewis, Randy S. . . . . (405) 744-9116
lewis@okstate.edu
Maddox, Robert N. (Emeritus) . . . . (405) 744-5280
rmaddox@okstate.edu
Rhinehart, R. Russell (Head) . . . . (405) 744-5280
rrr@okstate.edu
Robinson, Robert L. Jr. (Emeritus) . . . (405) 744-5280
rrobins@okstate.edu
Tree, D. Alan (Associate Dean) . . . . (405) 744-5280
tree@okstate.edu
Wagner, Jan . . . . (405) 744-5280
jwagner@okstate.edu

Associate Professors

High, Karen A. . . . . (405) 744-9112
high@okstate.edu
High, Martin S. . . . . (405) 744-9125
mhigh@okstate.edu
Whiteley, James R. . . . . (405) 744-9117
whitele@okstate.edu

Assistant Professors

Madihally, Sundararajan V. . . . . (405) 744-5280
svm@okstate.edu
Smay, James E. . . . . (405) 744-5280
smay@okstate.edu

Accredited by: ABET

Degrees granted 2003-2004:

B.S.: 30 M.S.: 12 Ph.D.: 9

Graduate advisor: Khaled A. M. Gasem

Undergraduate advisor: Randy S. Lewis

Student organization: AIChE
Advisor: Randy S. Lewis, Sundar V. Madihally

Department reports to: Karl N. Reid, Dean, College of Engineering, Architecture and Technology

Placement service: Cathy Southwick, Coordinator, Career Services

University of Tulsa

Chemical Engineering Department
600 South College
Tulsa, OK 74104-3189
Deliveries: 600 S. College

University phone . . . . (918) 631-2000
Department phone . . . . (918) 631-2226
Department FAX . . . . (918) 631-3268

Professors

Buthod, Paul (Emeritus)
Luks, Kraemer D. (918) 631-2974
kraemer-luks@utulsa.edu
Manning, Francis S. . . . . (918) 631-2977
frank-manning@utulsa.edu
Price, Geoffrey L. (Chair) . . . . (918) 631-2575
price@utulsa.edu
Sublette, Kerry L. (918) 631-3085
kerry-sublette@utulsa.edu
Thompson, Richard E. (Emeritus)
richardthompson@webzone.net

Associate Professors

Wisecarver, Keith D. . . . . (918) 631-2975
keith-wisecarver@utulsa.edu

Assistant Professors

Crunkleton, Daniel . . . . (918) 631-2644
daniel-crunkleton@utulsa.edu
Ford, Laura P . . . . (918) 631-2227
laura-ford@utulsa.edu

Other Faculty

Patton, Christi L. (918) 631-2978
christi-patton@utulsa.edu

Accredited by ABET

Degrees granted 2003-2004:

B.S.: 16 M.S.: 8

Graduate advisor: Geoffrey L. Price

Student organization: AIChE
Advisor: Laura P. Ford

Department reports to: Steve Bellovich (918) 631-2478, Dean of Engineering and Natural Science

Placement service: Mike Mills, Director of Placement

Oregon

Oregon State University

Chemical Engineering Department
103 Gleeson Hall
Corvallis, OR 97331-2702

Department phone .... (541) 737-4791
Department FAX .... (541) 737-4600

Professors

Kimura, Shoichi .... (541) 737-4831
kimuras@che.orst.edu
Knudsen, James G. (Emeritus) .... (541) 737-4791
knudsejg@che.orst.edu
Levenspiel, Octave (Emeritus) .... (541) 737-3618
levenspo@peak.org
McGuire, Joseph .... (541) 737-6306
mcguirej@engr.oregonstate.edu
Wicks, Charles E. (Emeritus) .... (541) 737-4797
wicksc@ucs.orst.edu
Williamson, Kenneth (Head) .... (541) 737-6836
kenneth.williamson@oregonstate.edu

Associate Professors

Bothwell, Michelle K. .... (541) 737-6313
bothwell@engr.orst.edu
Jovanovic, Goran .... (541) 737-3614
goran@che.orst.edu
Koretsky, Milo D. .... (541) 737-4591
koretsm@engr.orst.edu
Levien, Keith L. .... (541) 737-3155
levienk@engr.orst.edu
Peattie, Robert A. .... (541) 737-9359
peattie@che.orst.edu
Rochefort, Willie E. .... (541) 737-2408
rochefsk@engr.orst.edu
Rorrer, Gregory L. .... (541) 737-3370
rorrergl@che.orst.edu

Assistant Professors

Chang, Chih-hung (Alex) .... (541) 737-8548
changch@che.orst.edu
Yokochi, Alexandre .... (541) 737-9357
alex.yokochi@orst.edu

Other Faculty

Hackleman, David .... (541) 737-8988
david.hackleman@oregonstate.edu

Accredited by: ABET

Degrees granted 2003-2004:

B.S.: 36 M.S.: 3 Ph.D.: 1

Graduate advisor: Goran Jovanovic

Undergraduate advisor: W. E. Rochefort

Student organization: AIChE
Advisor: W. E. Rochefort

Department reports to: Ron Adams, Dean

Placement service: Tom Munnerlyn (541) 737-4085

Pennsylvania

Bucknell University

Chemical Engineering Department
Lewisburg, PA 17837

University phone .... (570) 577-1271
Department phone .... (570) 577-1114
Department FAX .... (570) 577-1141

Professors

Csernica, Jeffrey (Chair) .... (570) 577-1114
cernica@bucknell.edu
Hanyak, Michael E. Jr. .... (570) 577-3547
hanyak@bucknell.edu
King, William E. Jr. .... (570) 577-3403
wking@bucknell.edu
Pommersheim, James M. (Emeritus)
pommrshm@bucknell.edu
Prince, Michael J. .... (570) 577-1781
prince@bucknell.edu
Slonaker, Robert E. Jr. (Emeritus)
slonaker@bucknell.edu
Snyder, William J. .... (570) 577-1293
bsnyder@bucknell.edu

Associate Professors

Maneval, James E. .... (570) 577-1669
maneval@bucknell.edu

Assistant Professors

Cavanagh, Daniel P. .... (570) 577-3402
dcavanag@bucknell.edu
Jablonski, Erin L. .... (570) 577-1644
ejablons@bucknell.edu
Raymond, Timothy M. .... (570) 577-3192
traymond@bucknell.edu
Vigeant, Margot A.S. .... (570) 577-1646
mvigeant@bucknell.edu

Accredited by: ABET

Degrees granted 2003-2004:

B.S.: 25 M.S.: 3

Student organization: AIChE
Advisor: Timothy Raymond

Department reports to: James G. Orbison (570) 577-3711, Dean of Engineering

Placement service: Career Development (570) 577-1238, Director

Carnegie Mellon University

Chemical Engineering Department
Pittsburgh, PA 15213
Deliveries: Doherty Hall 1105, attn: Toni McIltrot

University phone .... (412) 268-2000
Department phone .... (412) 268-2230
Department FAX .... (412) 268-7139

Professors

Anderson, John L. (Adjunct) .... (412) 268-2537
johna@cmu.edu
Biegler, Lorenz T. .... (412) 268-2232
lb01@andrew.cmu.edu
Domach, Michael M. .... (412) 268-2246
md0q@andrew.cmu.edu
Gellman, Andrew J. (Head) .... (412) 268-3848
ag4b@andrew.cmu.edu
Grossmann, Ignacio E. .... (412) 268-2230
grossmann@cmu.edu
Jhon, Myung S. .... (412) 268-2233
mj3a@andrew.cmu.edu
Li, Kun (Emeritus) .... (412) 268-2226
kli@andrew.cmu.edu
Pandis, Spyros .... (412) 268-3531
spyros@andrew.cmu.edu
Powers, Gary J. .... (412) 268-3569
gp0c@andrew.cmu.edu
Prieve, Dennis C. .... (412) 268-2247
dcprieve@andrew.cmu.edu
Przybycien, Todd .... (412) 268-3857
todd@andrew.cmu.edu
Sides, Paul .... (412) 268-3846
ps7r@andrew.cmu.edu
Tilton, Robert .... (412) 268-1159
tilton@andrew.cmu.edu
Toor, Herbert L. (Emeritus) .... (412) 268-2225
ht16@andrew.cmu.edu
Westerberg, Arthur W. (Emeritus) .... (412) 268-2230
a.westerberg@cmu.edu
White, Lee .... (412) 268-2227
white@andrew.cmu.edu
Ydstie, B. Erik .... (412) 268-2235
ydstie@andrew.cmu.edu

Associate Professors

Hauan, Steinar .... (412) 268-4393
steinhau@andrew.cmu.edu
Schneider, James .... (412) 268-4314
jamess3@andrew.cmu.edu
Sholl, David .... (412) 268-4207
sholl@andrew.cmu.edu
Walker, Lynn .... (412) 268-3020
lwalker@andrew.cmu.edu

Assistant Professors

Donahue, Neil M. .... (412) 268-4415
nmd@andrew.cmu.edu

Accredited by: ABET

Degrees granted 2003-2004:
B.S.: 61 M.S.: 5 Ph.D.: 11

Graduate advisor: Paul Sides

Undergraduate advisor: Myung Jhon

Student organization: AIChE
Advisor: James Schneider

Department reports to: Andrew Gellman, Department Head, Chemical Engineering Department

Placement service: Lisa Dickter, Career Services

Drexel University

Chemical Engineering Department
Philadelphia, PA 19104
Deliveries: 34th and Ludlow St.

University phone . . . . (215) 895-2000
Department phone . . . . (215) 895-2227
Department FAX . . . . (215) 895-5837

Professors

Coughanowr, Donald R. (Emeritus) . . . . (215) 895-2233
Grossmann, Elihu D. . . . . (215) 895-2229
Mutharasan, Rajakannu . . . . (215) 895-2236
Soroush, Masoud . . . . (215) 895-1710
Weinberger, Charles B. . . . . (215) 895-2226

Associate Professors

Dan, Nily . . . . (215) 895-6624
Lowman, Anthony (Associate Dean) . . . (215) 895-2228
Palmese, Giuseppe (Chair) . . . . (215) 895-5814

Assistant Professors

Abrams, Cameron . . . . (215) 895-2231
Cairncross, Richard . . . . (215) 895-2230
Elabd, Yossef . . . . (215) 895-0986
Wrenn, Steven . . . . (215) 895-6694

Instructors

Meyer, Stephen P. (215) 895-1855
Rowell, George . . . . (215) 895-0987

Accredited by: ABET

Degrees granted 2003-2004:

B.S.: 44 M.S.: 14 Ph.D.: 4

Graduate advisor: M. Soroush

Undergraduate advisor: S. P. Meyer

Student organization: AIChE
Advisor: Cameron F. Abrams

Department reports to: S. Guceri, Dean of Engineering

Placement service: Aminta Hawkins-Breaux (215) 895-1630

Lafayette College

Acopian Engineering Center
Chemical Engineering Department
Easton, PA 18042-1775
Deliveries: Room 122, Acopian Engineering Center

University phone . . . . (610) 330-5000
Department phone . . . . (610) 330-5435
Department FAX . . . . (610) 330-5059

Professors

Martin, J. Ronald . . . . (610) 330-5430
martinj@lafayette.edu
Schaffer, James P. (Director) . . . . (610) 330-5406
schaffej@lafayette.edu
Uz, Mehmet . . . . (610) 330-5408
metmuz@lafayette.edu

Associate Professors

Piergiovanni, Polly . . . . (610) 330-5431
piergiop@lafayette.edu
Tavakoli, Javad (Head) . . . . (610) 330-5433
tavakoli@lafayette.edu

Assistant Professors

Bogaert-Alvarez, Ricardo . . . . (610) 330-5432
bogaertr@lafayette.edu
Ferri, James . . . . (610) 330-5820
ferrij@lafayette.edu
Morton, Samuel . . . . (610) 330-5417
mortons@lafayette.edu

Accredited by: ABET

Degrees granted 2003-2004:

B.S.: 25

Student organization: AIChE
Advisor: Javad Tavakoli

Department reports to: James P. Schaffer (610) 330-5403, Director of Engineering

Placement service: Linda Arra (610) 330-5116, Director, Career Services

Lehigh University

Chemical Engineering Department
Iacocca Hall, 111 Research Dr.
Bethlehem, PA 18015-4791

University phone .... (610) 758-3000
Department phone .... (610) 758-4260
Department FAX .... (610) 758-5057

Professors

Blythe, Philip A. .... (610) 758-3782
pab0@lehigh.edu
Caram, Hugo S. .... (610) 758-4259
hsc0@lehigh.edu
Charles, Marvin .... (610) 758-4268
mc02@lehigh.edu
Chaudhury, Manoj K. .... (610) 758-4471
mkc4@lehigh.edu
Chen, John C. .... (610) 758-4091
jcc0@lehigh.edu
Clump, Curtis W. (Emeritus) .... (610) 758-4260
cwc3@lehigh.edu
El-Aasser, Mohamed S. (Dean) .... (610) 758-5308
mse0@lehigh.edu
Hsu, James T. .... (610) 758-4257
jth0@lehigh.edu
Jagota, Anand .... (610) 758-4396
anj6@lehigh.edu
Klein, Andrew .... (610) 758-4219
ak04@lehigh.edu
Luyben, William L. .... (610) 758-4256
wll0@lehigh.edu
McHugh, Anthony J. (Chair) .... (610) 758-4470
ajm8@lehigh.edu
Schiesser, William (Emeritus) 610-758-4264
wes1@lehigh.edu
Sengupta, Arup K. .... (610) 758-3534
aks0@lehigh.edu
Silebi, Cesar A. .... (610) 758-4267
cas5@lehigh.edu
Sircar, Shivaji .... (610) 758-4469
shs3@lehigh.edu
Sperling, Leslie H. (Emeritus) .... (610) 758-3845
lhs0@lehigh.edu
Stein, Fred P. (Emeritus) .... (610) 758-6852
fps0@lehigh.edu
Stenger, Harvey G. Jr. .... (610) 758-4791
hgs0@lehigh.edu
Tuzla, Kemal (Associate Chair) .... (610) 758-4729
kt01@lehigh.edu
Wachs, Israel E. .... (610) 758-4274
iew0@lehigh.edu

Associate Professors

Kothare, Mayuresh V. .... (610) 758-6654
mvk2@lehigh.edu

Assistant Professors

Gilchrist, James F.
Laurenzi, Ian J.
Rajagopalan, Padma

Other Faculty

Matsen, John M. (Adjunct) .... (610) 758-4461
jom7@lehigh.edu
Simpkins, Peter G. (Adjunct) .... (610) 758-3782
pab0@lehigh.edu

Accredited by: ABET

Degrees granted 2003-2004:

B.S.: 47 M.S.: 15 Ph.D.: 9

Graduate advisor: James T. Hsu

Student organization: AIChE
Advisor: Harvey G. Stenger

Department reports to: Mohamed S. El-Aasser (610) 758-5308, Dean

Placement service: Donna Goldfeder (610) 758-3712

University of Pennsylvania

Department of Chemical and Biomolecular Engineering
311A Towne Building
220 South 33rd Street
Philadelphia, PA 19104-6393

University phone .... (215) 898-5000
Department phone .... (215) 898-8351
Department FAX .... (215) 573-2093

Professors

Churchill, Stuart W. (Emeritus) .... (215) 898-5579
churchil@seas.upenn.edu
Diamond, Scott L. .... (215) 573-5702
sld@seas.upenn.edu
Forsman, William C. (Emeritus) ...... (215) 898-7077
forsman@seas.upenn.edu
Glandt, Eduardo D. (Dean) .... (215) 898-7244
eglandt@seas.upenn.edu
Gorte, Raymond J. .... (215) 898-4439
gorte@seas.upenn.edu
Hammer, Daniel A. .... (215) 573-6761
hammer@seas.upenn.edu
Myers, Alan L. (Emeritus) .... (215) 898-7078
amyers@seas.upenn.edu
Perlmutter, Daniel D. (Emeritus) ...... (215) 898-8350
perlmutt@seas.upenn.edu
Quinn, John A. (Emeritus) .... (215) 898-8503
quinn@seas.upenn.edu
Seider, Warren D. .... (215) 898-7953
seider@seas.upenn.edu
Shieh, Wen K. .... (215) 898-4634
shieh@seas.upenn.edu
Vohs, John M. (Chair) .... (215) 898-6318
vohs@seas.upenn.edu

Associate Professors

Composto, Russell J. .... (215) 898-4451
composto@sol1.lrsm.upenn.edu
Discher, Dennis .... (215) 898-9950
discher@seas.upenn.edu
Graves, David J. .... (215) 898-7951
graves@seas.upenn.edu
Ungar, Lyle H. .... (215) 898-7449
ungar@central.cis.upenn.edu
Winey, Karen I. .... (215) 898-0593
winey@sol1.lrsm.upenn.edu

Assistant Professors

Boder, Eric T. .... (215) 898-5658
boder@seas.upenn.edu
Crocker, John C. .... (215) 898-9188
jcrocker@seas.upenn.edu
Sinno, Talid R. .... (215) 898-2511
talid@seas.upenn.edu

Other Faculty

Fabiano, Leonard A. (Adjunct) .... (215) 898-9879
lfabiano@seas.upenn.edu
Kuehl, Guenter H. (Adjunct) .... (215) 573-4201
kuehl@seas.upenn.edu
Levins, John M. (Adjunct)
McKenzie, Steven E. (Adjunct) .... (215) 955-7786
Olson, David H. (Adjunct) .... (215) 573-4201
dholson@seas.upenn.edu
Ramanarayanan, Trikur A. (Adjunct) .. (215) 573-0294
trikur@seas.upenn.edu

Accredited by: ABET

Degrees granted 2003-2004:

B.S.: 11 M.S.: 11 Ph.D.: 9

Graduate advisor: Dr. Eric T. Boder

Undergraduate advisor: Dr. David J. Graves

Student organization: AIChE
Advisor: Dr. David J. Graves

Department reports to: Eduardo D. Glandt (215) 898-7244, Dean

Placement service: Patricia Rose (215) 898-3208

Pennsylvania State University

Chemical Engineering Department
158 Fenske Laboratory
University Park, PA 16802

University phone . . . . . (814) 865-4700
Department phone . . . . . (814) 865-2574
Department FAX . . . . . (814) 865-7846

Professors

Ben-Jebria, Aziz . . . . . (814) 863-8049
axb23@psu.edu
Borhan, Ali . . . . . (814) 865-7847
borhan@psu.edu
Braun, Walter G. (Emeritus) . . . . . (814) 234-9888
wgb1@psu.edu
Curtis, Wayne R. . . . . . (814) 863-4805
wrc2@psu.edu
Danner, Ronald P. (Emeritus) . . . . . (814) 863-4814
rpd@psu.edu
Daubert, Thomas E. (Emeritus) . . . . . (814) 863-4816
ted@psu.edu
Duda, J. Larry . . . . . (814) 865-1640
jld@psu.edu
Engel, Alfred J. (Emeritus) . . . . . (814) 865-2574
aje@psu.edu
Fichthorn, Kristen A. . . . . . (814) 863-4807
fichthorn@psu.edu
Foley, Henry C. . . . . . (814) 865-2574
hcf2@psu.edu
Helfferich, Friedrich G. (Emeritus) . . . . . (814) 863-4675
fxh1@psu.edu
Kabel, Robert L. (Emeritus) . . . . . (814) 863-4815
r8k@psu.edu
Maranas, Costas . . . . . (814) 863-9958
costas@psu.edu
Nagarajan, R. . . . . . (814) 863-1973
rxn@psu.edu
Ultman, James S. . . . . . (814) 863-4802
jsu@psu.edu
Vannice, M. Albert . . . . . (814) 863-4803
mavche@engr.psu.edu
Vrentas, James S. . . . . . (814) 863-4808
jsv1@psu.edu
Zydney, Andrew L. (Interim Head) . . . . . (814) 863-7113
zydney@engr.psu.edu

Associate Professors

Matsoukas, Themistoklis . . . . . (814) 863-2002
matsoukas@psu.edu
Pishko, Michael V. . . . . . (814) 863-4810
mpishko@engr.psu.edu

Assistant Professors

Armaou, Antonios . . . . . (814) 863-5316
armaou@engr.psu.edu
Cirino, Patrick C. . . . . . (814) 865-2574
Hahm, Jong-in . . . . . (814) 863-4801
jhahm@engr.psu.edu
Kim, Seong Han . . . . . (814) 863-4809
shkim@engr.psu.edu
Maranas, Janna K. . . . . . (814) 863-6228
jmaranas@psu.edu
Nedwick, Robert . . . . . (814) 863-6229
nedwick@engr.psu.edu
Velegol, Darrell . . . . . (814) 865-8739
velegol@psu.edu

Other Faculty

Perez, Joseph M. . . . . . (814) 865-0340
jmp13@psu.edu

Accredited by: ABET

Degrees granted 2003-2004:

B.S.: 115 M.S.: 9 Ph.D.: 11

Graduate advisor: Ali Borhan

Undergraduate advisor: Themis Matsoukas

Student organization: AIChE
Advisor: Wayne Curtis

Department reports to: D. Wormley, Harold and Inge Marcus Dean of Engineering

Placement service: Career Development, Jack Rayman

University of Pittsburgh

Chemical and Petroleum Engineering Department
1249 Benedum Hall
Pittsburgh, PA 15261
Deliveries: 3700 O'Hara Street

University phone .... (412) 624-4141
Department phone .... (412) 624-9630
Department FAX .... (412) 624-9639

Professors

Ataai, Mohammad M. .... (412) 624-9648
ataai@pitt.edu
Balazs, Anna C. .... (412) 648-9250
balazsl@engr.pitt.edu
Beckman, Eric J. (Chair) .... (412) 624-9631
beckman@engr.pitt.edu
Borovetz, Harvey S. (Joint) .... (412) 624-4725
borovetzhs@msx.upmc.edu
Chiang, Shiao-Hung (Emeritus) .... (412) 624-9636
chiang@engr.pitt.edu
Enick, Robert M. .... (412) 624-9649
enick@engr.pitt.edu
Federspiel, William .... (412) 383-9499
federspielwj@upmc.edu
Holder, Gerald D. (Dean) .... (412) 624-9809
holder@engr.pitt.edu
Klinzing, George E. .... (412) 624-0784
klinzing@engr.pitt.edu
Kovalchuk, Vladimir .... (412) 624-4754
vkovalch@pitt.edu
Lindt, Jan T. .... (412) 624-9729
jtlindt@pitt.edu
Morsi, Badie I. .... (412) 624-9650
morsi@engr.pitt.edu
Russell, Alan J. (Joint) .... (412) 235-5109
russellaj@msx.upmc.edu
Tierney, John W. (Emeritus) .... (412) 624-9645
tierney@engr.pitt.edu
Wender, Irving .... (412) 624-9644
wender@engr.pitt.edu

Associate Professors

Beroes, Charles (Emeritus) .... (412) 648-8989
beroes@engr.pitt.edu
Cobb, James T. Jr. (Emeritus) .... (412) 624-7443
cobb@engr.pitt.edu
d'Itri, Julie L. .... (412) 624-9634
ditri@engr.pitt.edu
Johnson, J. Karl .... (412) 624-5644
karlj@engr.pitt.edu
McCarthy, Joseph .... (412) 624-7362
mccarthy@engr.pitt.edu
Wagner, William R. (Joint) .... (412) 647-8311
wagnerwr@msx.upmc.edu

Assistant Professors

Parker, Robert S. .... (412) 624-7364
rparker@engr.pitt.edu
Patzer, John F. II .... (412) 624-9819
patzer@engr.pitt.edu
Velankar, Sachin .... (412) 624-9984
velankar@pitt.edu
Veser, Goetz .... (412) 624-1042
gveser@pitt.edu

Accredited by: ABET

Degrees granted 2003-2004:

B.S.: 59 M.S.: 11 Ph.D.: 9

Graduate advisor: William J. Federspiel

Undergraduate advisor: Robert M. Enick

Student organization: AIChE
Advisor: Joseph McCarthy

Department reports to: Gerald D. Holder, Dean

Placement service: Chris Frankovic

Villanova University

Department of Chemical Engineering
800 Lancaster Avenue
Villanova, PA 19085

University phone .... (610) 519-4500
Department phone .... (610) 519-4950
Department FAX .... (610) 519-7354

Professors

Joye, Donald D. .... (610) 519-4966
donald.joye@villanova.edu
Kelly, C. Michael (Chair) .... (610) 519-4953
c.michael.kelly@villanova.edu
Myers, John A. (Emeritus)
Punzi, Vito L. .... (610) 519-4946
vito.punzi@villanova.edu
Rice, William J. (Emeritus)
Sweeny, Robert F. (Emeritus)
Yuen, Hsiu Huo (Emeritus)

Associate Professors

Muske, Kenneth R. .... (610) 519-6195
kenneth.muske@villanova.edu
Ritter, Edward R. .... (610) 519-4948
edward.ritter@villanova.edu
Skaf, Dorothy W. .... (610) 519-4952
dorothy.skaf@villanova.edu
Weinstein, Randy D. .... (610) 519-4954
randy.weinstein@villanova.edu

Assistant Professors

Kelly, William J. .... (610) 519-4947
william.j.kelly@villanova.edu

Accredited by: ABET

Degrees granted 2003-2004:

B.S.: 19 M.S.: 11

Graduate advisor: Vito L. Punzi

Student organization: AIChE
Advisor: Randy D. Weinstein

Department reports to: Dr. Barry Johnson, Dean of Engineering

Placement service: N. Dudak

Widener University

Chemical Engineering Department
Chester, PA 19013-5792
Deliveries: Kirkbride Hall, 17th and Walnut Sts.

University phone .... (610) 499-4000
Department phone .... (610) 499-4051
Department FAX .... (610) 499-4059

Professors

Maffia, Gennaro J. (Chair) .... (610) 499-4089
Gennaro.J.Maffia@widener.edu
McWilliams, Thomas G. Jr. (Emeritus) (610) 499-4608
tgm0001@widener.edu

Associate Professors

Chen, David H. T. .... (610) 499-4049
David.H.Chen@widener.edu
McNeil, Kenneth M. .... (610) 499-4056
Kenneth.M.McNeil@widener.edu
Nippert, Charles R. .... (610) 499-4050
Charles.R.Nippert@widener.edu

Other Faculty

Chitra, Surya (Adjunct) .... (610) 499-4051
Henderson, Louis (Adjunct) .... (610) 499-4051
Hoopes, John W. Jr. (Adjunct) .... (610) 499-4051
Pourki, Forouza (Adjunct) .... (610) 499-4051
Rowell, George (Adjunct) .... (610) 499-4051
Wu, Huey (Adjunct) ....(610)499-4051

Accredited by: ABET

Degrees granted 2003-2004:

B.S.: 5 M.S.: 9

Graduate advisor: John H. Dixon

Undergraduate advisor: G. J. Maffia

Student organization: AIChE
Advisor: K. M. McNeil

Department reports to: Dr. Fred Akl (610) 499-4036, Dean of Engineering

Placement service: Barbara Buckley (610) 499-4509

Puerto Rico

University of Puerto Rico

Department of Chemical Engineering
PO Box 9046
Mayaguez PR 00681-9046

University phone . . . . (787) 832-4040
Department phone . . . . (787) 832-4040 x2592
Department FAX . . . . (787) 834-3655

Professors

Aguayo, Guillermo . . . . x3605 †
gaguayo@uprm.edu

Benítez, Jaime x3209
jbenrod@caribe.net

Briano, Julio G. (Associate Chair) x2545
jbriano@uprm.edu

Cardona-Martínez, Nelson (Chair) x2592
ncardona@uprm.edu

Colucci-Rios, José A. x3749
jcolucci@uprm.edu

Colón, Guillermo x3727
gcolon@uprm.edu

Estévez, L. Antonio x2573
estevez@uprm.edu

Mandavilli, Satya N. x3589
s_mandavilli@rumac.uprm.edu

Ramírez, Carlos A. x2561
cramirez@uprm.edu

Rodríguez-Ramírez, Abraham x2581
arodrig@uprm.edu

Saliceti-Piazza, Lorenzo x2549
saliceti@uprm.edu

Sridhar, Lakshmi N. x3173
l_sridhar@rumac.uprm.edu

Villafañe-Ruiz, Gilberto x2569
gvilla@coqui.net

Associate Professors

Bogere, Moses N. x3607
mbogere@caribe.net

Suleiman-Rosado, David x2685
dsuleiman@uprm.edu

Torres-Lugo, Madeline x2585
madeline@ece.uprm.edu

Velázquez-Figueroa, Carlos (Associate Chair) . . . x2576
cvelazqu@uprm.edu

Assistant Professors

González-Pérez, Braulio x2473
braulio@uprm.edu

Hernández-Maldonado, Arturo J. x3748

Padrón, Federico x3606
fpadron@uprm.edu

Rinaldi-Ramos, Carlos Manuel 3585
crinaldi@uprm.edu

Instructors

Avilés-Molina, Misael O.
Córdova-Figueroa, Ubaldo M.
Martínez-Iñesta, María M.
Nazario-Acevedo, Aldo
Siberio-Pérez, Diana Y.

Other Faculty

Mehta, Narinder K. x3157
nmehta@uprm.edu

† Faculty numbers are extensions of (787) 832-4040.

Accredited by: ABET

Degrees granted 2003-2004:

B.S.: 87 M.S.: 7 Ph.D.: 1

Graduate advisor: Dr. Carlos M. Rinaldi-Ramos

Undergraduate advisor: Lourdes Fabregas

Student organization: AIChE and IIQPR
Advisor: Prof. Federico Padrón

Department reports to: Ramón Vásquez, Dean, College of Engineering

Placement service: Carmen Sol Ramírez, Director

Rhode Island

Brown University

Chemical Engineering Program
Division of Engineering, Box D
Providence, RI 02912
Deliveries: 182 Hope Street

University phone .... (401) 863-2677
Department phone .... (401) 863-2677
Department FAX .... (401) 863-1157

Professors

Breuer, Kenneth .... (401) 863-2870
kbreuer@brown.edu
Calo, Joseph M. (Contact) .... (401) 863-1421
Joseph_Calo@brown.edu
Caswell, Bruce (Emeritus) .... (401) 863-1448
Bruce_Caswell@brown.edu
Dobbins, Richard A. (Emeritus) .... (401) 863-2867
Richard_Dobbins@brown.edu
Hurt, Robert H. .... (401) 863-2685
Robert_Hurt@brown.edu
Karlsson, Sture K. F. (Emeritus) ...... (401) 863-3200
Sture_Karlsson@brown.edu
Liu, Joseph T. C. .... (401) 863-2654
Joseph_Liu@brown.edu
Richardson, Peter D. .... (401) 863-2687
Peter_Richardson@brown.edu
Sheldon, Brian W. .... (401) 863-2866
Brian_Sheldon@brown.edu
Sibulkin, Merwin (Emeritus) .... (401) 863-2867
Merwin_Sibulkin@brown.edu
Suuberg, Eric M. (Associate Dean) .... (401) 863-1420
Eric_Suuberg@brown.edu

Assistant Professors

Tripathi, Anubhav .... (401) 863-3063
Anubhav_Tripathi@brown.edu
Vlahovska, Petia (Visiting) .... (401) 863-2318
Petia_Vlahovska@brown.edu

Accredited by: ABET

Degrees granted 2003-2004:

B.S.: 4 M.S.: 2 Ph.D.: 4

Graduate advisor: Robert Hurt

Undergraduate advisor: Joseph M. Calo

Student organization: AIChE
Advisor: Joseph M. Calo

Department reports to: Clyde Briant (401) 863-2276, Dean of Engineering

Placement service: Sheila Curran (401) 863-3326, Director

University of Rhode Island

Department of Chemical Engineering
205 Crawford Hall
Kingston, RI 02881-0805

University phone .... (401) 874-1000
Department phone .... (401) 874-2655
Department FAX .... (401) 874-4689

Professors

Barnett, Stanley M. .... (401) 874-2443
barnett@egr.uri.edu
Bose, Arijit (Chair) .... (401) 874-2804
bose@egr.uri.edu
Brown, Richard A. .... (401) 874-2707
brown@egr.uri.edu
Gregory, Otto J. (Associate Dean) ..... (401) 874-2085
gregory@egr.uri.edu
Knickle, Harold N. .... (401) 874-2678
knickle@egr.uri.edu
Lucia, Angelo .... (401) 874-2814
lucia@egr.uri.edu
Rose, Vincent C. (Emeritus) .... (401) 874-5924
rose@egr.uri.edu

Associate Professors

Gray, Donald J. .... (401) 874-2651
gray@egr.uri.edu
Greenfield, Michael L. .... (401) 874-9289
greenfield@egr.uri.edu
Rivero-Hudec, Mercedes (Associate Dean) (401) 874-2271
rivero@egr.uri.edu

Other Faculty

Crisman, Everett .... (401) 874-5076
crisman@egr.uri.edu
Park, Eugene .... (401) 874-4303
epark@earthlink.net

Accredited by: ABET

Degrees granted 2003-2004:

B.S.: 19 M.S.: 4

Graduate advisor: M. L. Greenfield

Undergraduate advisor: S. M. Barnett

Student organization: AIChE
Advisor: D. J. Gray

Department reports to: Bahram Nassersharif (401) 874-2186, Dean, College of Engineering

Placement service: Roberta K. Koppel (401) 874-2311, Director

South Carolina

Clemson University

Department of Chemical Engineering
Box 340909
Clemson, SC 29634-0909
Deliveries: 127 Earle Hall

University phone .... (864) 656-3311
Department phone .... (864) 656-3055
Department FAX .... (864) 656-0784

Professors

Barron, Charles H. (Emeritus) .... (864) 656-5419
charles.barron@ces.clemson.edu
Beard, John N. (Emeritus)
Beckwith, William F. (Emeritus)
Edie, Danny D. .... (864) 656-4535
dan.edie@ces.clemson.edu
Gooding, Charles H. .... (864) 656-2621
charles.gooding@ces.clemson.edu
Goodwin, James G. (Chair) .... (864) 656-6614
james.goodwin@ces.clemson.edu
Melsheimer, Stephen S. (Associate Dean) (864) 656-5706
stephen.melsheimer@ces.clemson.edu
Mullins, Joseph C. (Emeritus) .... (864) 656-5426
Ogale, Amod A. .... (864) 656-5483
amod.ogale@ces.clemson.edu
Thies, Mark C. .... (864) 656-5424
mark.thies@ces.clemson.edu

Associate Professors

Bruce, David A. .... (864) 656-5425
david.bruce@ces.clemson.edu
Harcum, Sarah .... (864) 656-3055
harcum@clemson.edu
Hirt, Douglas E. .... (864) 656-0822
doug.hirt@ces.clemson.edu
Kilbey, S. Michael .... (864) 656-5423
mike.kilbey@ces.clemson.edu
Rice, Richard W. .... (864) 656-5428
richard.rice@ces.clemson.edu

Assistant Professors

Harrison, Graham M. .... (864) 656-6399
graham.harrison@ces.clemson.edu
Husson, Scott M. .... (864) 656-4502
scott.husson@ces.clemson.edu
Metters, Andrew .... (864) 656-3055
metters@clemson.edu

Accredited by: ABET

Degrees granted 2003-2004:

B.S.: 29 M.S.: 4 Ph.D.: 3

Graduate advisor: A.A. Ogale

Undergraduate advisor: S.M. Kilbey

Student organization: AIChE
Advisor: D. Bruce

Department reports to: T. M. Keinath (864) 656-3202, Dean of Engineering and Science

Placement service: Flora Riley (864) 656-2152

University of South Carolina

Department of Chemical Engineering
Swearingen Engineering Center
Columbia, SC 29208

Department phone .... (803) 777-4181
Department FAX .... (803) 777-0973

Professors

Amiridis, Michael D. (Chair) .... (803) 777-7294
amiridis@engr.sc.edu
Gibbons, Joseph H. .... (803) 777-8978
gibbons@engr.sc.edu
Matthews, Michael A. .... (803) 777-0556
matthews@engr.sc.edu
Ploehn, Harry J. (Associate Chair) .... (803) 777-7307
ploehn@engr.sc.edu
Popov, Branko N. .... (803) 777-7314
popov@engr.sc.edu
Ritter, James A. .... (803) 777-3590
ritter@engr.sc.edu
Van Brunt, Vincent .... (803) 777-3115
vanbrunt@engr.sc.edu
Van Zee, John W. .... (803) 777-2285
vanzee@engr.sc.edu
Weidner, John W. .... (803) 777-3207
weidner@engr.sc.edu
White, Ralph E. (Dean) .... (803) 777-3270
white@engr.sc.edu

Associate Professors

Gadala-Maria, Francis A. .... (803) 777-3182
gadala-m@engr.sc.edu
Jabbari, Esmaiel .... (803) 777-8022
jabbari@engr.sc.edu
Papathanasiou, Thanasis D. .... (803) 777-7219
papathan@engr.sc.edu

Assistant Professors

Bender, Jonathan W. .... (803) 777-5025
benderjw@engr.sc.edu
Gatzke, Edward P. .... (803) 777-1159
gatzke@engr.sc.edu
Moss, Melissa A. .... (803) 777-5604
mossme@engr.sc.edu
Stanford, Thomas G. .... (803) 777-4101
stanford@engr.sc.edu
Williams, Christopher .... (803) 777-0143
willia84@engr.sc.edu

Accredited by: ABET

Degrees granted 2003-2004:

B.S.: 33 M.S.: 5 Ph.D.: 11

Student organization: AIChE
Advisor: Christopher Williams

Department reports to: Ralph E. White, Dean

Placement service: Robert Rinehart, Engineering Career Services (803) 777-3972

South Dakota

South Dakota School of Mines and Technology

Chemistry/Chemical Engineering Department
501 E. Saint Joseph Street
Rapid City, SD 57701-3995

University phone .... (605) 394-2511
Department phone .... (605) 394-2421
Department FAX .... (605) 394-1232

Professors

Bauer, Larry G. (Emeritus)
Klemm, William A. (Emeritus)
Munro, James M. .... (605) 394-2422
James.Munro@sdsmt.edu
Puszynski, Jan A. (Dean) .... (605) 394-1230
Jan.Puszynski@sdsmt.edu
Sandvig, Robert L. (Emeritus)
Winter, Robb M. (Chair) .... (605) 394-1237
Robb.Winter@sdsmt.edu

Associate Professors

Dixon, David J. .... (605) 394-1235
David.Dixon@sdsmt.edu

Assistant Professors

Gilcrease, Patrick .... (605) 394-1239
Patrick.Gilcrease@sdsmt.edu

Instructors

Chian, Wei .... (605) 394-1231
wei.chian@sdsmt.edu
Smith, Stanley .... (605) 394-2421
stanley.smith@sdsmt.edu

Accredited by: ABET

Degrees granted 2003-2004:

B.S.: 12 M.S.: 11

Student organization: AIChE
Advisor: David J. Dixon

Department reports to: R.M. Winter, (605) 394-1237
Professor of Chemical Engineering, Chair

Placement service: Darrell Sawyer (605) 394-2667, Director

Tennessee

Christian Brothers University

Chemical and Biochemical Engineering Department
650 East Parkway South
Memphis, TN 38104-5581
Deliveries: Nolan Engineering Building

University phone . . . . (901) 321-3000
Department phone . . . . (901) 321-3401
Department FAX . . . . (901) 321-3402

Professors

Ray, Asit K . . . . (901) 321-3418
aray@cbu.edu

Associate Professors

Price, Randel M. . . . . (901) 321-3412
rprice@cbu.edu

Assistant Professors

Pourhashemi, Ali (Chair) . . . . (901) 321-3401
apourhas@cbu.edu

Instructors

Zhou, Youngquan (Adjunct) . . . . (901) 321-3405

Accredited by: ABET, SACS

Degrees granted 2003-2004:

B.S.: 6

Undergraduate advisor: Dr. Ali Pourhashemi

Student organization: AIChE
Advisor: Dr. Asit Ray

Department reports to: Dr. Siripong Malasari, Dean of Engineering

Placement service: Ms. Elizabeth McWillie, Director of Career Center

The University of Tennessee at Chattanooga

College of Engineering and Computer Science
Engineering: Chemical
615 McCallie Avenue
Chattanooga, TN 37403-2598
Deliveries: 430 EMCS, UTC, Chattanooga

University phone . . . . (423) 425-4411
Department phone . . . . (423) 425-5217
Department FAX . . . . (423) 425-5229

Professors

Cunningham, James . . . . (423) 425-4361
jim-cunningham@utc.edu

Henry, Jim . . . . (423) 425-4398
jim-henry@utc.edu

Associate Professors

Jones, Frank (Director) . . . . (423) 425-4366
frank-jones@utc.edu

Accredited by: ABET

Degrees granted 2003-2004:

B.S.: 4 M.S.: 2

Student organization: AIChE
Advisor: James Cunningham

Department reports to: Dr. Philip M. Kazemersky, Dean of Engineering

Placement service: Jean Dake (423) 425-4184, Director

University of Tennessee at Knoxville

Department of Chemical Engineering
The University of Tennessee
Knoxville, TN 37996-2200
Deliveries: 419 Dougherty Engineering Building

University phone .... (865) 974-1000
Department phone .... (865) 974-2421
Department FAX .... (865) 974-7076

Professors

Bienkowski, Paul R. .... (865) 974-1618
prb@utk.edu
Collier, John R. (Head) .... (865) 974-2421
collier@utk.edu
Counce, Robert M. .... (865) 974-5318
counce@utk.edu
Frazier, George C. (Emeritus)
Holmes, John M. (Emeritus)
Moore, Charles F. .... (865) 974-5339
cfmoore@utk.edu
Prados, John W. (Emeritus) .... (865) 974-6053
jprados@utk.edu
Watson, Jack S. (Adjunct)

Associate Professors

Ally, Moonis (Adjunct)
Arnold, John R. (Adjunct)
Bruns, Duane D. .... (865) 974-5317
dbruns@utk.edu
Davison, B.H. (Adjunct) .... (865) 576-8522
davisonbh@ornl.gov
Daw, C. Stuart (Adjunct)
DePaoli, David (Adjunct) .... (865) 574-6817
Edwards, Brian J. .... (865) 974-9596
bjedwards@chem.engr.utk.edu
Ferrada, J.J. (Adjunct) .... (865) 574-4998
ferradajj@ornl.gov
Frymier, Paul .... (865) 974-4961
pdf@ukt.edu
Keffer, David .... (865) 974-5322
dkeffer@utk.edu
Petrovan, Simioan .... (865) 974-5137
petrovan@chem.engr.utk.edu
Siirola, Jeffrey (Adjunct)
Wang, Tse-Wei .... (865) 974-6769
twang@utk.edu
Weber, Frederick E. .... (865) 974-6362
fweber@utk.edu

Assistant Professors

Cui, Shingting .... (865) 241-4896
scui@utk.edu

Other Faculty

Cummings, Peter T. (Adjunct)
Sheth, Atul .... (931) 393-7427
asheth@utsi.edu

Accredited by: ABET

Degrees granted 2003-2004:

B.S.: 24 M.S.: 5 Ph.D.: 5

Graduate advisor: Paul D. Frymier

Student organization: AIChE
Advisor: Paul R. Bienkowski

Department reports to: Way Kuo (865) 974-5321, Dean

Placement service: Robert Greenberg (865) 974-5435

Tennessee Technological University

Department of Chemical Engineering
P.O. Box 5013
Cookeville, TN 38505
Deliveries: 1020 Stadium Drive, Prescott Hall, Room 214

University phone . . . . (931) 372-3101
Department phone . . . . (931) 372-3297
Department FAX . . . . (931) 372-6352

Professors

Arce, Pedro E. (Chair) . . . . (931) 372-3189
parce@tntech.edu
Biernacki, Joseph J. . . . . (931) 372-3667
jbiernacki@tntech.edu
Holland, William D. (Emeritus) . . . . (931) 372-3297
Kerr, Clayton P. (Emeritus) . . . . (931) 372-3297
McGee, John C. (Emeritus) . . . . (931) 372-3297
Yarbrough, David W. (Emeritus) . . . . . . (931) 372-3297

Associate Professors

Booth, J. Richard (Adjunct) . . . . (931) 372-3495
rbooth@tntech.edu
Eliassen, John D. (Adjunct) . . . . (931) 372- 3297
jeliassen@tntech.edu
Whitmire, David R. . . . . (931) 372-3267
dwhitmire@tntech.edu

Assistant Professors

Dycus, Patricia J. M. . . . . (931) 372-3474
pdycus@tntech.edu
Subramanian, Venkat . . . . (931) 372-3494
vsubramanian@tntech.edu
Visco, Donald P. (931) 372-3606
dvisco@tntech.edu
Wang, Chunsheng . . . . (931) 372-3678
cswang@tntech.edu

Accredited by: ABET

Degrees granted 2003-2004:

B.S.: 15 M.S.: 7

Graduate advisor: Joseph J. Biernacki

Undergraduate advisor: Donald P. Visco, Jr.

Student organization: AIChE
Advisor: J. Richard Booth, Jr.

Department reports to: Glen Johnson (931) 372-3172, Dean of Engineering

Placement service: Alice Camuti (931) 372-3232, Director

Vanderbilt University

Chemical Engineering Department
VU Station B Box 351604
Nashville, TN 37235-1604
Deliveries: Olin Hall-107, 24th and Garland Avenues

Department phone . . . . (615) 322-2441
Department FAX . . . . (615) 343-7951

Professors

Bayuzick, Robert J. (Emeritus) . . . . (615) 322-7047
robert.j.bayuzick@vanderbilt.edu
Cummings, Peter T. . . . . (615) 322-8129
peter.cummings@vanderbilt.edu
Fort, Tomlinson (Emeritus) . . . . (615) 343-6992
tomlinson.fort@vanderbilt.edu
Godbold, Thomas M. (Emeritus)
LeVan, M. Douglas (Chair) . . . . (615) 322-2441
m.douglas.levan@vanderbilt.edu
Roth, John A. . . . . (615) 322-3517
john.roth@vanderbilt.edu
Schnelle, Karl B. Jr. . . . . (615) 322-3370
karl.b.schnelle@vanderbilt.edu
Tanner, Robert D. (615) 322-2061
robert.d.tanner@vanderbilt.edu

Associate Professors

Debelak, Kenneth A. . . . . (615) 322-2088
kenneth.a.debelak@vanderbilt.edu
Sharp, Julie E. . . . . (615) 322-3700
julie.e.sharp@vanderbilt.edu

Assistant Professors

Balcarcel, R. Robert . . . . (615) 322-2095
r.robert.balcarcel@vanderbilt.edu
Bowman, Frank M. . . . . (615) 343-7028
frank.m.bowman@vanderbilt.edu
Jennings, G. Kane . . . . (615) 322-2707
kane.g.jennings@vanderbilt.edu
McCabe, Clare . . . . (615) 322-6853
c.mccabe@vanderbilt.edu
Rogers, Bridget R. . . . . (615) 343-3269
bridget.rogers@vanderbilt.edu

Other Faculty

Prokop, Ales . . . . (615) 343-3515
ales.prokop@vanderbilt.edu

Accredited by: ABET

Degrees granted 2003-2004:

B.S.: 14 M.S.: 2 Ph.D.: 2

Graduate advisor: Bridget R. Rogers

Undergraduate advisor: Robert D. Tanner

Student organization: AIChE
Advisor: Frank M. Bowman

Department reports to: Kenneth Galloway, Dean

Texas

University of Houston

Chemical Engineering Department
S-222, Engineering Bldg. 1
Houston, TX 77204-4004
Deliveries: Engng. Bldg. 1, #S-222; 4800 Calhoun; Houston, TX 77004

University phone .... (713) 743-1000
Department phone .... (713) 743-4300
Department FAX .... (713) 743-4323

Professors

Amundson, Neal R. .... (713) 743-3492
NAmundson@aol.com
Balakotaiah, Vemuri .... (713) 743-4318
Bala@uh.edu
Bidani, Akhil (Adjunct) .... (713) 500-6826
akhil.bidani@uth.tmc.edu
Donnelly, Vincent M. .... (713) 743-4313
VMDonnelly@uh.edu
Economides, Michael J. .... (713) 743-4330
MJE@uh.edu
Economou, Demetre J. (Associate Chair) (713) 743-4320
Economou@uh.edu
Fleischer, Miguel T. (Adjunct) .... (713) 743-4388
MFleischer@uh.edu
Flumerfelt, Raymond W. (Dean) ...... (713) 743-4200
RWF@uh.edu
Fox, George (Joint) .... (713) 743-8363
Fox@uh.edu
Harold, Michael P. (Chair) .... (713) 743-4307
MHarold@uh.edu
Henley, Ernest J. (Emeritus) .... (713) 743-4326
EHenley@bayou.uh.edu
Jacobson, Allan J. (Joint) .... (713) 743-2785
AJJacob@uh.edu
Luss, Dan .... (713) 743-4305
DLuss@uh.edu
Mohanty, Kishore K. .... (713) 743-4331
Mohanty@uh.edu
Richardson, James T. .... (713) 743-4324
JTR@uh.edu
Rooks, Charles W. "Mickey" (Adjunct) (713) 743-4316
CWRooks@uh.edu
Willson, Richard C. .... (713) 743-4308
Willson@uh.edu

Associate Professors

Annapragada, Ananth (Adjunct) ...... (713) 743-4300
ananth.annapragada@uth.tmc.edu
Briggs, James M. (Joint) .... (713) 743-8366
JBriggs@uh.edu
Chellam, Shankar (Joint) .... (713) 743-4265
Chellam@uh.edu
Krishnamoorti, Ramanan .... (713) 743-4312
Ramanan@uh.edu
Lee, T. Randall (Joint) .... (713) 743-2724
TRLee@uh.edu
Marple, Stanley Jr. (Adjunct) .... (713) 743-4326
Nikolaou, Michael .... (713) 743-4309
Nikolaou@uh.edu
Perry, Scott S. (Joint) .... (713) 743-2715
Perry@uh.edu
Vekilov, Peter G. .... (713) 743-4315
Vekilov@uh.edu

Assistant Professors

Capitano, Adam T. .... (713) 743-4319
ACapitano@uh.edu
Tran, Veronique V. (Joint) .... (713) 743-4562
VTran@uh.edu

Other Faculty

Galkin, Oleg N. .... (713) 743-4344
OGalkin@uh.edu

Accredited by: ABET

Degrees granted 2003-2004:

B.S.: 38 M.S.: 38 Ph.D.: 13

Undergraduate advisor: Demetre Economou

Student organization: AIChE
Advisor: Richard Willson

Department reports to: Raymond Flumerfelt (713) 743-4200, Dean of Engineering

Placement service: David B. Small, (713) 743-5100

Lamar University

Chemical Engineering Department
P.O. Box 10053
Beaumont, TX 77710
Deliveries: 4400 Martin Luther King Dr., Beaumont, TX 77705

University phone .... (409) 880-7011
Department phone .... (409) 880-8784
Department FAX .... (409) 880-2197

Professors

Chen, Daniel H. .... (409) 880-8786
chendh@hal.lamar.edu
Cocke, David L. .... (409) 880-8372
cockdl@hal.lamar.edu
Ho, Thomas C. .... (409) 880-8790
hotc@hal.lamar.edu
Hopper, Jack R. (Dean) .... (409) 880-8785
hopperjr@hal.lamar.edu
Li, Ku-Yen (Chair) .... (409) 880-8789
liku@hal.lamar.edu
Yaws, Carl L. .... (409) 880-8787
yawscl@hal.lamar.edu

Assistant Professors

Gossage, John L. .... (409) 880-8788
gossagejl@hal.lamar.edu
Lou, Helen H. .... (409) 880-8207
louhh@hal.lamar.edu
Tadmor, Rafael .... (409) 880-7791
tadmorrx@hal.lamar.edu

Accredited by: ABET

Degrees granted 2003-2004:

B.S.: 14 M.S.: 24 Ph.D.: 2

Graduate advisor: Dr. Helen H. Lou

Undergraduate advisor: Mrs. Rebecca Caddy

Student organization: AIChE
Advisor: Dr. John L. Gossage

Department reports to: Jack R. Hopper, Dean of Engineering

Placement service: Dr. Jim Thomas

Prairie View A&M University

Chemical Engineering Department
P.O. Box 4229
Prairie View, TX 77446-4229
Deliveries: Preston and Obanion Streets

University phone .... (936) 857-3311
Department phone .... (936) 857-2427
Department FAX .... (936) 857-4540

Professors

Fotouh, Kamel H. .... (936) 857-2427
k_fotouh@pvamu.edu

Associate Professors

Gabitto, Jorge F. .... (936) 857-2427
jgabitto@aol.com
Osborne-Lee, Irvin W. (Head) .... (936) 857-2427
oslee@pvamu.edu

Assistant Professors

Aghara, Sukesh .... (936) 857-4606
sukesh@pvamu.edu
Gyamerah, Michael .... (936) 857-2427
michael_gyamerah@pvamu.edu
Nave, Felecia M. .... (936) 857-2427
felecia_nave@pvamu.edu

Accredited by: ABET

Degrees granted 2003-2004:

B.S.: 14 M.S.: 1

Graduate advisor: Irvin W. Osborne-Lee

Undergraduate advisor: Kamel H. Fotouh

Student organization: AIChE
Advisor: Irvin W. Osborne-Lee

Department reports to: Milton R. Bryant, Dean of College of Engineering

Placement service: Glenda Jones (936) 857-2120, Assistant Director

Rice University

Chemical Engineering Department
MS-362
P.O. Box 1892
Houston, TX 77251-1892
Deliveries: 6100 Main Street; Houston, TX 77005

University phone .... (713) 348-0000
Department phone .... (713) 348-4902
Department FAX .... (713) 348-5478

Professors

Akers, William W. (Emeritus) .... (713) 348-6055
wwakers@owlnet.rice.edu
Armeniades, Constantine .... (713) 348-3495
cda@rice.edu
Chapman, Walter G. .... (713) 348-4900
wgchap@rice.edu
Davis, Sam H. (Emeritus) .... (713) 348-3536
shdavis@rice.edu
Dyson, Derek C. (Emeritus) .... (713) 348-6055
dyson@rice.edu
Hellums, J. David (Emeritus) .... (713) 348-5116
jhellums@rice.edu
Hightower, Joe W. (Emeritus) .... (713) 348-5906
jhigh@owlnet.rice.edu
Hirasaki, George J. .... (713) 348-5416
gjh@rice.edu
Kobayashi, Riki (Emeritus) .... (713) 348-4901
lkobay@aol.com
McIntire, Larry V.
mcintire@rice.edu
Mikos, Antonios G. .... (713) 348-5355
mikos@rice.edu
Miller, Clarence A. .... (713) 348-4904
camill@pop.rice.edu
Robert, Marc A. .... (713) 348-3515
mrobert@rice.edu
San, Ka-Yiu .... (713) 348-5361
ksan@rice.edu
West, Jennifer .... (713) 348-5955
jwest@rice.edu
Zygourakis, Kyriacos (Chair) .... (713) 348-5208
kyzy@rice.edu

Associate Professors

Laibinis, Paul .... (713) 348-3539
pel@rice.edu

Assistant Professors

Mantzaris, Nikolas .... (713) 348-2955
nman@rice.edu
Pasquali, Matteo .... (713) 348-5830
mp@rice.edu
Wong, Michael .... (713) 348-3511
mswong@rice.edu

Instructors

Cox, Kenneth .... (713) 348-3529
krcox@rice.edu

Accredited by: ABET

Degrees granted 2003-2004:

B.S.: 16 M.S.: 2 Ph.D.: 4

Student organization: AIChE
Advisor: Walter G. Chapman

Department reports to: C. Sidney Burrus, Dean (713) 348-4009

Placement service: Cheryl Matherly, Director (713) 348-4055, Career Services Center

Texas A&M University

Chemical Engineering Department
3122 TAMU
College Station, TX 77843-3122

University phone . . . . (979) 845-3211
Department phone . . . . (979) 845-3361
Department FAX . . . . (979) 845-6446

Professors

Anthony, Rayford G. . . . . (979) 845-3370
rg-anthony@tamu.edu
Appleby, Anthony J. . . . . (979)845-5571
ajappleby@tamu.edu
Balbuena, Perla . . . . (979) 845-3375
balbuena@tamu.edu
Bukur, Dragomir B. (Associate Head) . . (979) 845-3401
d-bukur@tamu.edu
Bullin, Jerry A. (Emeritus)
J.Bullin@bre.com
Darby, Ronald (Emeritus) . . . . (979) 845-3301
R-Darby@tamu.edu
Davison, Richard R. (Emeritus) . . . . (979) 845-3391
rdavison@tamu.edu
Durbin, Leonel D. (Emeritus)
madurbin99@aol.com
El-Halwagi, Mahmoud M. (Assoc. Head) (979) 845-3484
el-halwagi@tamu.edu
Eubank, Philip T. . . . . (979) 845-3339
p-eubank@tamu.edu
Froment, Gilbert (Visiting) . . . . (979) 845-0406
g.froment@che.tamu.edu.edu
Glover, Charles J. (Associate Head) . . . (979) 845-3389
c-glover@tamu.edu
Hall, Kenneth R. (Head) . . . . (979) 845-3357
krhall@tamu.edu
Harris, William B. (Emeritus)
wbh5366@chemail.tamu.edu
Holland, Charles D. (Emeritus) . . . . (979) 845-3371
cdholland@tamu.edu
Holste, James C. (979) 845-3384
j-holste@tamu.edu
Holtzapple, Mark T. . . . . (979) 845-9708
m-holtzapple@tamu.edu
Kuo, Yue . . . . (979) 845-9807
yuekuo@tamu.edu
Mannan, Sam . . . . (979) 862-3985
mannan@tamu.edu
Seminario, Jorge . . . . (979) 845-3301
seminario@tamu.edu
Sevick-Muraca, Eva . . . . (979) 458-3206
sevick@che.tamu.edu
Slattery, John C. (409) 845-0407
Slattery@tamu.edu

Associate Professors

Ford, David M. . . . . (979) 862-4850
d-ford@chemail.tamu.edu
Hanson, Daniel T. . . . . (979) 845-3348
dthanson@tamu.edu

Assistant Professors

Bevan, Michael . . . . (979) 847-8766
mabevan@tamu.edu
Hahn, Juergen . . . . (979)845-3568
hahn@che.tamu.edu
Jayaraman, Arul
arul.jayaraman@chemail.tamu.edu
Shantz, Daniel F. . . . . (979) 845-3492
shantz@che.tamu.edu
Ugaz, Victor . . . . (979) 458-1002
ugaz@che.tamu.ed

Other Faculty

Baldwin, John T. (979) 845-9803
jt-baldwin@tamu.edu
Bradshaw, Jerry L. . . . . (979) 845-0610
j-bradshaw@tamu.edu
Isdale, Charles E. . . . . (979) 458-1168
c-isdale@tamu.edu
Yurttas, Lale . . . . (979) 847-9316
l-yurttas@tamu.edu

Accredited by: ABET

Degrees granted 2003-2004:

B.S.: 67 M.S.: 7 Ph.D.: 16

Graduate advisor: Mahmoud El-Halwagi

Undergraduate advisor: Lale Yurttas

Student organization: AIChE
Advisor: Daniel T. Hanson

Department reports to: G. K. Bennett, Vice Chancellor for Engineering, and Dean of Engineering, Wisenbaker Engineering Research Center

Placement service: Leigh Turner, Director of Placement, Rudder Tower

Texas A&M University-Kingsville

Chemical and Natural Gas Engineering Department
700 University Boulevard
MSC 193
Kingsville, TX 78363
Deliveries: Engineering Complex 917 W. Ave. B Suite 303

University phone . . . . (361) 593-2111
Department phone . . . . (361) 593-2002
Department FAX . . . . (361) 593-4026

Professors

Al-Saadoon, Faleh T. . . . . (361) 593-2380
KFFTA00@tamuk.edu
Heenan, William A. . . . . (361) 593-2002
W_HEENAN@tamuk.edu
Murphy, David (Visiting) . . . . (361) 593-4004
kfcdm00@tamuk.edu
Pilehvari, Ali (Chair) . . . . (361) 593-2089
A-PILEHVARI@tamuk.edu
Schruben, Dale L. (361) 593-2008
KFDLS00@tamuk.edu
Serth, Robert W. . . . . (361) 593-2093
R_SERTH@tamuk.edu

Associate Professors

Chisholm, John . . . . (361) 593-2092
CHISHOLM@tamuk.edu

Assistant Professors

Lee, Sang-Yong . . . . (361) 593-2629
S-Lee@tamuk.edu

Accredited by: ABET

Degrees granted 2003-2004:

B.S.: 22 M.S.: 25

Graduate advisor: R. W. Serth

Student organization: AIChE
Advisor: D. L. Schruben

Department reports to: William A. Heenan (361) 593-2001, Dean of Engineering

Placement service: Leticia Silquero (361) 593-2217

University of Texas at Austin

Chemical Engineering Department
1 University Station, C0400
Austin, TX 78712-1062
Deliveries: Speedway at E. Dean Keeton

University phone .... (512) 471-3434
Department phone .... (512) 471-5238
Department FAX .... (512) 471-7060

Professors

Allen, David T. .... (512) 471-0049
allen@che.utexas.edu
Bonnecaze, Roger T. .... (512) 471-1497
bonnecaze@che.utexas.edu
Brock, James R. (Emeritus) .... (512) 471-3348
chju344@uts.cc.utexas.edu
Edgar, Thomas F. .... (512) 471-3080
edgar@che.utexas.edu
Ekerdt, John G. (Chair) .... (512) 471-4689
ekerdt@che.utexas.edu
Fair, James R. (Emeritus) .... (512) 471-3689
fair@che.utexas.edu
Freeman, Benny D. .... (512) 232-2803
freeman@che.utexas.edu
Georgiou, George .... (512) 471-6975
gg@che.utexas.edu
Green, Peter F. .... (512) 471-3188
green@che.utexas.edu
Himmelblau, David M. (Emeritus) ..... (512) 471-7445
himmelblau@che.utexas.edu
Hougen, Joel O. (Emeritus)
Johnston, Keith P. .... (512) 471-4617
johnston@che.utexas.edu
Jose-Yacaman, Miguel .... (512) 232-9111
yacaman@che.utexas.edu
Lloyd, Douglas R. .... (512) 471-4985
lloyd@che.utexas.edu
McKetta, John J. (Emeritus) .... (512) 471-5227
mcketta@che.utexas.edu
Mullins, C. Buddie .... (512) 471-5817
mullins@che.utexas.edu
Paul, Donald R. .... (512) 471-5392
paul@che.utexas.edu
Peppas, Nicholas A. .... (512) 471-6644
peppas@che.utexas.edu
Qin, S. Joe .... (512) 471-4417
qin@che.utexas.edu
Rase, Howard F. (Emeritus) .... (512) 471-3251
rase@che.utexas.edu
Rochelle, Gary T. .... (512) 471-7230
rochelle@che.utexas.edu
Rossky, Peter J. .... (512) 471-3555
rossky@mail.utexas.edu
Sanchez, Isaac C. (Associate Chair) .... (512) 471-1020
sanchez@che.utexas.edu
Schechter, Robert S. (Emeritus) .... (512) 471-3245
robert_schechter@pe.utexas.edu
Steinfink, Hugo (Emeritus) .... (512) 471-5233
steinfink@che.utexas.edu
Stice, James E. (Emeritus) .... (512) 471-5527
stice@che.utexas.edu
Willson, C. Grant .... (512) 471-4342
willson@che.utexas.edu
Wissler, Eugene H. (Emeritus) .... (512) 471-5353
wissler@che.utexas.edu

Associate Professors

Korgel, Brian A. .... (512) 471-5633
korgal@che.utexas.edu
Schmidt, Christine .... (512) 471-1690
schmidt@che.utexas.edu

Assistant Professors

Ganesan, Venkat .... (512) 471-4856
ganesan@che.utexas.edu
Hwang, Gyeong S. .... (512) 471-4847
hwang@che.utexas.edu
Loo, Yueh Lin .... (512) 471-6300
lloo@che.utexas.edu
Truskett, Thomas M. .... (512) 471-6308
truskett@che.utexas.edu

Instructors

Eldridge, R. Bruce .... (512) 232-1407
rbeldr@che.utexas.edu
Poehl, Michael .... (512) 471-4438
poehlme@che.utexas.edu
Randall, D'Arcy .... (512) 471-8280
darcyr@mail.utexas.edu
Swinnea, J. Steven .... (512) 471-3173
swinnea@che.utexas.edu

Other Faculty

Heller, Adam .... (512) 471-8874
heller@che.utexas.edu
Sciance, Carroll T.
scscorp@earthlink.net

Accredited by: ABET

Degrees granted 2003-2004:

B.S.: 118 M.S.: 20 Ph.D.: 19

Graduate advisor: Charles B. (Buddie) Mullins

Undergraduate advisor: Benny D. Freeman

Student organization: AIChE
Advisor: Thomas M. Truskett

Department reports to: Ben G. Streetman (512) 471-1166, Dean of Engineering

Placement service: Nancy A. Evans (512) 471-7461, Director, Engineering Career Assistance Center, ECJ 2.400

Texas Tech University

Chemical Engineering Department
P. O. Box 43121
Lubbock, TX 79409-3121
Deliveries: 6th and Canton

University phone .... (806) 742-2011
Department phone .... (806) 742-3553
Department FAX .... (806) 742-3552

Professors

Bethea, Robert M. (Emeritus) .... (806) 742-4087
RBethea@coe.ttu.edu
Bradford, John R. (Emeritus)
Hoo, Karlene .... (806) 742-4079
Karlene.Hoo@coe.ttu.edu
Mann, Uzi .... (806) 742-4086
kmuzm@coe.ttu.edu
McKenna, Greg (Chair) .... (806) 742-4136
Greg.McKenna@coe.ttu.edu
Riggs, James B. .... (806) 742-1765
cpjbr@ttacs.ttu.edu
Simon, Sindee .... (806) 742-1763
Sindee.Simon@coe.ttu.edu
Tock, R. W. .... (806) 742-3998
tock@coe1.coe.ttu.edu

Associate Professors

Abbott, James .... (806) 742-3554
james.abbott@coe.ttu.edu
Wiesner, Theodore F. .... (806) 742-1448
twiesner@coe.ttu.edu

Assistant Professors

Dai, Lenore .... (806) 742-1757
lenore.dai@coe.ttu.edu
Leggoe, Jeremy .... (806) 742-1767
Jeremy.Leggoe@coe.ttu.edu
Siviniah, Easan .... (806) 742-3553
easan.siviniah@coe.ttu.edu
Vaughn, Mark .... (806) 742-5158
Mark.Vaughn@coe.ttu.edu

Accredited by: ABET

Degrees granted 2003-2004:

B.S.: 32 M.S.: 8 Ph.D.: 7

Graduate advisor: Karlene Hoo

Undergraduate advisor: Sindee L. Simon

Student organization: AIChE
Advisor: Uzi Mann

Department reports to: Pamela Eibeck, Dean of Engineering

Placement service: David Krauss

Utah

Brigham Young University

Chemical Engineering Department
350 Clyde Building
PO Box 24100
Provo, UT 84602
Deliveries: 350 Clyde Building

University phone .... (801) 422-1211
Department phone .... (801) 422-2586
Department FAX .... (801) 422-0151

Professors

Barker, Dee H. (Emeritus) .... (801) 422-4620
barkerd@et.byu.edu
Bartholomew, Calvin H. .... (801) 422-4162
bartc@et.byu.edu
Baxter, Larry L. .... (801) 422-8616
larry_baxter@byu.edu
Beckstead, Merrill W. (Associate Chair) (801) 422-6239
mwb@byu.edu
Fletcher, Thomas H. .... (801) 422-6236
tom_fletcher@byu.edu
Harb, John N. .... (801) 422-4393
jharb@et.byu.edu
Oscarson, John L. .... (801) 422-6243
oscarj@et.byu.edu
Pitt, William G. .... (801) 422-2589
pitt@byu.edu
Rowley, Richard L. .... (801) 422-2590
rowley@byu.edu
Smoot, L. Douglas (Emeritus) .... (801) 422-8930
lds@byu.edu
Solen, Kenneth A. .... (801) 422-6237
kenneth_solen@byu.edu
Terry, Ronald E. .... (801) 422-4297
ron_terry@byu.edu
Wilding, W. Vincent (Chair) .... (801) 422-2393
wildingv@et.byu.edu

Associate Professors

Hecker, William C. .... (801) 422-6235
hecker@byu.edu

Assistant Professors

Wheeler, Dean .... (801) 422-4126
dean_wheeler@byu.edu

Other Faculty

Hales, Hugh B. .... (801) 422-3749
hugh_hales@byu.edu

Accredited by: ABET

Graduate advisor: William G. Pitt

Undergraduate advisor: Richard L. Rowley

Student organization: AIChE
Advisor: William C. Hecker

Department reports to: Douglas M. Chabries, Dean, Engineering and Technology

Placement service: Lloyd Hawkins (801) 422-6932

University of Utah

Department of Chemical Engineering
50 S. Central Campus Drive (MEB)
Room 3290
Salt Lake City, UT 84112

University phone .... (801) 581-7200
Department phone .... (801) 581-6915
Department FAX .... (801) 585-9291

Professors

de Nevers, Noel H. .... (801) 581-6024
Noel.deNevers@m.cc.utah.edu
Deo, Milind D. .... (801) 581-7629
mddeo@eng.utah.edu
Hanson, Francis V. .... (801) 581-6591
Francis.Hanson@m.cc.utah.edu
Lighty, Jo Ann S. .... (801) 581-5763
jlighty@dean.eng.utah.edu
Massoth, Franklin E. .... (801) 581-5166
f.massoth@m.cc.utah.edu
Pershing, David W. .... (801) 581-6911
David.Pershing@dean.eng.utha.edu
Pugmire, Ronald J. .... (801) 581-7236
pug@vpres.adm.utah.edu
Ring, Terry A. .... (801) 585-5705
terry.ring@mse.utah.edu
Sarofim, Adel .... (801) 585-9258
sarofim@reaction-eng.com
Seader, J. D. .... (801) 581-6916
jseader@m.cc.utah.edu
Smith, Philip J. (Chair) .... (801) 585-3129
smith@opus.utah.edu

Associate Professors

Eddings, Eric .... (801) 585-3931
eddings@che.utah.edu
Fletcher, John V. .... (801) 585-6158
Jack.Fletcher@m.cc.utah.edu
Magda, Jules J. .... (801) 581-7536
jj.magda@m.cc.utah.edu
Peterson, Richard E. .... (801) 587-9058
rpeterson@eng.utah.edu
Silcox, Geoffrey D. (Associate Chair) .. (801) 581-8820
geoff@eng.utah.edu
Skliar, Mikhail .... (801) 581-6918
mikhail.skliar@utah.edu
Smith, Grant D. .... (801) 585-3381
gsmith@geoffrey.emro.utah.edu
Trujillo, Edward M. .... (801) 581-4460
edward.trujillo@m.cc.utah.edu
Tyler, Bonnie .... (801) 581-6915
rory.holley@m.cc.utah.edu

Assistant Professors

Krahenbuhl, Melinda .... (801) 581-4188
mpk@nuclear.utah.edu
Roper, Keith .... (801) 581-6915
rory.holley@m.cc.utah.edu
Solum, Mark .... (801) 581-7351
Spinti, Jennifer .... (801) 585-1241
spinti@crsim.utah.edu

Accredited by: ABET

Degrees granted 2003-2004:

B.S.: 27 M.S.: 3 Ph.D.: 4

Graduate advisor: Eric Eddings

Undergraduate advisor: P. Jeff Mulitalo

Student organization: AIChE
Advisor: Edward M. Trujillo

Department reports to: Richard B.Brown (801) 581-6911, Dean of Engineering

Placement service: Lisa Christensen

Virginia

Hampton University

Chemical Engineering Department
Hampton, VA 23668

University phone .... (757) 727-5000
Department phone .... (757) 727-5288
Department FAX .... (757) 727-5189

Professors

Adeyiga, Adeyinka A. .... (757) 727-5289
yinka@et.hamptonu.edu
Akyurtlu, Ates (Interim Chair) .... (757) 727-5599
ates@et.hamptonu.edu
Akyurtlu, Jale F. .... (757) 727-5589
jale@et.hamptonu.edu
Morgan, Morris H. III .... (757) 727-5583
morris@et.hamptonu.edu

Associate Professors

Shuming, Zheng .... (757) 727-5069
zheng@et.hamptonu.edu

Assistant Professors

Jothimurugesan, K. .... (757) 727-5817
jothi@et.hamptonu.edu

Accredited by: ABET

Degrees granted 1997-1998:
B.S.: 6

Undergraduate advisor: Department Chair

Student organization: AIChE
Advisor: Dr. Shuming Zheng

Department reports to: Samuel G. White, Jr., Dean, School of Engineering and Technology

Placement service: Bessie Willis (757) 727-5331

University of Virginia

Department of Chemical Engineering
School of Engineering and Applied Science
102 Engineers' Way
P.O. Box 400741
Charlottesville, VA 22904-4741
Deliveries: Chemical Engineering Building Room 117, McCormick Road

University phone .... (434) 924-0311
Department phone .... (434) 924-7778
Department FAX .... (434) 982-2658

Professors

Carta, Giorgio .... (434) 924-6281
gc@virginia.edu
Davis, Robert J. (Chair) .... (434) 924-6284
rjd4f@virginia.edu
Ford, Roseanne M. .... (434) 924-6283
rmf3f@virginia.edu
Gaden, Elmer L. Jr. (Emeritus) .... (434) 924-6276
elg@virginia.edu
Gainer, John L. .... (434) 924-6277
jlg@virginia.edu
Hudson, J. L. .... (434) 924-6275
hudson@virginia.edu
Kirwan, Donald J. .... (434) 924-6278
djk@virginia.edu
Laurencin, Cato .... (434) 923-0250
ctl3f@virginia.edu
Lilleleht, Lembit (Emeritus) .... (434) 924-6280
lul@virginia.edu
Neurock, Matthew .... (434) 924-6248
mn4n@virginia.edu
O'Connell, John P. .... (434) 924-3428
jpo2x@virginia.edu

Associate Professors

Fernandez, Erik J. .... (434) 924-1351
ejf3c@virginia.edu

Assistant Professors

Green, David L. .... (434) 924-1302
dlg9s@virginia.edu
Oberhauser, James P. .... (434) 924-7974
jpo8a@virginia.edu

Other Faculty

Espino, Ramon .... (434) 924-6279
re4n@virginia.edu

Accredited by: ABET

Degrees granted 2003-2004:
B.S.: 29 M.S.: 6 Ph.D.: 5

Graduate advisor: R.J. Davis

Undergraduate advisor: John L. Gainer

Student organization: AIChE
Advisor: John O'Connell

Department reports to: R. W. Miksad, Dean, School of Engineering and Applied Sciences

Placement service: C. J. Livesay, Assistant Dean and Director, Center for Engineering Career Development

Virginia Polytechnic Institute and State University/Virginia Tech

Chemical Engineering Department
0211
Blacksburg, VA 24061
Deliveries: 133 Randolph Hall

University phone .... (540) 231-6000
Department phone .... (540) 231-6631
Department FAX .... (540) 231-5022

Professors

Baird, Donald G. .... (540) 231-5998
dbaird@vt.edu
Demirel, Yasar (Visiting)
Durrill, Preston L. (Adjunct) .... (540) 231-6631
pdurrill@radford.edu
Kiran, Erdogan (Head) .... (540) 231-1375
ekiran@vt.edu
Liu, Y. A. .... (540) 231-7800
design@vt.edu
Oyama, S. Ted .... (540) 231-5309
oyama@vt.edu
Rony, Peter R. (Emeritus)
rony@vt.edu
Squires, Arthur M. (Emeritus) .... (540) 231-5972
verasqu@vt.edu
Sullivan, Joseph (Emeritus)
jtsull@vt.edu
Wilkes, Garth L. (Emeritus) .... (540) 231-5498
gwilkes@vt.edu
Wills, George B. (Emeritus)

Associate Professors

Cox, David F. .... (540) 231-6829
dfcox@vt.edu
Davis, Richey M. .... (540) 231-4578
rmdavis@vt.edu
Forsten Williams, Kimberly .... (540) 231-4851
kforsten@vt.edu
Marand, Eva .... (540) 231-8231
emarand@vt.edu
Michelsen, Donald L. (Emeritus)
Mischke, Roland A. (Emeritus)

Assistant Professors

Goldstein, Aaron .... (540) 231-3674
goldst@vt.edu
Sum, Amadeu .... (540) 231-7869
York, Steven (Adjunct) .... (540) 231-6555
syork@vt.edu

Accredited by: ABET

Degrees granted 2003-2004:

B.S.: 45 M.S.: 7 Ph.D.: 6

Graduate advisor: R. M. Davis

Undergraduate advisor: D. F. Cox

Student organization: AIChE
Advisor: Y. A. Liu

Department reports to: Hassan Aref, Dean of Engineering

Placement service: J. H. Malone

Washington

University of Washington

Chemical Engineering Department
Box 351750
Seattle, WA 98195-1750
Deliveries: Benson Hall, Room 105

University phone 206-543-2100
Department phone .... (206) 543-2250
Department FAX .... (206) 543-3778

Professors

Allan, G. Graham (Joint) .... (206) 543-1491
create@u.washington.edu
Babb, Albert L. (Emeritus) .... (206) 543-2754
mercury@u.washington.edu
Baneyx, François .... (206) 685-7659
baneyx@cheme.washington.edu
Berg, John C. .... (206) 543-2029
berg@cheme.washington.edu
Bowen, J. Ray (Emeritus) .... (206) 616-8128
bowen@engr.washington.edu
Castner, David G. .... (206) 543-8094
castner@nb.engr.washington.edu
David, Morton M. (Emeritus)
Davis, E. James .... (206) 543-0298
jdavis@cheme.washington.edu
Finlayson, Bruce A. .... (206) 685-1634
finlayson@cheme.washington.edu
Garlid, Kermit L. (Emeritus) .... (206) 543-4807
kgarlid@u.washington.edu
Gustafson, Richard (Adjunct) .... (206) 543-2790
pulp@u.washington.edu
Heideger, William J. (Emeritus)
Hodgson, Kevin (Adjunct) .... (206) 543-7346
hodgson@u.washington.edu
Hoffman, Allan S. (Joint) .... (206) 543-9423
hoffman@u.washington.edu
Horbett, Thomas A. (Joint) .... (206) 685-1392
horbett@uweb.engr.washington.edu
Jenekhe, Samson A. .... (206) 543-5525
jenekhe@cheme.washington.edu
Johanson, Lennart N. (Emeritus)
Lidstrom, Mary E. .... (206) 616-5282
lidstrom@u.washington.edu
McKean, Wm. T. (Joint) .... (206) 543-1626
wmckean@u.washington.edu
Pilat, Michael J. (Adjunct) .... (206) 543-4789
mpilat@u.washington.edu
Ratner, Buddy D. (Joint) .... (206) 685-1005
ratner@uweb.engr.washington.edu
Ricker, N. Lawrence .... (206) 543-8786
ricker@u.washington.edu
Schwartz, Daniel T. .... (206) 685-4815
dts@u.washington.edu
Seferis, James C. .... (206) 543-9371
seferis@cheme.washington.edu
Sleicher, Charles A. (Emeritus) .... (206) 543-4807
charles@sleicher.net
Stuve, Eric M. (Chair) .... (206) 543-2253
stuve@u.washington.edu
Woodruff, Gene L. (Emeritus)
woodruff@u.washington.edu
Yager, Paul (Adjunct) .... (206) 543-6126
yagerp@u.washington.edu

Associate Professors

Holt, Bradley R. .... (206) 543-0554
holt@cheme.washington.edu
Jiang, Shaoyi .... (206) 616-6509
sjiang@u.washington.edu
Krieger-Brockett, Barbara .... (206) 543-2216
krieger@cheme.washington.edu
Overney, Rene M. .... (206) 543-4353
roverney@u.washington.edu
Sarikaya, Mehmet (Adjunct) .... (206) 543-0724
sarikaya@u.washington.edu

Assistant Professors

Adler, Stuart .... (206) 543-2131
stuadler@u.washington.edu

Other Faculty

Baratuci, William B. .... (206) 543-2271
baratuci@u.washington.edu

Accredited by: ABET

Degrees granted 2003-2004:

B.S.: 54 M.S.: 9 Ph.D.: 12

Graduate advisor: N. Lawrence Ricker

Undergraduate advisor: Devota Madrano

Student organization: AIChE
Advisor: John C. Berg

Department reports to: Denice D. Denton (206) 543-0340, Dean of Engineering

Placement service: Susan Terry (206) 543-0535, Director

Washington State University

School of Chemical Engineering and Bioengineering
P.O. Box 642710
Pullman, WA 99164-2710
Deliveries: 118 Dana Hall, Spokane St.

University phone .... (509) 335-3564
Department phone .... (509) 335-4332
Department FAX .... (509) 335-4806

Professors

Campbell, Kenneth .... (509) 335-8011
cvselkbc@vetmed.wsu.edu
Davis, Denny C. .... (509) 335-7993
davis@wsu.edu
Ivory, Cornelius F. .... (509) 335-7716
ivory@mail.wsu.edu
Lee, James M. .... (509) 335-5252
jmlee@wsu.edu
Liddell, KNona C. .... (509) 335-3710
liddell@che.wsu.edu
Mahalingam, R. (Emeritus) .... (509) 335-1304
mahalingam@wsu.edu
Miller, Reid C. .... (509) 335-4001
millerrc@wsu.edu
Petersen, James N. .... (509) 335-4332
jn_petersen@wsu.edu
Thomson, William J. (Emeritus) ...... (509) 335-8580
thomson@che.wsu.edu
Van Wie, Bernard J. .... (509) 335-4103
bvanwie@che.wsu.edu
Zollars, Richard L. (Chair) .... (509) 335-4332
rzollars@che.wsu.edu

Associate Professors

Luedeking, Robert (Emeritus) .... (509) 335-4332
Peyton, Brent M. .... (509) 335-4002
bmpeyton@che.wsu.edu

Assistant Professors

Lin, David .... (509) 335-7534
davidlin@wsu.edu
Vasavada, Anita .... (509) 335-7533
vasavada@wsu.edu

Accredited by: ABET

Degrees granted 2003-2004:

B.S.: 17 M.S.: 8 Ph.D.: 2

Student organization: AIChE
Advisor: Richard L. Zollars

Department reports to: Anjan Bose, Dean, College of Engineering and Architecture

Placement service: Alton Jamison

West Virginia

West Virginia University

Department of Chemical Engineering
P.O. Box 6102
Morgantown, WV 26506-6102
Deliveries: 403 ESB

University phone .... (304) 293-0111
Department phone .... (304) 293-2111
Department FAX .... (304) 293-4139

Professors

Bailie, Richard C. (Emeritus)
Cho, Eung H .... (304) 293-2111x2433 †
Eung.Cho@mail.wvu.edu
Cilento, Eugene V. (Dean) ...... (304) 293-4821x2237
Eugene.Cilento@mail.wvu.edu
Dadyburjor, Dady B. (Chair) .... (304) 293-2111x2411
Dady.Dadyburjor@mail.wvu.edu
Galli, Alfred F. (Emeritus)
Gupta, Rakesh K. .... (304) 293-2111x2427
Rakesh.Gupta@mail.wvu.edu
Kono, Hisashi O. (Emeritus) ..... (304) 293-2111x2421
Hisashi.Kono@mail.wvu.edu
Kugler, Edwin L. .... (304) 293-2111x2414
Edwin.Kugler@mail.wvu.edu
Stiller, Alfred H. .... (304) 293-2111x2408
Alfred.Stiller@mail.wvu.edu
Turton, Richard .... (304) 293-2111x2415
Richard.Turton@mail.wvu.edu
Yang, Ray Y. K. .... (304) 293-2111x2419
Ray.Yang@mail.wvu.edu
Zondlo, John W. .... (304) 293-2111x2409
John.Zondlo@mail.wvu.edu

Associate Professors

Shaeiwitz, Joseph A. .... (304) 293-2111x2410
Joseph.Shaeiwitz@mail.wvu.edu
Stinespring, Charter .... (304) 293-2111x2425
Charter.Stinespring@mail.wvu.edu

Other Faculty

Doraiswamy, Deepak (Adjunct)
Drozdov, Aleksey .... (304) 293-2111x2434
Aleksey.Drozdov@mail.wvu.edu
Henry, Joseph D. (Adjunct)
Jaffe, Charles M. (Adjunct) ...... (304) 293-3435x4425
Charles.Jaffe@mail.wvu.edu
Keller, George (Adjunct)
Liang, Ruifeng .... (304) 293-2111x2623
Ruifeng.Liang@mail.wvu.edu
Olson, Fred (Adjunct)
Wallace, William E. (Adjunct)
Wildi, Robert (Adjunct)
Zhang, Wu .... (304) 293-2111x2422
Wu.Zhang@mail.wvu.edu

† Faculty numbers are extensions of (304) 293-2111

Accredited by: ABET

Degrees granted 2003-2004:

B.S.: 13 M.S.: 5 Ph.D.: 3

Graduate advisor: Charter D. Stinespring

Undergraduate advisor: Joseph A. Shaeiwitz

Student organization: AIChE
Advisor: Dady Dadyburjor

Department reports to: Dean, College of Engineering and Mineral Resources

Placement service: Robert Kent, Director (304) 293-2221

Wisconsin

University of Wisconsin-Madison

Department of Chemical and Biological Engineering
1415 Engineering Drive
Madison, WI 53706-1691

University phone . . . . (608) 262-1234
Department phone . . . . (608) 262-1092
Department FAX . . . . (608) 262-5434

Professors

Abbott, Nicholas L. . . . . (608) 265-5278
abbott@engr.wisc.edu
Bird, R. Byron (Emeritus) . . . . . . (608) 262-5920
bird@chewi.che.wisc.edu
Chapman, Thomas W. (Emeritus) . . . . . (608) 263-1979
chapman@engr.wisc.edu
de Pablo, Juan J. . . . . (608) 262-7727
depablo@engr.wisc.edu
Duffie, John A. (Emeritus) . . . . . . (608) 263-1587
jaduffie@engr.wisc.edu
Dumesic, James A. . . . . (608) 262-1095
dumesic@engr.wisc.edu
Graham, Michael D. . . . . (608) 265-3780
graham@engr.wisc.edu
Hill, Charles G. Jr. . . . . (608) 263-4593
hill@engr.wisc.edu
Kuech, Thomas F. (Chair) . . . . . . (608) 263-2922
kuech@engr.wisc.edu
Langer, Stanley H. (Emeritus) . . . . . (608) 262-1190
langer@engr.wisc.edu
Lightfoot, Edwin N. (Emeritus) . . . . . (608) 262-6934
lightfoot@engr.wisc.edu
Murphy, Regina M. (Associate Chair) . . (608) 262-1587
murphy@engr.wisc.edu
Rawlings, James B. . . . . (608) 263-5859
jbraw@che.wisc.edu
Ray, W. Harmon (Emeritus) . . . . . . (608) 263-4732
ray@chewi.che.wisc.edu
Sather, Glenn A. (Emeritus) . . . . . . (608) 262-1092
gasather@facstaff.wisc.edu
Stewart, Warren E. (Emeritus) . . . . . (608) 262-0188
stewart@chewi.che.wisc.edu
Yin, John . . . . (608) 265-3779
yin@engr.wisc.edu

Associate Professors

Klingenberg, Daniel J. . . . . (608) 262-8932
klingenberg@engr.wisc.edu
Nealey, Paul F. . . . . (608) 265-8171
nealey@engr.wisc.edu
Root, Thatcher W. . . . . (608) 262-8999
root@chewi.che.wisc.edu
Swaney, Ross E. . . . . (608) 262-3641
swaney@engr.wisc.edu

Assistant Professors

Lynn, David M. . . . . (608) 262-1086
dlynn@engr.wisc.edu
Maravelias, Christos T. . . . . (608) 262-1092
maravelias@engr.wisc.edu
Mavrikakis, Manos . . . . (608) 262-9053
manos@engr.wisc.edu
Palecek, Sean P. . . . . (608) 262-8931
palecek@engr.wisc.edu
Shusta, Eric V. . . . . (608) 262-1092
shusta@engr.wisc.edu

Accredited by: ABET

Degrees granted 2003-2004:

B.S.: 74 M.S.: 5 Ph.D.: 17

Graduate advisor: Paul F. Nealey

Undergraduate advisor: Regina M. Murphy

Student organization: AIChE
Advisor: Eric Shusta/David Lynn

Department reports to: Paul Peercy, Dean of Engineering

Placement service: Sandra L. Arnn (608) 262-3471, Director

Wyoming

University of Wyoming

Department of Chemical and Petroleum Engineering
Dept 3295, 1000 E University Avenue
Laramie, Wyoming 82071
Deliveries: 16th and Gibbon Street

University phone .... (307) 766-5160
University FAX .... (307)-766-4042
Department phone .... (307) 766-2500
Department FAX .... (307) 766-6777

Professors

Ackerman, John (Adjunct) .... (307) 766-2566
wyoming1@uwyo.edu
Cha, Chang Yul (Emeritus) .... (307) 766-2500
cha@uwyo.edu
Deans, Harry A. (Emeritus) .... (307) 766-2500
Evers, John F. (Emeritus)
Harris, H. Gordon .... (307) 766-6558
harrishg@uwyo.edu
Haynes, Henry W. (Emeritus) .... (307) 766-2500
haynes@uwyo.edu
Mason, Geoffrey (Adjunct)
Morrow, Norman R. .... (307) 766-2838
morrownr@uwyo.edu
Radosz, Maciej .... (307) 766-2500
radosz@uwyo.edu
Sharma, Mrityunjai P. .... (307) 766-6317
sharma@uwyo.edu
Takamura, Koichi (Adjunct)

Associate Professors

Bell, David A. .... (307) 766-5769
davebell@uwyo.edu
Towler, Brian F. (Chair) .... (307) 766-2189
towler@uwyo.edu

Assistant Professors

Argyle, Morris .... (307) 766-2500
mdargyle@uwyo.edu
Rosinski, Matthew .... (307) 766-2500
mrosinski@uwyo.edu
Shen, Youqing .... (307) 766-2500

Accredited by: ABET

Degrees granted 2003-2004:

B.S.: 13 M.S.: 6 Ph.D.: 1

Student organization: AIChE
Advisor: Morris D. Argyle

Department reports to: O. A. (Gus) Plumb, Dean of Engineering

Placement service: Jo Chytka, Director

United States Engineering Technical Schools (Two-year programs leading to Associate Degrees)

Tennessee

Pellissippi State Technical Community College

Engineering and Media Technologies Department
10915 Hardin Valley Road
P.O. Box 22990
Knoxville, TN 37933-0990

University phone .... (865) 694-6400
Department phone .... (865) 694-6483
Department FAX .... (865) 694-6426

Associate Professors

Driver, E. Ely .... (865) 694-6506
edriver@pstcc.cc.tn.us
Scott, Robert (Adjunct)
Stephens, Frederick .... (865) 694-6507
fstephens@pstcc.cc.tn.us

Accredited by: ABET

Department reports to: Dr. Margaret Ann Jeffries

Placement service: Carolyn Carson (423) 694-6559, Director of Placement

Argentina

Instituto Tecnologico de Buenos Aires

Departamento de Ingenieria Quimica
Av. E. Madero 399
1106 Buenos Aires

University phone . . . . . (54-11)43147778
Department phone . . . . . (54-11)43147778 Ext 260
Department FAX . . . . . (54-11)43147778 Ext 558

Professors

Cao, Eduardo G.
Costa, Jose L.
Kammerer, Jorge R.
Marques, Dardo (Head) . . . . . Ext 242
dmarques@itba.edu.ar
Orce, Luis V.
Paya, Miguel A.
mpaya@itba.edu.ar
Pontiggia, Claudio J.
Soler, Susana M.
Zuazo, Beatriz N.

Associate Professors

Peralta, Carmen E.
Pirk, Tomas
Rausei, Diego N.
Tudino, Mabel

Assistant Professors

Aude Luppi, Vergenie
Bertini, Liliana M.
Coppari, Norberto R.
Fernandez, Beatriz E.
Gagey, Susana M.
Iglesias, Graciela S.
Iñon, Fernando A.
Pizarro, Ramon A.
Popik, Pablo R.
Sanchez, Lidia P.
Testa Fernandez, Juan J.

Accredited by: CONEAU

Degrees granted 2003-2004:

B.S.: 13

Undergraduate advisor: Dardo Marques

Department reports to: Ing. Jose Luis Roces (Vicerrector)

Universidad Nacional de la Plata

Facultad de Ingenieria
Departmento Ingenieria Quimica
1 esq. 47-1900 La Plata

Department phone 54 21 82-1600
Department FAX 54 21 25-9471

Professors

Barreto, Guillermo F.
Cunningham, Roberto E.
Farina, Italo H.
Ferretti, Osmar A.
Iglesias, Omar A.
Martinez, Osvaldo M.
Mascheroni, Rodolfo H.
Pessacq, Raul A (Head)
Vilche, Jorge R.
Zaritzky, Noemi E.

Assistant Professors

Bevilacqua, Alicia E.
Caminos, Rolando A.
Canale, Guillermo J.
D. Sorbo, C. L.
Duarte, Susana I. P.
Elsner, Cecilia I.
Gervasi, Claudio A.
Giner, Sergio A.
Graieb, Jorge A.
Grattoni, Carlos A.
Mazza, German D.
Navarro, Agustin F.
Pesarini, Carlos E.
Polito, Oscar A.
Salvadori, Viviana O.
Santana, Ines M
Siri, Guillerno J.

Department reports to: R. A. Pessacq, Head

Universidad Nacional del Litoral

Inst. de Desarrollo Tecnol. para la Industria Quimica - INTEC
Guemes 3450
S3000GLN - Santa Fe - Santa Fe

University phone . . . . . (54-342) 4571110
Department phone . . . . . (54-342) 4532965
Department FAX . . . . . (54-342) 4532965

Professors

Alfano, Orlando M.
alfano@intec.unl.edu.ar
Baltanas, Miguel A.
tderliq@ceride.gov.ar
Campanella, Enrique A.
tquique@ceride.gov.ar
Cassano, Alberto E. (Director) 4532965
acassano@ceride.gov.ar
Cerda, Jaime
jcerda@intec.unl.edu.ar
Chiovetta, Mario G.
mchiove@intec.unl.edu.ar
Costanza, Vicente
tsinoli@ceride.gov.ar
Deiber, Julio A.
treoflu@ceride.gov.ar
Grau, Ricardo J.
cqfina@ceride.gov.ar
Gugliotta, Luis M.
lgug@intec.unl.edu.ar
Henning, Gabriela
ghenning@intec.unl.edu.ar
Irazoqui, Horacio A.
hirazo@ceride.gov.ar
Luna, Julio
jluna@ceride.gov.ar
Marchetti, Jacinto L.
jlmarch@ceride.gov.ar
Meira, Gregorio R.
gmeira@ceride.gov.ar

Rubiolo, Amelia C. — tceidal@ceride.gov.ar

Saita, Fernando A. — fasaita@ceride.gov.ar

Associate Professors

Bortolozzi, Raul A. — rabor@ceride.gov.ar

Brandi, Rodolfo J. — rbrandi@ceride.gov.ar

Cabrera, María I. — cqfina@ceride.gov.ar

Estenoz, Diana A. — destenoz@ceride.gov.ar

Giavedoni, Maria D. — madelia@ceride.gov.ar

Isla, Miguel A. — misla@ceride.gov.ar

Martin, Carlos A. — cmartin@ceride.gov.ar

Pozzo, Roberto L. — rpozzo@intec.unl.edu.ar

Vega, Jorge R. — jvega@ceride.gov.ar

Zorrilla, Susana E. — zorrilla@intec.unl.edu.ar

Assistant Professors

Berli, Claudio L. — cberli@ceride.gov.ar

Dondo, Rodolfo G. — rdondo@ceride.gov.ar

Accredited by:

Degrees granted 2003-2004:

Ph.D.: 2

Department reports to: Mario Barletta, Rector de la Universidad Nacional del Litoral

Universidad Nacional del Salta

Facultad de Ingenieria
Buenos Aires 177
4400 Salta
Deliveries: Jorge Almazan

University phone 54 387 432-0563
Department phone 54 387 425-5420
Department FAX 54 387 425-5351

Professors

Armada, Margarita (Chair) .... 425 5362 †
armadam@unsa.edu.ar

Borla, Ricardo J. (Chair) 425 5351
rborla@epassaporte.com

Cuevas, Carlos M. (Chair) 425 5362
ccuevas@unsa.edu.ar

Finetti, Julián E. (Chair) 425 5412
finetti@unsa.edu.ar

Flores, Horacio R. (Chair) 254 5412
hrflores@unsa.edu.ar

Gonzo, Elio E. (Chair) 425 5410
gonzo@unsa.edu.ar

Gottifredi, Juan C. (Chair) 425 5410
gottifre@unsa.edu.ar

Mercado, Lorgio (Associate Dean) 425 5343
lmercado@unsa.edu.ar

Pocoví, Rubens E. (Chair) 425 5412
inbemi@unsa.edu.ar

Quiroga, Oscar D. (Chair) 425 5409
oquiroga@unsa.edu.ar

Associate Professors

Abán, Francisco E. (Chair) 425 5351
faban@unsa.edu.ar

Capretto, María E. (Chair) 425 5326
castillo@unsa.edu.ar

Destéfanis, Hugo A. (Chair) 425 5391
hdestefa@unsa.edu.ar

Macoritto, Alberto M. (Chair) 425 5350
amacori@unsa.edu.ar

Romero, Luis César (Chair) 425 5343
lcromero@unsa.edu.ar

† Faculty numbers have prefix 54 387.

Accredited by: CONEAU

Undergraduate advisor: Hugo Destéfanis

Student organization: Centro Estudiantes de Ingeniería

Department reports to: Jorge Almazán, Facultad de Ingenieria UNSa

Placement service: Facultad de Ingeniería, Universidad Nacional de Salta

Universidad Nacional del Sur

Departamento de Ingeniería Química
Avenida Alem 1253
8000 Bahía Blanca
Buenos Aires

University phone 54 291 459-5101
Department phone 54 291 459-5170
Department FAX 54 291 459-5171

Professors

Bessone, Jorge B. .... 4595170-3616 †
jbessone@criba.edu.ar

Brignole, Esteban A. 4861700-231
ebrignole@plapiqui.edu.ar

Capiati, Numa J. 4861700-205
ncapiati@plapiqui.edu.ar

Crapiste, Guillermo H. 4861700-203
gcrapiste@plapiqui.edu.ar

Damiani, Daniel E. 4861700-211
ddamiani@plapiqui.edu.ar

Gigola, Carlos E. 4861700-204
cgigola@plapiqui.edu.ar

Lozano, Jorge E. (Head) 4595170-3602
jlozano@plapiqui.edu.ar

Mayer, Carlos E. (Emeritus) 4595170-3610
cmayer@criba.edu.ar

Porras, José A. 4538605
jporras@criba.edu.ar

Urbicain, Martín J. (Emeritus) 4861700-201
urbicain@criba.edu.ar

Valles, Enrique M. 4861700-228
valles@plapiqui.edu.ar

Associate Professors

Bandoni, José A. 4861700-254
abandoni@plapiqui.edu.ar

Bottini, Susana B. 4861700-213
sbottini@plapiqui.edu.ar
Duarte, Marta M. 4595170-3618
mduarte@criba.edu.ar
Echarte, Roberto E. 4861700-248
recharte@plapiqui.edu.ar
Eliceche, Ana M. 4861700-252
meliceche@plapiqui.edu.ar
Errazu, Alberto F. 4861700-269
aerrazu@plapiqui.edu.ar
Salinas, Daniel R. 4595170-3614
dsalinas@criba.edu.ar

Assistant Professors

Aduriz, Rodolfo 4861700-260
raduriz@plapiqui.edu.ar
Borio, Daniel O. 4861700-208
dborio@plapiqui.edu.ar
Brandolin, Adriana 4861700-272
abrandolin@plapiqui.edu.ar
Bucalá, Verónica 4861700-265
vbucala@plapiqui.edu.ar
Diaz, María S 4861700-243
sdiaz@plapiqui.edu.ar
Quinzani, Lidia M. 4861700-282
poquinza@criba.edu.ar
Sanchez, Mabel C. 4861700-249
msanchez@plapiqui.edu.ar
Sarmoria, Claudia 4861700-273
csarmoria@plapiqui.edu.ar
Villar, Marcelo A. 4595170-3603
mvillar@plapiqui.edu.ar

† Phone code 54 291 precedes each number.

Graduate advisor: Susana B. Bottini

Undergraduate advisor: Adriana Brandolin

Department reports to: Dr. Jorge E. Lozano - Head

Australia

University of Adelaide

School of Chemical Engineering
University of Adelaide
Adelaide SA 5005

University phone 61 8 8303 4455
Department phone 61 8 8303 5455
Department FAX 61 8 8303 4373

Professors

Agnew, John A. (Emeritus) 8303 4118
jagnew@chemeng.adelaide.edu.au
King, Keith D. (Chair) 8303 5448
kking@chemeng.adelaide.edu.au

Associate Professors

Davey, Ken R. 8303 5457
kdavey@chemeng.adelaide.edu.au
Mullinger, Peter J. 8303 5085
pmulling@chemeng.adelaide.edu.au
Nguyen, Q. D. 8303 5456
dnguyen@chemeng.adelaide.edu.au
O'Neill, Brian K. (Associate Chair) 8303 4647
boneill@chemeng.adelaide.edu.au

Assistant Professors

Alwahabi, Zeyad T. 8303 3768
zalwahab@chemeng.adelaide.edu.au
Ashman, Peter J. 8303 5072
pashman@chemeng.adelaide.edu.au
Colby, Chris B. 8303 5846
cbcolby@chemeng.adelaide.edu.au
Lewis, David M. 8303 5503
dlewis@chemeng.adelaide.edu.au
Ngothai, Yung 8303 5445
yngothai@chemeng.adelaide.edu.au

Accredited by: Institute of Engineers (Australia), IChemE (UK)

Degrees granted 2003-2004:

B.S.: 41 M.S.: 1

Graduate advisor: Dr Chris Colby

Undergraduate advisor: Dr Yung Ngothai

Department reports to: Professor Peter Dowd, Executive Dean of Engineering

Placement service: Ms Amanda Phillis, Careers Administrator

Curtin University of Technology

Department of Chemical Engineering
GPO Box U1987
Perth WA 6845
Deliveries: Kent Street, Bentley, WA 6102

University phone .... (618) 9266-9266
Department phone .... (618) 9266-7581
Department FAX .... (618) 9266-3554

Professors

Smith, Terence N (Emeritus)
Tade, Moses O. (Head) 618 9266 7704
tadem@che.curtin.edu.au
Zhang, Dong-ke 618 9266 7581
dkzhang@che.curtin.edu.au

Associate Professors

Ang, Ha M. 618 9266 7894
angm@che.curtin.edu.au
Ray, Martyn 618 9266 7702
M.Ray@exschange.curtin.edu.au

Assistant Professors

Hongwei, Wu 618 9266 7592
Wuh@vesta.curtin.edu.au
Li, Qin 618 9266 7704
Pareek, Vishnu 618 9266 4687
pareekv@vesta.curtin.edu.au
Vuthaluru, Hari B 618 9266 4685
haribabu@che.curtin.edu.au
Wang, Shaobin 618 9266 3776
wangshao@vesta.curtin.edu.au
Zhu, Jianing 618 9266 1137
jnzhu@vesta.curtin.edu.au
Zhu, John 618 9266 3777
zhuz@vesta.curtin.edu.au

Accredited by: Institution of Engineers Australia, Institution of Chemical Engineers (UK)

Degrees granted 2003-2004:

B.S.: 64 Ph.D.: 2

Graduate advisor: H.M. Ang

Undergraduate advisor: H Vuthaluru

Student organization: ACES

Department reports to: Professor Peter Lee, Executive Dean, Engineering and Science

University of Melbourne

Department of Chemical and Biomolecular Engineering
Victoria 3010
Deliveries: Monash Road, Parkville 3052

University phone .... (61 3) 8344-4000
Department phone .... (61 3) 8344 6631
Department FAX .... (61 3) 8344 4153

Professors

Boger, David V. (Director) 8344-7440
dvboger@unimelb.edu.au

Caruso, Frank 8344-3461
fcaruso@unimelb.edu.au

Davidson, Malcolm R. 8344-6615
m.davidson@unimelb.edu.au

Dunstan, David E. 8344-8261
davided@unimelb.edu.au

Gray, Neil B. 8344-6639
n.gray@unimelb.edu.au

Healy, Thomas W (Emeritus) 8344-7624
tomhealy@unimelb.edu.au

Pamment, Neville B. 8344-6627
nbp@unimelb.edu.au

Scales, Peter J (Head) 8344-6480
peterjs@unimelb.edu.au

Shallcross, David C. 8344-6614
dcshal@unimelb.edu.au

Siemon, S. R. (Emeritus)

Solomon, David H. 8344-8200
davids@unimelb.edu.au

Stevens, Geoffrey W. 8344-6621
gstevens@unimelb.edu.au

van Deventer, Jannie S.J. (Dean) 8344-6619
jannie@unimelb.edu.au

Wood, David G. (Emeritus) 8344-3800
dgwood@unimelb.edu.au

Associate Professors

Connor, Michael A 8344-6875
maconnor@unimelb.edu.au

Kentish, Sandra E. 8344-6682
sandraek@unimelb.edu.au

O'Connor, Andrea J. 8344-8962
andreajo@unimelb.edu.au

Qiao, Greg G. 8344-3683
gregghq@unimelb.edu.au

Yeow, Y. L. 8344-6613
yly@unimelb.edu.au

Accredited by: IChemE (UK), Institute of Engineers (Australia)

Degrees granted 2003-2004:

B.S.: 98 M.S.: 1 Ph.D.: 11

Graduate advisor: A/Prof David Shallcross

Undergraduate advisor: Dr Michael Connor

Student organization: MUCESS
Advisor: Mr Mike Parksinson

Department reports to: Jannie van Deventer 8344-6619, Dean of Engineering

Monash University

Chemical Engineering Department
PO Box 36
Clayton, Victoria 3800
Deliveries: Building 35, Room 226, Wellington Road

University phone 61 3 990 54000
Department phone 61 3 990 53450
Department FAX 61 3 990 59649

Professors

Brisk, Mike L. (Emeritus) .... 990 52592 †
mike.brisk@eng.monash.edu.au

Johnston, Bob E. (Emeritus) 990 52592
R.E.Johnston@eng.monash.edu.au

Lawson, Frank (Honorary) 990 53426
frank.lawson@eng.monash.edu.au

Mathews, Joseph F. (Honorary) 990 53425
joe.mathews@eng.monash.edu.au

Potter, Owen E. (Emeritus) 990 53420

Prince, Ian G. 990 53449
ian.prince@eng.monash.edu.au

Rhodes, Martin J. (Head) 990 53445
martin.rhodes@eng.monash.edu.au

Seale, Alan (Adjunct) 990 59620
alan.seale@eng.monash.edu.au

Sridhar, Tamarapu (Dean) 990 53427
tam.sridhar@eng.monash.edu.au

Tiu, Carlos (Honorary) 990 53423
carlos.tiu@eng.edu.au

Toner, Mark (Adjunct) 990 59620
mark.toner@eng.monash.edu.au

Associate Professors

Andrews, John R.G. 990 53428
john.andrews@eng.monash.edu.au

Batchelor, Warren J. 990 53452
warren.batchelor@eng.monash.edu.au

Hoadley, Andrew F. 990 53421
andrew.hoadley@eng.monash.edu.au

Jagadeeshan, Ravi P. 990 53274
ravi.jagadeeshan@eng.monash.edu.au

Jeffrey, Matthew I. 990 51873
matthew.jeffrey@eng.monash.edu.au

Li, Chun-Zhu 990 59623
chun-zhu.li@eng.monash.edu.au

Nguyen, Kien L. 990 53429
loi.nguyen@eng monash.edu.au

Olbrich, W. E. 990 53436
erich.olbrich@eng monash.edu.au

Parker, Ian H. 990 55078
ian.parker@eng.monash.edu.au

Shen, Wei 990 53447
wei.shen@eng.monash.edu.au

Webley, Paul A. 990 51874
paul.webley@eng.monash.edu.au

† Numbers have 61 3 prefix.

Accredited by: IChemE (U.K.), Inst. Engrs. (Aust.), RACI

Degrees granted 2003-2004:

B.S.: 25 M.S.: 1 Ph.D.: 8

Graduate advisor: Dr Paul A Webley

Undergraduate advisor: Prof Martin J Rhodes

Student organization: SMUCE
Advisor: Dr W. Erich Olbrich

Department reports to: Prof Tamarupu Sridhar, Dean of Engineering

Placement service: Ms Gilda Moss, 61 3 990 54170

University of New South Wales

School of Chemical Engineering and Industrial Chemistry
Sydney, New South Wales, 2052

University phone .... (61) 02-9385 1000
Department phone .... (61) 02-9385 4315
Department FAX .... (61) 02-9385 5966

Professors

Brungs, Michael P. (Head) 9385-4306
m.brungs@unsw.edu.au
Burford, Robert P. 9385-4308
r.burford@unsw.edu.au
Chaplin, Rod 9385-4309
r.chaplin@unsw.edu.au
Davis, Tom 9385-4371
t.davis@unsw.edu.au
Fane, Anthony G. 9385-4315
a.fane@unsw.edu.au
Foster, Neil R. 9385-4341
n.foster@unsw.edu.au
Fowler, Robert (Emeritus) 9385-4422
Skyllas-Kazacos, Maria 9385-4335
m.kazacos@unsw.edu.au
Trimm, David (Emeritus) 9385-4340
d.trimm@unsw.edu.au

Associate Professors

Adesina, Adesoji 9385-5268
a.adesina@unsw.edu.au
Amal, Rose 9385-4361
r.amal@unsw.edu.au
Pham, Tuan 9385-5267
tuan pham@unsw.edu.au
Sheikholeslami, Roya 9385-4343
r.sheikholeslami@unsw.edu.au
Stubington, John F. 9385-4338
j.stubington@unsw.edu.au
Tran, Tam 9385-4448
t.tran@unsw.edu.au

Assistant Professors

Chen, Vicky 9385-4813
v.chen@unsw.edu.au
Crisp, Phillip T. 9385-4447
p.crisp@unsw.edu.au
Lucien, Frank 9385-4302
f.lucien@unsw.edu.au
Wiley, Dianne E. 9385-4304
d.wiley@unsw.edu.au

Accredited by: IChemE (UK), Institution of Engineers (Australia)

Department reports to: Mark Wainwright, Dean Fac. of Engineering

University of Newcastle

Chemical Engineering Department
New South Wales, NSW 2308

University phone .... (02) 4921-5000
Department phone .... (02) 4921-6180
Department FAX .... (02) 4921-6920

Professors

Evans, Geoff .... (02) 4921-5897
Geoffrey.Evans@newcastle.edu.au
Jameson, Graeme J. .... (02) 4921-6181
Graeme.Jameson@newcastle.edu.au
Wall, Terry F. .... (02) 4921-6179
Terry.Wall@newcastle.edu.au

Associate Professors

Dlugogorski, Bogdan .... (02) 4921-6176
Bogdan.Dlugogorski@newcastle.edu.au
Galvin, Kevin .... (02) 4921-6194
Kevin.Galvin@newcastle.edu.au
Gupta, Raj .... (02) 4921-6195
Rajender.Gupta@newcastle.edu.au
Kennedy, Eric .... (02) 4921-6177
Eric.Kennedy@newcastle.edu.au
Lucas, John .... (02) 4921 6193
John.Lucas@newcastle.edu.au

Assistant Professors

Franks, George .... (02) 4921-5889
George.Franks@newcaslte.edu.au
Moghtaderi, Behdad .... (02) 4921-6183
Behdad.Moghtaderi@newcastle.edu.au

Accredited by: Institution of Engineers (Australia), IChemE (UK)

Degrees granted 2003-2004:

B.S.: 21 Ph.D.: 3

Graduate advisor: B Dlugogorski

Undergraduate advisor: B Moghtaderi

Department reports to: A. Page (02) 4921-6025, Pro Vice-Chancellor

Placement service: Susan Eade

University of Queensland

Division of Chemical Engineering
St. Lucia, Brisbane
Queensland 4072
Deliveries: Building 74 College Road

University phone 61 7 3365 1111
Department phone 61 7 3365 3708
Department FAX 61 7 3365 4199

Professors

Bhatia, Suresh 3365 4263
sureshb@cheque.uq.edu.au

Cameron, Ian T. 3365 4261
ianc@donald.cheque.uq.edu.au

Do, Duong D. 3365 4154
duongd@cheque.uq.edu.au

James, David 3365 3708
davidj@cheque.uq.edu.au

Kavanagh, Lydia 3365 7517
lydiak@cheque.uq.edu.au

Keller, Jurg 3365 4727
j.keller@cheque.uq.edu.au

Litster, Jim D. (Head) 3365 3616
j.litster@cheque.uq.edu.au

Lu, G. Q. (Max) 3365 3735
maxlu@cheque.uq.edu.au

Middelberg, Anton (Director) 3366 8784
a.middelberg@uq.edu.au

Rudolph, Victor 3365 4171
victorr@cheque.uq.edu.au

White, E.T. (Ted) (Emeritus) 3365 4153
tedw@cheque.uq.edu.au

Associate Professors

Bell, Peter 3365 3801
p.bell@cheque.uq.edu.au

Clarke, Bill 3365 6464
billc@cheque.uq.edu.au

Cooper-White, Justin 3346 8715
j.cooperwhite@uq.edu.au

Crosthwaite, Caroline 3365 4264
carolc@cheque.uq.edu.au

Halley, Peter 3365 4158
p.halley@cheque.uq.edu.au

Howes, T. 3365 4262
tonyh@cheque.uq.edu.au

Lant, Paul 3365 4728
paull@cheque.uq.edu.au

Nielsen, Lars 3365 6960
larsn@cheque.uq.edu.au

Wiles, Robert J. (Gus) (Hon) 3365 3708
rjwiles@pocketmail.com.au

Assistant Professors

da Costa, Joe 3365 4218
joedac@cheque.uq.edu.au

Guo, Qipeng 3365 3661
QipengG@cheque.uq.edu.au

Hardin, Matt 3365 3708
m.hardin@auckland.ac.nz

Martin, Darren 3365 4152
darrenm@cheque.uq.edu.au

Masserotto, Paul 3365 4152
paulm@cheque.uq.edu.au

O'Brien, Kate 3365 3534
k.obrien@mailbox.uq.edu.au

Reid, Steve 3365 4001
stever@cheque.uq.edu.au

Sopade, Peter 3365 3931
p.sopade@cheque.uq.edua.u

Wang, F-Y. 3365 4552
fuyang@cheque.uq.edu.au

Accredited by: Institute of Engineers (Australia), IChemE (UK)

Degrees granted 2003-2004:

B.S.: 52 M.S.: 4 Ph.D.: 14

Graduate advisor: Tony Howes

Undergraduate advisor: Caroline Crosthwaite

Department reports to: John Simmons, Head, School of Engineering

Placement service: Marilyn Wilckens

Royal Melbourne Institute of Technology

School of Civil and Chemical Engineering
GPO Box 2476V
Melbourne, Victoria 3001

University phone .... (61 3) 9925-2000
Department phone .... (61-3) 9925-2080
Department FAX .... (61 3) 9639 0138

Professors

Bhattacharya, Sati N. 9925-2086
satinath.bhattacharya@rmit.edu.au

Roddick, Felicity (Head) 9925-2080
felicity.roddick@rmit.edu.au

Associate Professors

Swinbourne, Doug 9925-2201
douglas.swinbourne@rmit.edu.au

Accredited by: Institution of Engineers (Australia), IChemE (UK)

Undergraduate advisor: Assoc Prof Roger Hadgraft

Department reports to: Prof. Robert Snow 61 3 9925-2523, Dean

Placement service: Joanne Tyler 61 3 9925-2078, Careers and Appointments

University of Sydney

Chemical Engineering Department
Sydney, New South Wales 2006

University phone .... (02) 9351-2222
Department phone .... (02) 9351-2455
Department FAX .... (02) 9351-2854

Professors

Haynes, Brian S. (Chair) 9351-3435
haynes@chem.eng.usyd.edu.au

Petrie, Jim G.. (Head) 9351-4115
petrie@chem.eng.usyd.edu.au

Prince, Rolf G. H. (Emeritus) 9351-2354
prince@chem.eng.usyd.edu.au

Romagnoli, Jose A (Chair) 9351-4794
jose@chem.eng.usyd.edu.au

Associate Professors

Barton, Geoff W. 9351-3780
barton@chem.eng.usyd.edu.au

Langrish, Tim A. G. 9351-4568
timl@chem.eng.usyd.edu.au

Assistant Professors

Fletcher, David (Adjunct) 9351-4147
davidf@chem.eng.usyd.edu.au
Furzer, Ian A. 9351-2926
furzer@chem.eng.ce.su.oz.au
Gomes, Vincent G. 9351-4868
vgomes@chem.eng.usyd.edu.au
McNevin, Dennis 9351-4994
mcnevin@chem.eng.usyd.edu.au
See, Howard 9351-3832
howards@chem.eng.usyd.edu.au
Valix, Marjorie 9351-4995
mvalix@chem.eng.usyd.edu.au

Accredited by: IChemE (UK), Institute of Engineers (Australia)

Graduate advisor: T A G Langrish

Undergraduate advisor: G W Barton

Student organization: S U C E S

Department reports to: Dean of Engineering and ViceChancellor

Placement service: The Secretary, Appointments Board

Bangladesh

Bangladesh University of Engineering and Technology

Chemical Engineering Department
BUET, Dhaka 1000
Bangladesh

University phone 880-2-9665650-80
University FAX 880-2-9665622
Department phone 880-2-9665609

Professors

Ahmed, Nooruddin .... 7347 †
nahmed@che.buet.ac.bd
Ali, M. S. 7309
ali@buet.ac.bd
Begum, Dil A. 7346
afroza@buet.ac.bd
Hossain, Ijaz (Head) 7329
pmrebuet@bangla.net
Mahmud, Iqbal (Emeritus) 7341
imahmed@bangla.net
Quader, A.K.M.A. 7308
quader@buet.ac.bd

Associate Professors

Islam, M. S. 7559
seraj@buet.ac.bd
Khan, Sirajul H. 7326
shkhan@buet.ac.bd
Mondal, Harendra N. 7245
hmondal@buet.ac.bd

Assistant Professors

Kazi, M F. K. 7894
Syeda, Sultana R. 7893
syedasrazia@che.buet.ac.bd

† Numbers are extensions of the university number.

Accredited by:

Degrees granted 2003-2004:
B.S.: 62 M.S.: 2

Graduate advisor: Feroz Kabir

Undergraduate advisor: Syeda Sultana Razia

Student organization: Chemical Engineering Association
Advisor: M. Sabder Ali

Department reports to: A S w Kurny, Dean of Engineering

Placement service: M. Fazlul Bari, Director of Students Welfare

Belgium

University of Gent

Laboratorium voor Petrochemische Techniek
Krijgslaan 281 S5
9000 Gent
Deliveries: Krijgslaan 281, Building S5 B-9000

Department phone 32 9 264 45 16
Department FAX 32 9 264 49 99

Professors

Heynderickx, Geraldine
Marin, Guy B. (Director)
Menon, P. G. (Visiting)
Reyniers, Marie-Francoise
Yablonsky, Gregory (Visiting)

Associate Professors

Dumez, Francis.
Olea, Maria
van den Berg, Henk

Department reports to: P. Kiekens, Dean, Faculty of Applied Science

University of Liege

Group of Applied Chemistry
and Chemical Engineering
Institut de Chimie (B6)
4000 Sart Tilman—Liege

Department phone ....(32) 4 366 3541
Department FAX ....(32) 4 366 3545

Professors

Crine, M. ....(32) 4 366 3556
M.Crine@ulg.ac.be
Germain, A. H. ....(32) 4 366 3547
Albert.Germain@ulg.ac.be
Heyen, G. ....(32) 4 366 3527
G.Heyen@ulg.ac.be
Kalitventzeff, B. (Emeritus) ....(32) 4 366 3521
B.Kalitventzeff@ulg.ac.be
L'Homme, G. A. ....(32) 4 366 3541
Guy.LHomme@ulg.ac.be

Lecloux, A.
Marchot, P. ....(32) 4 366 3513
Pierre.Marchot@ulg.ac.be
Pirard, J. P. ....(32) 4 366 3558
Jean-Paul.Pirard@ulg.ac.be

Assistant Professors

Heinrichs, B. ....(32) 4 366 35 05
B.Heinrichs@ulg.ac.be
Toye, D. ....(32) 4 366 35 09
Dominique.Toye@ulg.ac.be

Accredited by:

Degrees granted 2003-2004:
M.S.: 5 Ph.D.: 3

Brazil

Universidade Estadual de Campinas (UNICAMP)

School of Chemical Engineering - FEQ
Av. Albert Einstein, 500
Cidade Universitaria, C.P. 6066
13081-970, Campinas SP

University phone ....(55) 019-788 3900
Department FAX ....(55) 019 788 3910

Professors

Rocha Pereira, J. A. F. da — pereira@desq.unicamp.br
Rocha, Sandra C.dos S. — rocha@feq.unicamp.br
Bittencourt, Edison — edsonbit@dtp.feq.unicamp.br
Francesconi, Artur Z. — francesconi@desq.feq.unicamp.br
Maciel Filho, Rubens — maciel@feq.unicamp.br
Maciel, Maria R. W. — wolf@feq.unicamp.br
Mei, Lúcia H. I. — lumeibit@turing.unicamp.br
Mendes, Mario de J. — mendes@desq.feq.unicamp.br
Mori, Milton (Dean) — mori@feq,unicamp.br
Santana, Cesar C. — santana@feq.unicamp.br

Associate Professors

Campos, Joao S. de C. — siniezio@dtp.feq.unicamp.br
Cruz, Sandra L. da — cruz@desq.feq.unicamp.br
Bueno, Sônia M. A. — sonia@feq.unicamp.br
Cremasco, Marco A. — cremasco@feq.unicamp.br
Franco, Telma T. — franco@feq.unicamp.br
Fratini, Ana M. F. — fattini@desq.feq.unicamp.br
Guirardello, Reginaldo — guira@feq.unicamp.br
Jordão, Elisabete — bete@desq.feq.unicamp.br
Kieckbusch, Theo G. — theo@feq.unicamp.br
Krähenbühl, Maria A. — mak@feq.unicamp.br
Lona, Liliane M.F. — liliane@feq.unicamp.br
Miranda, Everson A. — everson@feq.unicamp.br
Mohamed, Rahoma S. — mohamed@feq.unicamp.br
Moraes, Ângela M. — ammoraes@feq.unicamp.br
Ravagnani, Sergio P. — ravag@dtp.feq.unicamp.br
Ravagnani, Teresa M. K. — kakuta@desq.feq.unicamp.br
Santana, Maria H. A. — lena@feq.unicamp.br
Silva, Maria A. — cida@feq.unicamp.br
Silva, Meuris G. C. da — meuris@bla.feq.unicamp.br
Tannous, Kátia — katia@feq.unicamp.br
Taranto, Osvaldir P. (Associate Dean) — val@feq.unicamp.br
Zemp, Roger J. — zemp@desq.feq.unicamp.br

Assistant Professors

Aznar, Martin — maznar@feq.unicamp.br
Bartoli, Julio R. — bartoli@feq.unicamp.br
Beppu, Marisa M. — beppu@feq.unicamp.br
Bufo, Moacir J. — bufo@feq.unicamp.br
Cobo, Antonio J. G. — acobo@bla.feq.unicamp.br
Lisboa, Antonio C. L. — lisboa.feq.unicamp.br
Morales, Ana R. — morales@dtp.feq.unicamp.br
Oliveira, Wagner S. — oliveira@dtp.feq.unicamp.br
Peres, Leila — lperes@dtp.feq.unicamp.br
Rodrigues, Maria T. M. — maite@desq.feq.unicamp.br
Sprung, Renato — sprung@bla.feq.unicamp.br
Tomaz, Edson — etomaz@feq.unicamp.br
Valenca, Gustavo P. — gustavo@feq.unicamp.br

Department reports to: Milton Mori, Dean

Placement service: Marisa Augusto Cavalhaes

Universidade Estadual de Maringa

Departamento de Engenharia Quimica
Av. Colombo, 5790, BL D-90
CEP 87020-900 - MARINGA

University phone (55 44) 261 4000
Department phone (55 44) 261 44755
Department FAX (55 44) 263 3440

Professors

de Moraes, Flavio F. 261 4754 †
flavio@deq.uem.br
Pereira, Nehemias C. (Head) 261 4760
nehemias@deq.uem.br
Zanin, Gisella M. (Dean) 261 4754
gisellazanin@deq.uem.br

Associate Professors

Bergamasco, Rosangela 261 4763
rosangela@deq.uem.br
dos Santos, Onelia A. A. (Associate Head) 261 4756
onelia@deq.uem.br
Gimenes, Marcelino L. 261 4761
marcelino@deq.uem.br
Machado, Nadia R. C. F. 261 4747
nadia@deq.uem.br
Mendes, Elisabete S. (Dean) 261 4756
elisabete@deq.uem.br
Neitzel, Ivo 261 4761
ibneitzel@deq.uem.br
Ravagnani, Mauro A. 261 4746
ravag@deq.uem.br
Tavares, Celia R. G. 261 4746
celia@deq.uem.br

Assistant Professors

Andrade, Cid M. G. 261 4752
cid@deq.uem.br
Arroyo, Pedro A. 261 4758
arroyo@deq.uem.br
Barros, Maria A. D. 261 4758
angelica@deq.uem.br
Canassa, Edson M. 261 4752
edson@deq.uem.br
Cardozo Filho, Lucio 261 4749
lucio@deq.uem.br
Cossich, Eneida S. 261 4748
eneida@deq.uem.br
Costa, Alexando M. de S. 2614755
costa@deq.uem.br
da Rocha Pietrobon, Carmen L. 261 4762
carmen@deq.uem.br
de Barros Jr, Carlos 261 4780
carlos@deq.uem.br
de Barros, Sueli T. D. (Associate Dean) 261 4762
sueli@deq.uem.br
Gianotto, Valter R. 261 4753
valter@deq.uem.br
Higa, Márcio 2614755
higa@deq.uem.br
Jorge, Luiz M. M. 261 4747
lmmj@deq.uem.br
Mafra, Luciana I. 261 4749
luciana@deq.uem.br
Motta Lima, Oswaldo C. da 261 4759
oswaldo@deq.uem.br
Muller, José M. 261 4759
jmu@deq.uem.br
Olivo, Jose E. 261 4753
olivo@deq.uem.br
Paraiso, Paulo R. 261 4752
paulo@deq.uem.br
Povh, Nancy P. 261 4748
nancy@deq.uem.br
Ugri, Miriam C. B. A. 261 4749
miriam@deq.uem.br

† Faculty numbers have prefix 55 44.

Accredited by: Brazilian Ministry of Education

Degrees granted 2003-2004:

B.S.: 73 M.S.: 13 Ph.D.: 4

Graduate advisor: Gisella Maria Zanin

Undergraduate advisor: Oswaldo Curty da Motta Lima

Student organization: CAEQ
Advisor: Oswaldo Curty da Motta Lima

Department reports to: Nehemias Curvelo Pereira

Placement service: Jose Eduardo Olivo

Universidade Federal de Minas Gerais

Departamento de Engenharia Quimica
Rua Espirito Santo 35, sala 613
Belo Horizonte, Minas Gerais
CEP 30160-030

University phone (5531) 238-1780
Department phone (5531) 238-1780
Department FAX (5531) 238-1789

Professors

de Souza Freitas, Roberto F. (5531) 238-1780
Passos, Maria L. A. (5531) 238-1783
mlpassos@ig.com.br

Associate Professors

Andrade, Maria H. C. (5531) 238-1779
cano@deq.ufmg.br
Cardoso, Marcelo (5531) 238-1786
mcardoso@deq.ufmg.br
Castro, Maria M. R. (5531) 238-1788
deia@deq.ufmg.br
Coclho, Solange V. (5531) 238-1780
solange@deq.ufmg.br
da Silva, Sergio D. C. (5531) 238-1780
ssilva@magnesita.com.br
de Oliveira, Leandro S. (5531) 238-1777
leandro@deq.ufmg.br
Franca, Adriana S. (5531) 238-1777
franca@deq.ufmg.br
Lins, Vanessa F. C. (5531) 238-1775
vlins@deq.ufmg.br
Miranda, Tania L. S. (Head) (5531) 238-1779
tania@deq.ufmg.br
Oliveira, Eder D. (5531) 238-1785
eder@deq.ufmg.br
Rocha, Sonia D. F. (5531) 3238-1775
sdrocha@deq.ufmg.br
Salum, Adriane (5531) 238-1757
adrsalum@gold.com.br
Silva, Maria E. S. R. (5531) 238-1778
elisa@deq.ufmg.br

Souza, Ricardo G. .... (5531) 238-1785
rick@deq.ufmg.br

Assistant Professors

Castro, Roberto M. Z. .... (5531) 238-1788
rzcastro@deq.ufmg.br

Machado, Heloisa H. V. .... (5531) 238-1780
fisiqui@pucminas.br

Mansur, Marcelo B. .... (5531) 238-1778
mbmansur@ig.com.br

Torquetti, Zuleika S. C. .... (5531) 238-1780
ztorquetti@aol.com

Undergraduate advisor: Adriane Salum

Department reports to: Léo Heller, Director, School of Engineering

Placement service: Sônia Denise Ferreira Rocha

Universidade Federal Do Rio De Janeiro

Escola de Quimica
Centro de Tecnologia-Bloco E
Cidade Universitaria
21949-900, Rio de Janeiro, RJ

University phone .... (5521) 2562-7037
University FAX .... (5521) 2562-7567
Department phone .... (5521) 2562-7040
Department FAX .... (5521) 2562-7567

Professors

Coutinho, Jorge de A. (Emeritus)

Gentil, Vicente (Emeritus) .... (5521) 2562-7595

Martelli, Hebe H. L. (Emeritus)

Mascarenhas, Bernardo J. G. (Emeritus)

Moritz, Vitalis (Emeritus)

Perlingeiro, Carlos A. G. (Emeritus) .... (5521) 2562-7532
per@eq.ufrj.br

Rajagopal, Krishnaswamy .... (5521) 2562-7654
raja@eq.ufrj.br

Schmal, Martin
schmal@peq.coppe.ufrj.br

Seidl, Peter .... (5521) 2562-7586
pseidl@eq.ufrj.br

Telles, Affonso C.S da S. .... (5521) 2562-7653
affonso@eq.ufrj.br

Valdman, Belkis (Director) .... (5521) 2562-7037
diretora@eq.ufrj.br

Associate Professors

Aguiar, Eduardo F. de S.
falabella@cenpes.petrobras.com

Almeida, Valéria C. de .... (5521) 2562-7595
valeria@eq.ufrj.br

Antunes, Adelaide M. de S. .... (5521) 2562-7426
adelaide@eq.ufrj.br

Aranda, Donato A. G. .... (5521) 2562-7657
donato@eq.ufrj.br

Araujo, Ofelia de Q. F. .... (5521) 2562-7637
ofelia@eq.ufrj.br

Barreto, Daniel W. .... (5521) 2562-7475
dbarreto@eq.ufrj.br

Borschiver, Suzana .... (5521) 2562-7426
suzana@eq.ufrj.br

Brasil, Simone L. D.C. .... (5521) 2562-7642
simone@eq.ufrj.br

Calado, Verônica M. de A. .... (5521) 2562-7533
calado@eq.ufrj.br

Cammarota, Magali C. .... (5521) 2562-7568
christe@eq.ufrj.br

Carvalho, Denize D. de. .... (5521) 2562-7566
denize@eq.ufrj.br

Castier, Marcelo .... (5521) 2562-7607
castier@eq.ufrj.br

Castro, Mário Sérgio O. de .... (5521) 2562-7584
msoc@eq.ufrj.br

Chrisman, Erika C. A. N. .... (5521) 2562-7582
enunes@eq.ufrj.br

Coelho, Maria A. Z. .... (5521) 2562-7572
alice@eq.ufrj.br

Couto, Maria A. P.G. .... (5521) 2562-7577
gimenes@eq.ufrj.br

Cunha, Osvaldo G. C. da .... (5521) 2562-7641
osvaldo@eq.ufrj.br

d'Avila, Luiz A. .... (5521) 2562-7041
davila@eq.ufrj.br

Dweck, Jo .... (5521) 2562-7594
dweck@eq.ufrj.br

Folly, Rossana O. M. .... (5521) 2562-7652
rossana@eq.ufrj.br

Guimarães, Maria J. de O. .... (5521) 2562-7597
zeze@eq.ufrj.br

Leite, Selma G. F. .... (5521) 2562-7578
selma@eq.ufrj.br

Leão, Maria H. M. R. .... (5521) 2562-7580
mhrl@eq.ufrj.br

Maia, Maria C. A. .... (5521) 2562-7576
antun@eq.ufrj.br

Margarit, Isabel C. P. .... (5521) 2562-8550
margarit@eq.ufrj.br

Martins, José Vítor B. .... (5521) 2562-7610
vitor@eq.ufrj.br

Medeiros, José L. de .... (5521) 2562-7535
jlm@eq.ufrj.br

Medronho, Ricardo de A. .... (5521) 2562-7635
medronho@eq.ufrj.br

Milfont Jr., Wilson de N. .... (5521) 2562-7585
milfont@eq.ufrj.br

Mothé, Cheila G. .... (5521) 2562-7587
cheila@eq.ufrj.br

Nicolaiewsky, Eliôni M. de A. .... (5521) 2562-7634
elioni@eq.ufrj.br

Penteado Filho, Alberto de F. .... (5521) 2562-7537
penteado@eq.ufrj.br

Pereira Jr., Nei .... (5521) 2562-7639
nei@eq.ufrj.br

Pessoa, Fernando L. P. .... (5521) 2562-7603
pessoa@eq.ufrj.br

Peçanha, Ricardo P. .... (5521) 2562-7633
pecanha@eq.ufrj.br

Queiroz, Eduardo M. .... (5521) 2562-7603
mach@eq.ufrj.br

Salgado, Andréa M. .... (5521) 2562-7579
andrea@eq.ufrj.br

Servulo, Eliana F. C. .... (5521) 2562-7600
servulo@eq.ufrj.br

Silva, Monica A. P. da .... (5521) 2562-7606
monica@eq.ufrj.br

Souza Jr., Maurício B. de .... (5521) 2562-7636
mbsj@eq.ufrj.br

Souza, Mariana de M. V.M. .... (5521) 2562-7640
mmattos@eq.ufrj.br

Tavares, Frederico W. .... (5521) 2562-7650
tavares@eq.ufrj.br
Valle, Maria L. M. .... (5521) 2562-7581
murta@eq.ufrj.br
Yokoyama, Lidia .... (5521) 2562-7640
lidia@eq.ufrj.br
Zakon, Abraham .... (5521) 2562-7643
zakon@eq.ufrj.br

Assistant Professors

Alhadeff, Eliana M. .... (5521) 2562-7647
ema@eq.ufrj.br
Andrade, Hubmaier L. B. de .... (5521) 2206-9229
hubmaier.andrade@br.bureauveritas.com
Andrade, José E. P.de .... (5521) 2562-7717
joseedu@eq.ufrj.br
Awerianow, Cláudia de M. J. .... (5521) 2598-1715
jardim@eq.ufrj.br
Campello, Ana E. F. .... (5521) 2562-7638
campello@eq.ufrj.br
Gomes, Alexandre C. L. .... (5521) 2562-7574
aleiras@eq.ufrj.br
Silva, Sílvia M. C. da .... (5521) 2562-7651
sebrao@eq.ufrj.br
Vieira, Pedro A. P. .... (5521) 2562-7536
papv@eq.ufrj.br

Accredited by: ABEQ, ABQ, CRQ, MEC-SESU, CAPES

Degrees granted 2003-2004:

B.S.: 80 M.S.: 40 Ph.D.: 20

Undergraduate advisor: Prof. Eduardo Mach Queiroz

Student organization: DAEQ

Department reports to: Profa. Belkis Valdman

Placement service: Prof. Pedro Antonio Peixoto Vieira

Universidade de Sao Paulo USP

Escola Politecnica da USP
Department of Chemical Engineering
Av. Prof. Luciano Gualberto, travessa 3, n. 380
05508-900 Sao Paulo, SP, Brasil

Department phone 55 11 3091-2280
Department FAX 55 11 3031-3020

Professors

Buchler, Pedro M. .... 3091-2225 †
pbuchler@usp.br
Facciotti, Maria C. R. 3091-2234
mcrfacci@usp.br
Giudici, Reinaldo (Head) 3091-2254
rgiudici@usp.br
Guardani, Roberto 3091-2246
guardani@usp.br
Nascimento, Claudio A. O. 3091-2246
oller@usp.br
Odloak, Darci 3091-2237
odloak@usp.br

Associate Professors

Kilikian, Beatriz V. 3091-2232
kilikian@usp.br
LeRoux, Galo A.C. 3091-2237
galoroux@usp.br
Neves, Jose M. 3091-2280
jmango@ipt.br
Pinto, Jose M. 3091-2237
jompinto@usp.br
Silva, Gil A. 3091-2213
ganderis@usp.br
Tadini, Carmen C. 3091-2258
catadini@usp.br
Terron, Luiz R. 3091-2240
lrterron@usp.br

Assistant Professors

Antunha, Andre G. 3091-2244
aantunha@usp.br
Aoki, Idalina V. 3091-2274
idavaoki@usp.br
Camacho, Jose L. P. 3091-2243
jlpcam@usp.br
Gombert, Andreas K. 3091-2229
andreas.gombert@poli.usp.br
Guedes, Isabel C. 3091-2223
icguedes@usp.br
Loureiro, Luiz V. 30912237
valcov@usp.br
Maeda, Masazi 3091-2233
masmaeda@usp.br
Matai, Patricia 3091-2224
pmatai@usp.br
Melo, Hercilio G. 3091-2231
hgdemelo@usp.br
Neiva, Augusto C. 3091-2228
augusto.neiva@poli.usp.br
Pacheco, Claudio R. F. 3091-2265
crfp@usp.br
Paiva, Jose L. 3091-2262
jolpaiva@usp.br
Park, Song W. 3091-2246
sonwpark@usp.br
Pessoa, Pedro A. 3091-2282
pedro.pessoa@poli.usp.br
Salvagnini, Wilson M. 3091-2285
jackwil@usp.br
Sayer, Claudia 3091-2237
csayer@lscp.pqi.ep.usp.br
Shimizu, Priscila A. 3091-2242
pashimiz@usp.br
Song, Tah W. 3091-2245
tahwsong@usp.br
Taqueda, Maria E. S. 3091-2252
mtaqueda@usp.br
Tonso, Aldo 3091-2283
aldtonso@usp.br

† Dial with prefix 55 11.

Accredited by:

Student organization: AEQ

Universidade Federal de Uberlandia

Federal University of Uberlândia
School of Chemical Engineering
Av. João Naves de Ávila, 2121
P. O. Box 593
38400-902 Uberlândia-MG/Brazil

University phone 55 34 3239-4292
Department phone 55 34 3239-4292
Department FAX 55 34 3239-4188/4249

Professors

Damasceno, João J. R. (Director) .... ext. 213/235 †
jjrdamasceno@ufu.br

de Araújo, Euclides H. .... ext. 234
euclides@ufu.br

Fehr, Manfred .... ext. 227
mfehr@ufu.br

Limaverde, José R. .... ext. 206
jrlimaverde@ufu.br

Ribeiro, Eloízio J. .... ext. 239
ejriberio@ufu.br

Associate Professors

Ataíde, Carlos H. .... ext. 211
chataide@ufu.br

Burjaili, Mauro M. .... ext. 236
mmburjaili@ufu.br

Cardoso, Vicelma L. .... ext. 220
vicelma@ufu.br

Coelho, Márcia G. .... ext. 232
mgcoelho@ufu.br

de Assis, Adilson J. .... ext. 233
ajassis@ufu.br

de Souza Barrozo, Marcos A. .... ext. 203
masbarrozo@ufu.br

Franco, Moilton R. .... ext. 218
moilton@ufu.br

Henrique, Humberto M. .... ext. 217
humberto@ufu.br

Hori, Carla E. .... ext. 212
cehori@ufu.br

Murata, Valéria V. .... ext. 229
valeria@ufu.br

Nascimento, Alvimar F. .... ext. 228
alvimar@ufu.br

Oliveira Lopes, Luís C. .... ext. 230
lcol@ufu.br

Oliveira, Daniel T. .... ext. 237
tostes@ufu.br

Romanielo, Lucienne L. .... ext. 226
lucienne@ufu.br

Soares, Ricardo R. .... ext. 209
rrsoares@ufu.br

Assistant Professors

Sesso, Maria das G. V. .... ext. 236
mgvas@ufu.br

† Faculty numbers are extensions of 55 34-3239-4189

Accredited by: Brazilian Ministry of Education

Degrees granted 2003-2004:

B.S.: 60 M.S.: 18

Placement service: copev@ufu.br

Canada

University of British Columbia

Department OF Chemical and Biological Engineering
2216 Main Mall
Vancouver, British Columbia
V6T 1Z4
Deliveries: Mr. Horace Lam

University phone .... (604) 822-2211
Department phone .... (604) 822-3238
Department FAX .... (604) 822-6003

Professors

Bowen, Bruce D. .... (604) 822-3198
Branion, Richard (Emeritus) .... (604) 822-8752
Chieng, Sie-Tan .... (604) 822-4426
Duff, Sheldon .... (604) 822-9485
Englezos, Peter .... (604) 822-6184
Epstein, Norman (Emeritus) .... (604) 822-2815
Grace, John R. .... (604) 822-3121
Hatzikiriakos, Savvas .... (604) 822-3107
Kerekes, Richard J. .... (604) 224-8561
Lielmezs, Janis (Emeritus)
Lim, C. J. .... (604) 822-4871
Lo, K. V. .... (604) 822-4880
Oloman, Colin W. .... (604) 822-4345
Pinder, Kenneth L. (Emeritus) .... (604) 822-2583
Piret, James M. .... (604) 822-5835
Smith, Kevin J. (Head) .... (604) 822-3601
Thompson, D. W. (Emeritus) .... (604) 822-3486
Watkinson, A. P. .... (604) 822-2741

Associate Professors

Baldwin, Susan .... (604) 822-1973
Bennington, Chad .... (604) 822-8573
Bi, Xiaotao .... (604) 822-4408
Haynes, Charles .... (604) 822-5136
Kwok, K. E. .... (604) 822-3238
Lau, Anthony .... (604) 822-3476
Petrell, Royanne .... (604) 822-3475

Assistant Professors

Ellis, Naoko .... (604) 822-1243
Gyenge, Elod .... (604) 822-3217
Martinez, Mark .... (604) 822-2693
Mohseni, Madjid .... (604) 822-0047
Taghipour, Fariborz .... (604) 822-1902

Accredited by: Canadian Engineering Accreditation Board

Graduate advisor: S. Hatizkiriakos

Undergraduate advisor: J. Grace, S. Duff & L. Creagh

Student organization: AIChE and CSChE

Department reports to: M. Isaacson(604) 822-6413, Dean of Applied Science

Dalhousie University

Chemical Engineering Department
P.O. Box 1000
Halifax, Nova Scotia B3J 2X4
Deliveries: 1360 Barrington Street, Room F201, B3J 1Z1

University phone .... (902) 494-2211
Department phone .... (902) 494-3953
Department FAX .... (902) 420-7639

Professors

Al Taweel, Adel M. .... (902) 494-3992
Al.Taweel@dal.ca
Amyotte, Paul R. .... (902) 494-3976
Paul.Amyotte@dal.ca
Chen, Bin-hwa (Emeritus) .... (902) 494-3953
Fels, Mort .... (902) 494-3254
Mort.Fels@dal.ca
Gomaa, Hassan (Adjunct) .... (902) 494-3953
Gupta, Yash P. (Head) .... (902) 494-3948
Yash.Gupta@dal.ca
Mintz, Kenneth (Adjunct) .... (902) 494-3953
Pegg, Michael J. .... (902) 494-3225
Michael.Pegg@dal.ca
Thibault, Paul (Adjunct) .... (902) 494-3953
Woo, Stephen S. (Adjunct) .... (902) 494-3953

Associate Professors

Ghanem, Amyl .... (902) 494-3225
Amyl.Ghanem@dal.ca
Kuzak, Stephen G. .... (902) 494-3253
Stephen.Kuzak@dal.ca

Assistant Professors

Yuet, Pak K. .... (902) 494-3213
Pak.Yuet@dal.ca

Accredited by: Canadian Engineering Accreditation Board

Degrees granted 2003-2004:

B.S.: 32 M.S.: 6 Ph.D.: 1

Graduate advisor: A.M. Al Taweel

Undergraduate advisor: S. G. Kuzak

Student organization: CSChE
Advisor: S. G. Kuzak

Department reports to: W. Caley (902) 494-6055, Dean of Engineering

Placement service: A. M. Coolen

University of Alberta

Department of Chemical and Materials Engineering
University of Alberta
Edmonton, Alberta, T6G 2G6
Deliveries: Room 536, 116 St. and 92 Ave.

University phone .... (780) 492-3111
Department phone .... (780) 492-3321
Department FAX .... (780) 492-2881

Professors

Burrell, Robert .... (780) 492-8111
rburrell@ualberta.ca
Chuang, Karl T. .... (780) 492-4676
karlt.chuang@ualberta.ca
Dalla Lana, Ivo G. (Emeritus) .... (780) 492-3391
dalla.lana@ualberta.ca
Eadie, Reginald L. .... (780) 492-2858
reg.eadie@ualberta.ca
Elliott, Janet A. W. .... (780) 492-7963
janet.elliott@ualberta.ca
Etsell, Thomas H. .... (780) 492-5594
tom.etsell@ualberta.ca
Fisher, D. G. (Emeritus) .... (780) 492-3301
grant.fisher@ualberta.ca
Forbes, J. F. (Chair) .... (780) 492-0873
fraser.forbes@ualberta.ca
Gray, Murray R. .... (780) 492-7965
murray.gray@ualberta.ca
Hayes, Robert E. .... (780) 492-3571
bob.hayes@ualberta.ca
Henein, Hani .... (780) 492-7304
hani.henein@ualberta.ca
Huang, Biao .... (780) 492-9016
biao.huang@ualberta.ca
Ivey, Douglas G .... (780) 492-2957
doug.ivey@ualberta.ca
Kresta, Suzanne M. .... (780) 492-9221
suzanne.kresta@ualberta.ca
Kuznicki, Steven M. .... (780) 492-8819
kuznicki@ualberta.ca
Luo, Jingli .... (780) 492-2232
jingli.luo@ualberta.ca
Lynch, David T. (Dean) .... (780) 492-3596
dave.lynch@ualberta.ca
Masliyah, Jacob H. .... (780) 492-4673
jacob.masliyah@ualberta.ca
Mather, Alan E. (Emeritus) .... (780) 492-3957
alan.mather@ualberta.ca
McCaffrey, William .... (780) 492-6733
william.mccaffrey@ualberta.ca
Nandakumar, Kumar .... (780) 492-5810
kumar.nandakumar@ualberta.ca
Otto, Fred D. (Emeritus) .... (780) 492-5963
fred.otto@ualberta.ca
Patchett, Barry M. (Emeritus) .... (780) 492-2604
barry.patchett@ualberta.ca
Rao, Ming .... (780) 492-8250
ming.rao@ualberta.ca
Ryan, James T. (Emeritus)
jim.ryan@ualberta.ca
Shah, Sirish L. .... (780) 492-5162
sirish.shah@ualberta.ca
Shaw, John .... (780) 492-8236
jmshaw@ualberta.ca
Sundararaj, Uttandaraman .... (780) 492-1044
u.sundararaj@ualberta.ca
Wanke, Sieghard E. .... (780) 492-3817
sieg.wanke@ualberta.ca
Wayman, Michael L. (Emeritus) .... (780) 492-3418
mike.wayman@ualberta.ca
Williams, Michael C. (Emeritus) .... (780) 492-3962
mike.williams@ualberta.ca
Wood, Reginald K. (Emeritus) .... (780) 492-5963
reg.wood@ualberta.ca
Xu, Zhenghe .... (780) 492-7667
zhenghe.xu@ualberta.ca

Associate Professors

Chen, Weixing .... (780) 492-7706
weixing.chen@ualberta.ca

Choi, Phillip .... (780) 492-9018
phillip.choi@ualberta.ca
Li, Dongyang .... (780) 492-6750
dongyang.li@ualberta.ca
Liu, Qi .... (780) 492-8628
qi.liu@ualberta.ca
Uludag, Hasan .... (780) 492-0988
hasan.uludag@ualberta.ca
Yeung, Anthony .... (780) 492-3669
tony.yeung@ualberta.ca

Assistant Professors

Bhushan, Mani .... (780)492-7651
mbhushan@ualberta.ca
Liu, Shijie .... (780) 492-0981
shijie.liu@ualberta.ca
Meadows, E. S. .... (780) 492-3181
scott.meadows@ualberta.ca
Mitlin, David .... (780) 492-1542
dmitlin@ualberta.ca
Nelson, Alan .... (780) 492-7380
alan.nelson@ualberta.ca

Accredited by: Canadian Engineering Accreditation Board

Degrees granted 2003-2004:

B.S.: 176 M.S.: 20 Ph.D.: 11

Undergraduate advisor: U.T. Sundararaj

Student organization: CSChE and AIChE
Advisor: Anthony Yeung

Department reports to: J. Fraser Forbes, Department Chair

Placement service: Wendy L. Coffin (780) 492-4291

University of Calgary

Department of Chemical and Petroleum Engineering
2500 University Drive N.W.
Calgary, Alberta, T2N 1N4

Department phone .... (403) 220-5751
Department FAX .... (403) 284-4852

Professors

Aguilera, Roberto (Adjunct)
Badakhshan, Amir (Emeritus)
Behie, Leo A. .... (403) 220-6692
behie@ucalgary.ca
Bennion, Douglas W. (Emeritus)
Berruti, Franco (Adjunct)
Bishnoi, P. R. .... (403) 220-6695
bishnoi@ucalgary.ca
Butler, Roger M. (Emeritus)
Chakma, Amitabha (Adjunct)
Donnelly, John K. (Adjunct)
Exall, Douglas (Adjunct)
Farouq Ali, S. M. (Honorary)
Goberdhansingh, E. (Adjunct)
Gregory, Garry A. (Adjunct)
Harding, Thomas (Head)
Heidemann, Robert A. (Emeritus) .... (403) 220-8755
heideman@ucalgary.ca
Jeje, Ayodeji A. (Associate Head) .... (403) 220-5753
jeje@ucalgary.ca
Kantzas, Apostolos .... (403) 220-5752
akantzas@ucalgary.ca
Maini, Brij B. .... (403) 220-8777
bmaini@ucalgary.ca
Mattar, Luis (Adjunct)
Mehrotra, Anil K. .... (403) 220-7406
mehrotra@ucalgary.ca
Mehta, S.A. (Raj) .... (403) 220-4804
mehta@ucalgary.ca
Mohtadi, Matt F. (Emeritus)
Moore, R. G. .... (403) 220-5750
moore@ucalgary.ca
Mungan, Nick (Honorary)
Pereira, Pedro .... (403) 220-4799
ppereira@ucalgary.ca
Pruden, Barry B. (Emeritus)
Satyro, M. (Adjunct)
Sayegh, Selim (Adjunct)
Schramm, Laurie (Adjunct)
Settari, A. (Tony) .... (403) 220-7133
asettari@ucalgary.ca
Stanislav, Jaroslav F. (Emeritus)
Svrcek, William Y. (Associate Head) .... (403) 220-5755
svrcek@ucalgary.ca
Thusoo, Autar (Adjunct)
Tollefson, Eric L. (Emeritus)
Towson, Donald (Adjunct)
Trebble, Mark A. .... (403) 220-4823
trebble@ucalgary.ca
Wichert, Edward (Adjunct)
Wong, Dale (Adjunct)
Younger, A. (Andy) (Adjunct)
Zanzotto, Ludo .... (403) 220-8918
lzanzott@ucalgary.ca

Associate Professors

Adegbesan, Kenny (Adjunct)
Aikman, Michael (Adjunct)
Azaiez, Jalel .... (403) 220-7526
azaiez@ucalgary.ca
Baker, Richard (Adjunct)
Bellehumeur, Celine T. .... (403) 220-8804
cbellehu@ucalgary.ca
Monnery, Wayne D. (Adjunct)
Pooladi-Darvish, Mehran .... (403) 220-8779
pooladi@ucalgary.ca
Yarranton, Harvey W. .... (403) 220-6529
hyarrant@ucalgary.ca
Young, Brent R. .... (403) 220-8751
byoung@ucalgary.ca

Assistant Professors

Farrell, Patrick J. (Adjunct)
Hill, Josephine .... (403) 210-9488
jhill@ucalgary.ca
Husein, Maen .... (403) 220-6691
mhusein@ucalgary.ca
Kallos, Michael S. .... (403) 220-7447
mskallos@ucalgary.ca
Sen, Arindom .... (403) 210-9452
asen@ucalgary.ca
Srinivasan, Sanjay (Adjunct)

Accredited by: Canadian Engineering Accreditation Board

Degrees granted 2003-2004:

B.S.: 62 M.S.: 15 Ph.D.: 1

Undergraduate advisor: A. A. Jeje

Department reports to: S. C. Wirasinghe, Dean (403) 220-5731 Faculty of Engineering

Ecole Polytechnique-University of Montreal

Department of Chemical Engineering
Campus de l'Universite de Montreal
P.O. Box 6079, Station Centre-ville
Montreal, Quebec H3C 3A7
Deliveries: 2550 Chemin Polytechnique, Montreal H3T 1J4

University phone .... (514) 340-4711
Department phone .... (514) 340-4613
Department FAX .... (514) 340-4159

Professors

Ajersch, Franck .... 4533 †
franck.ajersch@polymtl.ca
Bale, Christopher 4535
christopher.bale@polymtl.ca
Buschmann, Michael D. 4931
mike@grbb.polymtl.ca
Carreau, Pierre 4924
pierre.carreau@polymtl.ca
Chaouki, Jamal 4034
jamal.chaouki@polymtl.ca
Favis, Basil D. 4527
basil.favis@polymtl.ca
Guy, Christophe (Dean) 4526
christophe.guy@polymtl.ca
Klvana, Danilo 4927
danilo.klvana@polymtl.ca
Lafleur, Pierre G. (Dean) 4618
pierre.lafleur@polymtl.ca
Legros, Robert (Head) 4922
robert.legros@polymtl.ca
Pelton, Arthur 4531
arthur.pelton@polymtl.ca
Perrier, Michel 4130
michel.perrier@polymtl.ca
Samson, Rejean 4898
Samson@biopro.polymtl.ca
Tanguy, Philippe A. 4017
Tanguy@urpei.polymtl.ca

Associate Professors

Stuart, Paul 4384
paul.stuart@polymtl.ca

Assistant Professors

Bertrand, Francois 5773
bertrand@urpei.polymtl.ca
Deschenes, Louise 5974
ldeschenes@site.polymtl.ca
Dubois, Charles 4893
charles.dubois@polymtl.ca
Heuzey, Marie-Claude 5930
mcheuzey@polymtl.ca
Jolicoeur, Mario 4525
mario.jolicoeur@polymtl.ca

† Faculty numbers are extensions of (514) 340-4711.

Accredited by: Engineering Institute of Canada

Student organization: CEGCh

Department reports to: Robert Papineau, Dean of Engineering

Placement service: Maryse Deschenes (514) 340-4374

Lakehead University

Chemical Engineering Department
955 Oliver Road
Thunder Bay, Ontario P7B 5E1

University phone .... (807) 343-8110
Department phone .... (807) 343-8170
Department FAX .... (807) 343-8928

Professors

Garred, Laurie J. .... (807) 343-8415
Laurie.Garred@lakeheadu.ca
Gilbert, Allan F. (Chair) .... (807) 343-8583
Allan.Gilbert@lakeheadu.ca
Nirdosh, Inderjit .... (807) 343-8343
INirdosh@gale.lakeheadu.ca
Puttagunta, V. R. (Emeritus)
Venugopala.Puttagunta@lakeheadu.ca

Associate Professors

Catalan, Lionel .... (807) 343-8573
Lionel.Catalan@lakeheadu.ca
Cooper, David .... (807) 343-8697
David.Cooper@lakeheadu.ca

Assistant Professors

Liao, Baoqiang .... (807) 343-8437
baoqiang.liao@lakeheadu.ca

Accredited by: Canadian Council of Professional Engineers

Graduate advisor: K. Liu

Undergraduate advisor: A. Gilbert

Student organization: CSChE
Advisor: I. Nirdosh

Department reports to: H.T. Saliba, (807) 343-8509, Dean, Faculty of Engineering

Placement service: John DeGiacomo, Manager, Student Placement and Cooperative Education Centre

Universite Laval

Chemical Engineering Department
Pouliot Building (3550)
Faculte des Sciences et de Genie
Cite Universitaire
Quebec, Quebec, G1K 7P4

University phone .... (418) 656-2131
Department phone .... (418) 656-3375
Department FAX .... (418) 656-5993

Professors

Bousmina, Mostapha .... (418) 656-2769
mosto.bousmina@gch.ulaval.ca
Cholette, Albert (Emeritus)
Grandjean, Bernard (Chair) .... (418) 656-2859
bernard.grandjean@gch.ulaval.ca
Kaliaguine, Serge .... (418) 656-2708
serge.kaliaguine@gch.ulaval.ca
Larachi, Faical .... (418) 656-3566
faical.larachi@gch.ulaval.ca

LeDuy, Anh .... (418) 656-2634
anh.leduy@gch.ulaval.ca
Ramalho, Rubens S. (Emeritus)
Roy, Christian .... (418) 656-7406
christian.roy@gch.ulaval.ca

Associate Professors

Garnier, Alain .... (418) 656-3106
alain.garnier@gch.ulaval.ca
Rodrigue, Denis (Associate Chair) .... (418) 656-2903
denis.rodrigue@gch.ulaval.ca

Assistant Professors

Do, Trong-On .... (418) 656-3774
Trong-On.Do@gch.ulaval.ca
Duchesne, Carl .... (418) 656 5184
carl.duchesne@gch.ulaval.ca
Mighri, Frej .... (418) 656-2241
Frej.Mighri@gch.ulaval.ca

Accredited by: Canadian Engineering Accreditation Board

Degrees granted 2003-2004:

B.S.: 21 M.S.: 4 Ph.D.: 5

Undergraduate advisor: Denis Rodrigue

Student organization: CSChE
Advisor: Denis Rodrigue

Department reports to: Jean Serodes (418) 656-2354, Dean, Faculty of Science and Engineering

Placement service: Micheline Grenier (418) 656-6877

McGill University

Department of Chemical Engineering
M.H. Wong Building
Room 3060, Floor 3A
3610 University Street
Montreal, Quebec, Canada
H3A 2B2

University phone .... (514) 398-4455
Department phone .... (514) 398-4494
Department FAX .... (514) 398-6678

Professors

Cooper, David G. .... (514) 398-4278
dcoope5@po-box.mcgill.ca
Dealy, John M. .... (514) 398-4264
jdealy@po-box.mcgill.ca
Douglas, W. J. M. (Emeritus) .... (514) 398-6186
mdougl2@po-box.mcgill.ca
Kamal, Musa R. .... (514) 398-4262
mkamal@po-box.mcgill.ca
Munz, Richard J. (Chair) .... (514) 398-4277
rmunz@po-box.mcgill.ca
Rey, Alejandro D. .... (514) 398-4196
arey@po-box.mcgill.ca
Vera, Juan H. .... (514) 398-4274
jvera@po-box.mcgill.ca
Volesky, Bohumil .... (514) 398-4276
bvoles@po-box.mcgill.ca
Weber, Martin E. .... (514) 398-4269
mweber7@po-box.mcgill.ca

Associate Professors

Berk, Dimitrios .... (514) 398-4271
dberk@po-box.mcgill.ca
Charrier, Jean-Michel (Emeritus) .... (514) 398-4265
jcharr@po-box.mcgill.ca
Meunier, Jean-Luc .... (514) 398-8331
jmeuni1@po-box.mcgill.ca
Simandl, Jana .... (514) 398-8308
jsiman@po-box.mcgill.ca

Assistant Professors

Brown, Wayne .... (514) 398-4275
wayne.brown@mcgill.ca
Omanovic, Sasha .... (514) 398-4273
TBA

Accredited by: Canadian Engineering Accreditation Board

Degrees granted 2003-2004:

B.S.: 63 M.S.: 16 Ph.D.: 5

Graduate advisor: Sylvain Coulombe

Undergraduate advisor: Jean-Luc Meunier

Student organization: CSChE
Advisor: Reghan Hill

Department reports to: J.E. Gruzleski, Dean of Engineering

Placement service: Gregg Blachford, Director

McMaster University

Chemical Engineering Department
Hamilton, Ontario L8S 4L7

University phone .... (905) 525-9140
Department phone .... (905) 525-9140x24957
Department FAX .... (905) 521-1350

Professors

Baird, Malcolm H. (Emeritus) .... x24945 †
bairdmhi@mcmaster.ca
Crowe, Cameron M. (Emeritus) x24947
crowe@mcmaster.ca
Dickson, James x24948
dickson@mcmaster.ca
Feuerstein, Irwin A. (Emeritus) x23118
feuerst@mcmaster.ca
Hamielec, Alvin E. (Emeritus) x24950
hamielec@mcmaster.ca
Hrymak, Andrew N. (Chair) x23136
hrymak@mcmaster.ca
MacGregor, John F. x24951
macgreg@mcmaster.ca
Marlin, Thomas E. x27125
marlint@mcmaster.ca
Pelton, Robert H. x27045
peltonrh@mcmaster.ca
Shemilt, Leslie W. (Emeritus) x24735
shemiltl@mcmaster.ca
Taylor, Paul A. x24952
taylorpa@mcmaster.ca
Vlachopoulos, John x24954
vlachopj@mcmaster.ca
Wood, Philip E. x24755
woodpe@mcmaster.ca
Woods, Donald R. (Emeritus) x27339
woodsdr@mcmaster.ca

Wright, Joseph D. (Adjunct) .... (514) 630-4102
wright@paprican.ca
Zhu, Shiping x24962
zhuship@mcmaster.ca

Associate Professors

Kourti, Theodora (Adjunct) x27035
kourtit@mcmaster.ca
Sheardown, Heather x24794
sheardow@mcmaster.ca
Swartz, Christopher x27945
swartzc@mcmaster.ca

Assistant Professors

Crossley, Lisa x26027
crossll@mcmaster.ca
Filipe, Carlos x27278
filipec@mcmaster.ca
Ghosh, Raja x27415
rghosh@mcmaster.ca
Jones, Kim x26333
kjones
Kostanski, Kris (Adjunct) x27018
kostans@mcmaster.ca
Quinn, Shannon (Adjunct) .... (905) 548-7200x6648
Shannon_quinn@dofasco.ca
Samyudia, Yudi x23273
samyudi@mcmaster.ca
Thompson, Michael R. x23213
mthomps@mcmaster.ca

† Faculty numbers are extensions of (905) 525-9140.

Accredited by: Canadian Engineering Accreditation Board

Degrees granted 2003-2004:

B.S.: 31 M.S.: 8 Ph.D.: 4

Student organization: CSChE
Advisor: J.M. Dickson

Department reports to: M. Elbestawi (905) 525-9140x24288, Dean, Faculty of Engineering

Placement service: Anne Markey (905) 525-9140x22571

University of Ottawa

Department of Chemical Engineering
161, Louis Pasteur, Room A-412
PO Box 450, Station A
Ottawa, Ontario
K1N 6N5

University phone .... (613) 562-5800
Department phone .... (613) 562-5772
Department FAX .... (613) 562-5172

Professors

Hornof, Vladimir (Emeritus) .... x6104 †
hornof@eng.uottawa.ca
Kennedy, Kevin (Joint) x6133
kennedy@genie.uottawa.ca
Kozicki, William (Emeritus) x6103
kozicki@eng.uottawa.ca
Lu, Benjamin C.-Y. (Emeritus) x6097
lu@eng.uottawa.ca
Mann, Ranveer S. (Emeritus) x6118
mann@eng.uottawa.ca
Matsuura, Takeshi x6114
matsuura@eng.uottawa.ca
McLean, David D. x6110
mclean@eng.uottawa.ca
Neale, Graham H. x6107
neale@eng.uottawa.ca
Talbot, Frank D. F. (Emeritus)
Thibault, Jules (Chair) .... (613) 562-5920
thibault@genie.uottawa.ca

Associate Professors

Dubé, Marc x6105
dube@eng.uottawa.ca
Taylor, David G. x6098
taylor@eng.uottawa.ca
Tezel, F. H. x6099
tezel@eng.uottawa.ca
Tremblay, Andre-Yves x6094
ayt@eng.uottawa.ca

Assistant Professors

Kruczek, Boguslaw x6302
kruczek@eng.uottawa.ca

† Faculty numbers are extensions of (613) 562-5800.

Accredited by: Canadian Engineering Accreditation Board

Student organization: CSChE
Advisor: Boguslaw Kruczek

Department reports to: Tyseer Aboulnasr, Dean of Engineering

Placement service: Yolaine Ruel

University of New Brunswick

Chemical Engineering Department
Post Office Box 4400
Fredericton, New Brunswick
E3B 5A3
Deliveries: 15 Dineen Drive, Head Hall, Room B-14

University phone .... (506) 453-4666
Department phone .... (506) 453-4520
Department FAX .... (506) 453-3591

Professors

Bendrich, Guida .... (506) 453-5019
bendrich@unb.ca
Chaplin, Robin A. .... (506) 453-4656
rchaplin@unb.ca
Couturier, Michel F. (Associate Dean) .... (506) 453-4690
cout@unb.ca
Eic, Mladen .... (506) 453-4689
meic@unb.ca
Lister, Derek H. (Chair) .... (506) 453-3299
dlister@unb.ca
Ni, Yonghao (Director) .... (506) 453-4547
yonghao@unb.ca
Picot, Jules J. C. (Emeritus) .... (506) 453-3542
jjcp@unb.ca

Associate Professors

Li, Kecheng .... (506) 451-6861
kecheng@unb.ca
Lowry, Brian .... (506) 453-4691
bjl@unb.ca

Xiao, Huining .... (506) 453-3532
hxiao@unb.ca

Zheng, Ying .... (506) 447-3329
yzheng@unb.ca

Accredited by: Canadian Engineering Accreditation Board

Degrees granted 2003-2004:

B.S.: 21 M.S.: 3 Ph.D.: 3

Graduate advisor: K. Li

Undergraduate advisor: B. Lowry

Student organization: CSChE
Advisor: B. Lowry

Department reports to: Dr. D. Coleman, Dean of Engineering

Placement service: Lois Clowater, Student Employment Service

Queen's University

Department of Chemical Engineering
Dupuis Hall
Kingston, ON
K7L 3N6
Deliveries: Dupuis Hall, Room 201

University phone .... (613) 533-6000
Department phone .... (613) 533-2765
Department FAX .... (613) 533-6637

Professors

Bacon, David W. (Emeritus) .... (613) 533-2781
bacond@post.queensu.ca

Becker, Henry A. (Emeritus) .... (613) 533-2767
beckerha@chee.queensu.ca

Code, Russel K. (Emeritus)

Cunningham, Michael F. .... (613) 533-2782
cunning@chee.queensu.ca

Daugulis, Andrew J. .... (613) 533-2784
daugulis@chee.queensu.ca

Downie, John (Emeritus)
downiej@post.queensu.ca

Grandmaison, Edward W. .... (613) 533-2771
grandmai@chee.queensu.ca

Harris, Thomas J. (Dean) .... (613) 533-2056
harrist@post.queensu.ca

Hsu, Cheng C. (Emeritus) .... (613) 533-2786
hsuc@chee.queensu.sa

McAuley, Kimberley B. .... (613) 533-2768
mcauleyk@chee.queensu.ca

McLellan, P. J. .... (613) 533-6343
mclellnj@chee.queensu.ca

Neufeld, Ronald J (Head) .... (613) 533-2785
neufeld@chee.queensu.ca

Wojciechowski, Bohdan W. (Emeritus)
wojciech@post.queensu.ca

Associate Professors

Amsden, Brian G. .... (613) 533-3093
amsden@chee.queensu.ca

Guay, Martin .... (613) 533-2788
guaym@chee.queensu.ca

Hutchinson, Robin A. .... (613) 533-3097
hutchra@chee.queensu.ca

Parent, J. S. .... (613) 533-6266
parent@chee.queensu.ca

Ramsay, Juliana A. .... (613) 533-2770
ramsayj@chee.queensu.ca

Assistant Professors

Docoslis, Aris .... (613) 533-6949
docoslis@chee.queensu.ca

Karan, Kunal .... (613) 533-3095
karan@chee.queensu.ca

Kontopoulou, Marianna .... (613) 533-3079
kontop@chee.queensu.ca

Accredited by: Canadian Engineering Accreditation Board

Degrees granted 2003-2004:

B.S.: 43 M.S.: 8 Ph.D.: 2

Graduate advisor: Dr. R.A. Hutchinson

Undergraduate advisor: Dr. E.W. Grandmaison

Department reports to: Thomas J. Harris, Dean of the Faculty of Applied Science

Placement service: J. Kelly, Career Planning and Placement

Royal Military College of Canada

Chemistry and Chemical Engineering Department
P. O. Box 17000 Station Forces
Kingston, Ontario, K7K 7B4

Department phone .... (613) 541-6000x6271
Department FAX .... (613) 542-9489

Professors

Amphlett, John C (Head) .... (613) 541-6000x6272
amphlett-j@rmc.ca

Bennett, Leslie G. I. .... (613) 541-6000x6614
bennett-l@rmc.ca

Bonin, Hugues W. .... (613) 541-6000x6613
bonin-h@rmc.ca

Bui, V. T. .... (613) 541-6000x6621
bui-v@rmc.ca

Lewis, Brent J. .... (613) 541-6000x6611
lewis-b@rmc.ca

Mann, Ronald F. (Emeritus) .... (613) 541-6000x6055
mann-r@rmc.ca

Roberge, Pierre .... (613) 541-6000x6485
roberge-p@rmc.ca

Thompson, William T. .... (613) 541-6000x6081
thompson-w@rmc.ca

Weir, Ronald D. (Dean) .... (613) 541-6612x6612
weir-r@rmc.ca

Associate Professors

Andrews, William S. .... (613) 541-6000x6052
andrews-w@rmc.ca

Bates, P. J. .... (613) 541-6609x6609
bates-p@rmc.ca

Peppley, Brant A .... (613) 541-6000x6702
peppley-b@rmc.ca

Assistant Professors

Pagé, Danny YSD .... (613) 541-6000x6353
page-d@rmc.ca

Pierre, J. M. .... (613) 541-6000x6713
pierre-j@rmc.ca

Accredited by: Canadian Engineering Accreditation Board

Student organization: CSChE
Advisor: H. W. Bonin

Department reports to: J. C. Amphlett, Head, Department of Chemistry and Chemical Engineering

Ryerson University

Department of Chemical Engineering
350 Victoria Street
Toronto, Ontario M5B 2K3

University phone .... (416) 979-5000
Department phone .... (416) 979-5157
Department FAX .... (416) 979-5083

Professors

Cuenca, Manuel A. .... x6346 †
mcuenca@ryerson.ca
Fayed, Muhammad x5217
mfayed@ryerson.ca
Lohi, Ali (Chair) x7028
alohi@ryerson.ca
Turcotte, Ginette x7312
gturcott@ryerson.ca

Associate Professors

Chan, Philip (Associate Chair) x6960
p4chan@ryerson.ca
Dhib, Ramdhane x6343
rdhib@ryerson.ca
Doan, Huu (Associate Chair) x6341
hdoan@ryerson.ca
Mehrvar, Mehrab x6555
mmehrvar@ryerson.ca
Wu, Jiangning x6549
j3wu@ryerson.ca

Assistant Professors

Ein-Mozaffari, Farhad
Upreti, Simant R. x6344
supreti@ryerson.ca

† Faculty numbers are extensions of (416) 979-5000

Accredited by: Canadian Engineering Accreditation Board

Degrees granted 2003-2004:

B.S.: 42 M.S.: 8

Graduate advisor: Dr. Philip Chan

Undergraduate advisor: Dr. Huu Doan

Student organization: CSChE
Advisor: Dr. Huu Doan

Department reports to: Stalin Boctor (416) 979-5000x7224, Dean, Faculty of Engineering and Applied Science

Placement service: John Easton (416) 979-5000x 6589

University of Saskatchewan

Department of Chemical Engineering
Research Annex
105 Maintenance Road
Saskatoon, Saskatchewan S7N 5C5

University phone .... (306) 966-4342
Department phone .... (306) 966-4760
Department FAX .... (306) 966-4777

Professors

Bakhshi, Narendra N. (Emeritus) .... (306) 966-4763
bakhshi@engr.usask.ca
Dalai, Ajay .... (306) 966-4771
dalai@engr.usask.ca
DeCoursey, W. J. (Emeritus) .... (306) 966-7854
Hill, Gordon (Head) .... (306) 966-4765
hill@engr.usask.ca
Macdonald, Douglas G. (Emeritus) .... (306) 966-2174
macdonal@engr.usask.ca
Peng, D.-Y. .... (306) 966-4767
pengd@engr.usask.ca
Postlethwaite, John (Emeritus) .... (306) 966-4768
jack@engr.usask.ca

Associate Professors

Evitts, Richard .... (306) 966-4766
Richard_W_Evitts@engr.usask.ca
Lin, Yen-Han .... (306) 966-4764
linyen@engr.usask.ca
Nemati, Mehdi .... (306) 966-4769
Mehdi.Nemati@usask.ca
Phoenix, Aaron .... (306) 966-4190
phoenix@engr.usask.ca
Pugsley, Todd .... (306) 966-4761
pugsley@engr.usask.ca
Sumner, Robert .... (306) 966-4775
sumner@engr.usask.ca

Assistant Professors

Wang, Hui .... (306) 966-2685
hui.wang@usask.ca

Accredited by: Canadian Engineering Accreditation Board

Degrees granted 2003-2004:

B.S.: 29 M.S.: 9

Graduate advisor: Y.-H. Lin

Undergraduate advisor: A. Phoenix

Student organization: CSChE
Advisor: R. Evitts

Department reports to: C. Lague, Dean of Engineering

Placement service: Sheila Taylor, Canada Manpower Centre

Universite de Sherbrooke

Chemical Engineering Department
Faculty of Engineering
2500, Blvd Universite
Sherbrooke
Quebec J1K2R1

University phone .... (819) 821-8000
Department phone .... (819) 821-7171
Department FAX .... (819) 821-7955

Professors

Abatzoglou, Nicolas .... (819) 821 7904[†]
nicolas.abatzoglou@USherbrooke.ca
Boulos, Maher I. .... (819) 821 7168
maher.boulos@USherbrooke.ca
Broadbent, Arthur D. x2172
authur.broadbent@USherbrooke.ca
Chornet, Esteban .... (819) 821 7170
Esteban.Chornet@USherbrooke.ca
Gitzhofer, Francois .... (819) 821 7841
Francois.Gitzhofer@USherbrooke.ca
Gravelle, Denis x2177
Denis.Gravelle@USherbrooke.ca
Heitz, Michele x2827
Michele.Heitz@USherbrooke.ca
Jones, J. P. (Chair) x2165
Peter.Jones@USherbrooke.ca
Marcos, Bernard x2166
Bernard.Marcos@USherbrooke.ca
Proulx, Pierre x2173
Pierre.Proulx@USherbrooke.ca
Soucy, Gervais x2167
Gervais.Soucy@USherbrooke.ca
Therien, Normand x2169
Normand.Therien@USherbrooke.ca

Associate Professors

Jurewicz, Jerzy .... (819) 821 7178
Jerzy.Jurewicz@USherbrooke.ca

Assistant Professors

Vermette, Patrick x2826
Patrick.Vermette@USherbrooke.ca

[†] Extensions have prefix (819) 821-8000.

Accredited by: Canadian Engineering Accreditation Board

Graduate advisor: J. Peter Jones

Undergraduate advisor: A.D. Broadbent

Student organization: Societe canadienne de genie chimique
Advisor: Gervais Soucy

Department reports to: Richard Marceau, Dean (819) 821-7111

Placement service: Renald Mercier, Director, Service des stages et du placement

University of Toronto

Department of Chemical Engineering and Applied Chemistry
200 College Street
Toronto, Ontario, M5S 3E5

University phone .... (416) 978-2011
Department phone .... (416) 978-3063
Department FAX .... (416) 978-8605

Professors

Allen, D. G. (Associate Chair) .... (416) 978-8517
allendg@chem-eng.utoronto.ca
Balke, Stephen T. .... (416) 978-7495
balke@chem-eng.utoronto.ca
Barham, David (Emeritus) .... (416) 978-3063
Basmadjian, Diran (Emeritus) .... (416) 978-5621
Bidleman, Terry F. (Adjunct) .... (416) 739-5730
terry.bidleman@ec.gc.ca
Boocock, David G.B. .... (416) 978-4020
boocock@chem-eng.utoronto.ca
Brook, Jeffrey (Adjunct) .... (416) 978-3063
Burgess, William H. (Emeritus) .... (416) 978-5827
Chaffey, Charles E. (Emeritus) .... (416) 978-3067
chaffey@chem-eng.utoronto.ca
Charles, Michael E. (Emeritus) .... (416) 978-5890
charles@chem-eng.utoronto.ca
Cheng, Yu-Ling .... (416) 978-5500
ylc@chem-eng.utoronto.ca
Cluett, William R. .... (416) 978-6697
cluett@chem-eng.utoronto.ca
Colcleugh, David W. (Adjunct) .... (416) 978-3063
Cormack, Donald E. .... (416) 978-4074
cormack@chem-eng.utoronto.ca
Cox, Brian (Joint) .... (416) 978-2127
cox@ecf.utoronto.ca
Davies, John E. (Joint) .... (416) 978-1471
davies@ecf.utoronto.ca
Diamond, Miriam L. (Joint) .... (416) 978-1749
diamond@geog.utoronto.ca
Diosady, Levente L. .... (416) 978-4137
diosady@chem-eng.utoronto.ca
Evans, Greg J. (Associate Dean) .... (416) 978-1821
evans@chem-eng.utoronto.ca
Foulkes, Frank R. .... (416) 978-6432
foulkes@chem-eng.utoronto.ca
Goodfellow, Howard D. (Adjunct) .... (416) 978-3063
Grace, Thomas M. (Adjunct) .... (416) 978-3063
Graydon, William F. (Emeritus) .... (416) 978-3063
Hummel, Richard L. (Emeritus) .... (416) 978-3063
James, David F. (Joint) .... (416) 978-3049
james@mie.utoronto.ca
Jervis, Robert E. (Emeritus) .... (416) 978-3071
robert.jervis@utoronto.ca
Jones, Andrew (Adjunct) .... (416) 978-3063
Kawaji, Masahiro .... (416) 978-3064
kawaji@chem-eng.utoronto.ca
Kirk, Donald W. .... (416) 978-7406
kirk@chem-eng.utoronto.ca
Kortschot, Mark T. .... (416) 978-8926
kortsch@chem-eng.utoronto.ca
Kuhn, David C.S. (Associate Chair) .... (416) 978-3065
kuhn@chem-eng.utoronto.ca
Luus, Rein (Emeritus) .... (416) 978-5200
luus@chem-eng.utoronto.ca
Lyne, M. B. (Adjunct)
Mackay, Donald (Emeritus) .... (416) 978-3063
Meresz, Otto (Adjunct) .... (416) 978-3063
Mims, Charles A. .... (416) 978-4575
mims@chem-eng.utoronto.ca
Missen, Ronald W. (Emeritus) .... (416) 978-3063
Napier, Douglas H. (Emeritus) .... (416) 978-3063
Newman, Roger C. .... (416) 946-0604
newman@chem-eng.utoronto.ca
Ojha, Matadial (Joint) .... (416) 978-7584
Papangelakis, Vladimiros G. .... (416) 978-1093
papange@chem-eng.utoronto.ca
Paradi, Joseph C. .... (416) 978-3714
paradi@mie.utoronto.ca
Phillips, M. J. (Emeritus) .... (416) 978-5872
mjp@chem-eng.utoronto.ca
Piggott, Michael R. (Emeritus) .... (416) 978-4745
michael.piggot@ecf.utoronto.ca
Reeve, Douglas W. (Chair) .... (416) 946-7939
reeve@chem-eng.utoronto.ca
Rizvi, Syed S.H. (Adjunct) .... (607) 255-7913
ssr3@cornell.edu

Sandler, Samuel (Emeritus) .... (416) 978-3063
Santerre, Paul (Joint) .... (416) 979-4903 x4341
paul.santerre@utoronto.ca
Saville, Bradley A. .... (416) 978-7745
saville@chem-eng.utoronto.ca
Sefton, Michael V. .... (416) 978-3088
sefton@chem-eng.utoronto.ca
Shoichet, Molly S. .... (416) 978-1460
molly@chem-eng.utoronto.ca
Smith, James W. (Emeritus) .... (416) 978-4906
smithjw@chem-eng.utoronto.ca
Smith, William R. (Adjunct) .... (416) 978-3063
Spinner, Irving H. (Emeritus) .... (416) 978-3063
Stanford, William L. (Joint) .... (416) 946–8379
william.stanford@utoronto.ca
Szabo, Paul (Adjunct) .... (416) 978-3063
Tran, Honghi .... (416) 978-8585
tranhn@chem-eng.utoronto.ca
Trass, Olev (Emeritus) .... (416) 978-6901
trass@chem-eng.utoronto.ca
Tremaine, Peter (Adjunct) .... (519) 823-4120
tremain2@uoguelph.ca
Utigard, Torstein A. (Joint) .... (416) 978-6267
utigard@ecf.utoronto.ca
Wania, Frank (Joint) .... (416) 287-7225
frank.wania@utoronto.ca
Woo, Stephen S (Adjunct) .... (905) 940-1859
sswoo@rogers.com
Woodhams, Raymond T. (Emeritus) ...... (416) 978-3063
Woodhouse, Kimberly A .... (416) 978-3060
kas@chem-eng.utoronto.ca

Associate Professors

Beller, Harry R. (Adjunct) .... (925) 422-0081
beller2@llnl.gov
Bialkowski, Wojciech (Adjunct) .... (416) 978-3063
Cooper, Paul A. (Joint) .... (416) 946-5078
p.cooper@utoronto.ca
Coyle, Tom W. .... (416) 978-5647
coyle@ecf.utoronto.ca
Edwards, Elizabeth A. .... (416) 978-3506
edwards@chem-eng.utoronto.ca
Fulthorpe, Roberta (Joint) .... (416) 287-7221
fulthorpe@scar.utoronto.ca
Jia, Charles Q. .... (416) 946-3097
cqjia@chem-eng.utoronto.ca
Kumacheva, Eugenia (Joint) .... (416) 978-3576
ekumache@alchemy.chem.utoronto.ca
Liss, Steven N (Adjunct) .... (416) 979-5000 x6357
sliss@ryerson.ca
McKague, Bruce (Adjunct) .... (416) 978-7297
mckague@chem-eng.utoronto.ca
Oshinowo, Toks (Adjunct) .... (416) 978-6948
toks@chem-eng.utoronto.ca
Sain, Mohini (Joint) .... (416) 946-3191
m.sain@utoronto.ca
Sayad, Saed (Adjunct) .... (416) 978-3063
Sodhi, Rana (Adjunct) .... (416) 978-5381
sodhi@chem-eng.utoronto.ca
Thomson, Murray J. (Joint) .... (416) 978-1827
thomson@mie.utoronto.ca
Yip, Chris M. .... (416) 978-7853
christopher.yip@utoronto.ca
Zandstra, Peter (Joint) .... (416) 978-8888
peter.zandstra@utoronto.ca

Assistant Professors

Acosta, Edgar J. .... (416) 946-0742
acosta@chem-eng.utoronto.ca
Allen, Christine (Joint) .... (416) 946-8594
callen@adm.utoronto.ca
Audet, Julie (Joint) .... (416) 946-0209
julie.audet@utoronto.ca
Farnood, Ramin R. .... (416) 946-7525
farnood@chem-eng.utoronto.ca
Gong, Sunling (Adjunct) .... (416) 739-5749
Sunling.Gong@ec.gc.ca
Yan, Ning (Joint) .... (416) 946-8070
ning.yan@utoronto.ca

Accredited by: Canadian Engineering Accreditation Board

Degrees granted 2003-2004:

B.S.: 82 M.S.: 37 Ph.D.: 11

Graduate advisor: D.G. Allen

Undergraduate advisor: D.C.S. Kuhn

Department reports to: A Venetsanopoulos, Dean, Faculty of Applied Science and Engineering

Placement service: Ms. Rivi Frankle

University of Waterloo

Chemical Engineering Department
200 University Ave. W.
Waterloo, Ontario, N2L 3G1

University phone......................... (519) 885-1211
Department phone .... (519) 888-4567x2404
Department FAX .... (519) 746-4979

Professors

Budman, Hector .... x6980 †
hbudman@uwaterloo.ca
Chatzis, Ioannis x3306
ichatzis@uwaterloo.ca
Douglas, Peter L. x2913
pdouglas@uwaterloo.ca
Duever, Thomas (Chair) x2540
tduever@uwaterloo.ca
Huang, Robert (Emeritus) x3409
ryhuang@uwaterloo.ca
Hudgins, Robert R. (Emeritus) x2092
rhudgins@uwaterloo.ca
Legge, Raymond L. x6728
rllegge@uwaterloo.ca
Moo-Young, Murray (Emeritus) x4006
mooyoung@uwaterloo.ca
Ng, Flora T. T. x3979
fttng@uwaterloo.ca
Pal, Rajinder x2985
rpal@uwaterloo.ca
Penlidis, Alexander x6634
penlidis@uwaterloo.ca
Pritzker, Mark x2542
pritzker@uwaterloo.ca
Rempel, G. L. x2702
grempel@uwaterloo.ca
Scharer, Jeno M. (Emeritus) x2703
jscharer@uwaterloo.ca
Silveston, Peter (Emeritus) x2092
silvesto@uwaterloo.ca
Tzoganakis, Costas (Associate Chair) x3442
ctzogan@uwaterloo.ca

Associate Professors

Anderson, William A. (Associate Chair) x5011
wanderson@uwaterloo.ca

Chen, Pu x5586
p4chen@uwaterloo.ca

Chou, Chih-Hsiung 3310
cpchou@uwaterloo.ca

Croiset, Eric x6472
ecroiset@uwaterloo.ca

Feng, Xianshe x6555
xfeng@uwaterloo.ca

Ioannidis, Marios x2914
mioannidis@uwaterloo.ca

Jervis, Eric x3928
ejervis@uwaterloo.ca

Moresoli, Christine x5254
cmoresoli@uwaterloo.ca

Pan, Qinman x7111
qpan@uwaterloo.ca

Soares, Joao B. P. x3436
jsoares@uwaterloo.ca

Assistant Professors

Elkamel, Ali x7157
aelkamel@uwaterloo.ca

Folwer, Michael x3415
mfowler@uwaterloo.ca

Simon, Leonardo x3301
lsimon@uwaterloo.ca

† Faculty numbers are extensions of (519) 888-4567.

Accredited by: Canadian Engineering Accreditation Board

Degrees granted 2003-2004:

B.S.: 98 M.S.: 26 Ph.D.: 12

Graduate advisor: Wendy Irving

Undergraduate advisor: William A. Anderson

Student organization: CSChE
Advisor: T.A. Duever

Department reports to: A. Sedra, Dean of Engineering

Placement service: J. F. Westlake

University of Western Ontario

Chemical and Biochemical Engineering
London, Ontario N6A 5B9

University phone .... (519) 661-2111
Department phone .... (519) 661-2131
Department FAX .... (519) 661-3498

Professors

Bergougnou, Maurice (Emeritus) .... (519) 661-2143 †
Berruti, Franco (Dean) .... (519) 661-2128
berruti@eng.uwo.ca

Briens, Cedric (Associate Dean) .... (519) 661-2145
cbriens@eng.uwo.ca

de Lasa, Hugo .... (519) 661-2144
hdelasa@eng.uwo.ca

Jutan, Arthur .... (519) 661-2111x88322
ajutan@uwo.ca

Margaritis, Argyrios .... (519) 661-2146
amarg@uwo.ca

Nakhla, George .... (519) 661-5470
nakhla@eng.uwo.ca

Rohani, Sohrab (Chair) .... (519) 661-4116
rohani@eng.uwo.ca

Zhu, Jesse .... (519) 661-3807
jzhu@uwo.ca

Associate Professors

Bassi, Amarjeet S. .... (519) 661-2111x88324
abassi@eng.uwo.ca

Karamanev, Dimitre .... (519) 661-2111x88230
dkaramanev@eng.uwo.ca

Prakash, Anand .... (519) 661-2111x88528
aprakas2@uwo.ca

Wan, WanKei .... (519) 661-2111x88440
wkwan@uwo.ca

Assistant Professors

Briens, Lauren .... (519) 661-2111x88849
lbriens@eng.uwo.ca

Charpentier, Paul .... (519) 661-3466
pcharpentier@eng.uwo.ca

Rizkalla, Amin .... (519) 661-2111x82212
arizkalla@eng.uwo.ca

† Numbers are extensions of (519) 661-2111.

Accredited by: Canadian Engineering Accreditation Board, APEO Canadian Society, ChE/CIC

Degrees granted 2003-2004:

B.S.: 59 M.S.: 17 Ph.D.: 3

Graduate advisor: A. Bassi

Undergraduate advisor: A. Bassi

Student organization: CIC-CSChE
Advisor: L. Briens

Department reports to: F. Berruti (519) 661-2128, Dean of Engineering

Placement service: Lesley Mounteer, Career Development Officer

Chile

Universidad de Concepcion

Departamento de Ingenieria Quimica
Facultad de Ingenieria
Casilla 160-C, Correo 3
Concepcion
Deliveries: Campus Universitario, Edificio Gustavo Pizarro Castro

University phone .... (56) 41 20 4000
Department phone .... (56) 41 20 4534
Department FAX .... (56) 41 24 7491

Professors

Borquez, Rodrigo (Chair) .... 3654 †
rborquez@diq.udec.cl

Canales, Edgardo 4669
ecanales@diq.udec.cl

Gordon, Alfredo L. 4534
agordon@diq.udec.cl

Melo, Roberto (Emeritus) 4762
rmelo@diq.udec.cl

Mendez, Julio 3660
jmendez@diq.udec.cl
Reich, Ricardo 3297
rreich@mail.udec.cl
Roeckel, Marlene 4987
mroeckel@diq.udec.cl
Toledo, Pedro G. 3658
pgtoledo@diq.udec.cl
Zaror, Claudio A. 3659
czaror@diq.udec.cl

Associate Professors

Aspe, Estrella 4376
easpe@diq.udec.cl
Garcia, Ximena 3661
xgarcia@diq.udec.cl
Klattenhoff, Dieter 4234
Marquez, Fernando 4755
fmarquez@diq.udec.cl
Melo, Diogenes 4722
dmelo@diq.udec.cl
Segura, Hugo 3662
hsegura@diq.udec.cl
Tippmann, Carlos 4534

† Dial 41 20 and extension.

Accredited by: CONAP(Chile)

Graduate advisor: Diogenes L. Melo

Undergraduate advisor: Claudio A. Zaror

Department reports to: Joel Zambrano, Decano, Facultad de Ingenieria

Pontificia Universidad Catolica de Chile

Departamento de Ingeniería Química y Bioprocesos
Vicuña Mackenna 4860,
C.P. 6904411
Santiago

University phone .... (562) 354-4198
Department phone .... (562) 354-4254
Department FAX .... (562) 686-5803

Professors

Agosin, Eduardo (Associate Dean) .... 354-4253 †
agosin@ing.puc.cl
Aguilera, José M. 354-4256
jmaguile@ing.puc.cl

Associate Professors

De la Barra, Fernando 354-4237
del Valle, José M. 354-4418
delvalle@ing.puc.cl
Jorquera, Héctor I. (Head) 354-4421
jorquera@ing.puc.cl
Olivares, Marcela 354-4234
maoliva@ing.puc.cl
Pérez, José R. 354-4258
perez@ing.puc.cl
San Martín, Ricardo 354-4263
sanmarti@ing.puc.cl
Schmidt, Cristian 354-4237

Assistant Professors

Bouchon, Pedro 354-4927
pbouchon@ing.puc.cl
Sáez, César A. 354-4257
csaez@ing.puc.cl

† Numbers above have prefix 562.

Accredited by: CONAP, CONICYT

Graduate advisor: Ricardo Perez

Undergraduate advisor: Loreto Valenzuela

Department reports to: Aldo Cipriano, Dean, School of Engineering

Placement service: Rosa Salas, Assistant Director, Students Affairs Office, (562) 686-4200

Universidad de Santiago de Chile

Facultad de Ingenieria
Depto. Ingenieria Quimica
Avda. Libertador Bernardo
O'Higgins 3363
P.O. Box 10233
Santiago

University phone 56 2 681-1100
Department phone 56 2 681-2398
Department FAX 56 2 681-1422

Professors

Alvarez, Ivan (Head)
Correa, Horacio
Romo, Claudio
Vega, Ricardo

Associate Professors

Bravo, Luis
Blasco, Ramon
Carvajal, Cynthia
Castro, Jose
Correa, Duberlis
Herrera, Luis
Levy, Isaac
Moyano, Pedro
Retamales, Lautaro
Reyes, Alejandro
Thoma, Enrico
Vega, Rolando

Assistant Professors

Contreras, Elsa
Cubillos, Francisco
Diaz, Georgina
Marquardt, Fritz-Hans
Palominos, Pedro

Accredited by: Ministry of Education, Colegio de Ing. de Chile A.G.

Department reports to: Bernd Schulz, Decano, Facultad de Ingenieria

Placement service: Elsa Contreras

Universidad Tecnica Federico

Santa Maria
Department de Procesos Quimicos, Biotecnologicos y Ambientales
Casilla 110-V
Valparaiso

University phone 56 032 654258
Department phone 56 032 654258
Department FAX 56 032 797505

Professors

Bergh, Luis
Livingston, Alejandro (Emeritus)
Riquelme, Gerardo D.
Yianatos, Juan B.

Associate Professors

Alkalay, Daniel (Director)
Nunez, Patricio

Assistant Professors

Almonacid, Sergio
Borzone, Luis (Head)
De la Fuente, Juan
Guerrero, Lorna
Hansen, Henrik
Rojo, Adrián
Simpson, Ricardo

Undergraduate advisor: Adrián Rojo

Universidad Catolica de Valparaiso

Escuela de Ingenieria Quimica
Ingeniero Civil Quimico
Casilla 4059
Valparaiso
Deliveries: Avda. Brasil 2950

University phone .... (032) 273000
Department phone .... (032) 273721
Department FAX .... (032) 273807

Professors

Aros, Horacio M. .... (032) 273737
Dondero, Amelia C. .... (032) 273727
Meyer, Eduardo A. .... (032) 273726
Proust, Patricio C. .... (032) 273730
Santana, C. J. .... (032) 273733
Soto, Horacio A. .... (032) 273733
Torres, Jose T. .... (032) 273732

Associate Professors

Fernandez, Jaime C. (Director) .... (032) 273720
Garcia, M. V. .... (032) 273725
Vega, Luis A. .... (032) 273722

Department reports to: Feliciano Tomarelli Z., Decano de Ingenieria

Peoples Republic of China

Beijing University of Chemical Technology

Chemical Engineering Department
15 KeiSanhuan East Road
ChaoYang District
Beijing 100029

University phone 886-10 6421-2091
Department phone 86-10-64434754
Department FAX 86-10-64423610

Professors

Chen, Biaohua .... 64429057 †
chenbh@mail.buct.edu.cn
Guo, Fen 64448808
guof@mail.buct.edu.cn
Guo, Kai 64448808
guok@163bj.com
Hai, Reti 64427356
hjzhx@mail.buct.edu.cn
Li, Chengyue 64436787
licy@mail.buct.edu.cn
Li, Chunxi 64444912
Licx@mail.buct.edu.cn
Liu, Hui 64433695
hliu@mail.buct.edu.cn
Ma, Runyu 64448919
r.ma@mail.buct.edu.cn
Wang, Jianhong 64411065
wjhmaster@263.net
Wang, Wenchuan 64444905
wangwc@mail.buct.edu.cn
Zhang, Zeting 64434775
zhangzt@mail.buct.edu.cn
Zheng, Danxing 64416406
dxzh@mail.buct.edu.cn
Zhong, Chongli 64419862
zhongcl@mail.buct.edu.cn

Associate Professors

Chen, Xiaochun 64434783
chenxc@mail.buct.edu.cn
Feng, Lui 64427356
chenql@btamail.net.cn
Gao, Zhengming 64418267
gaozm@mail.buct.edu.cn
Huang, Xiongbin 64418267
huangxb@mail.buct.edu.cn
Li, Jianwei 64433695
lijw@mail.buct.edu.cn
Li, Qunsheng 64449695
qsli@mail.buct.edu.cn
Li, Xiujin 64427356
lxiujin@hotmail.com

† Faculty numbers preceded by 86-10

Accredited by: State Education Commission of China

Department reports to: Xiangzhi Wu, Dean

South China University of Technology

Chemical Engineering Research Center
Guangzhou, 510640 P. R. China

University phone .... (86) 20 87110000
Department phone .... (86) 20 87113871

Professors

Chen, Huanqin 87112093
cehqchen@scut.edu.cn
Deng, Xianhe 87111814
xianhedeng@263.net
Hua, Ben 87113744
cehuaben@scut.edu.cn
Li, Zhong (Chair) 87113501
cezhli@scut.edu.cn
Lu, Enxi 87113744
ceenxilu@scut.edu.cn
Qiu, Xueqin 87110201
adxqqiu@scut.edu.cn
Tu, Weiping 87113486
cewptu@scut.edu.cn
Yang, Xiaoxi 87114341
cexxyang@scut.edu.cn
Yang, Zhuorn 87112057
cezhyang@scut.edu.cn
Yin, Qinghua 87113614
ceqhyin@scut.edu.cn
Zhu, Dongsheng 87112043
cedshzhu@scut.edu.cn

Associate Professors

Chen, Qinglin 87113614
ceqlchen@scut.edu.cn
Cheng, Jiang 87112047
cejcheng@scut.edu.cn
Lan, Renhuan 87112047
cerhlan@scut.edu.cn
Liu, Xiaohong 83305432
cexhliu@scut.edu.cn
Xi, Hongxia 87113501
hongxia@public.guangzhou.gd.cn
Xiao, Xinyan 87112074
cexyxiao@scut.edu.cn

Assistant Professors

Jiang, Yanbin 87112046
cebjiang@scut.edu.cn
Li, Xiuxi 87112046
cexxli@scut.edu.cn
Lu, Shushen 87113614
cesslu@scut.edu.cn
Qian, Yu 87110491
ceyuqian@scut.edu.cn

Accredited by: State Education Ministry of China

Department reports to: Liu, Huanbin

Placement service: Li, Zhong

Tianjin University

School of Chemical Engineering and Technology
Tianjin
P. R. China 300072

University phone 011 86 22-27405374
Department phone 011 86 22-27404770
Department FAX 011 86 22-27403389

Professors

Bai, Peng 86 22 27404154
baip@eyou.com
Chen, Hong-fang 86 22 27406043
hfchen@tju.edu.cn
Dong, Xiaoyan 86 22 27406590
ysun@tju.edu.cn
Feng, Yaqing 86 22 87894220
fengyaqing@eyou.com
Gan, Yiru 86 22 27409598
yrgan@tju.edu,cn
Han, Jinyu 86 22 27491156
jyhan@tju.edu.cn
He, Zhimin 86 22 27403154
zhe@tju.edu.cn
Hu, Jinbang 86 22 8354744
jbhu@eyou.com
Hu, Zong-ding 86 22 27404196
zdhu@public.tpt.tj.cn
Jia, Shaoyi 86 22 27405803
shaoyiJ@eyou.com
Jiang, Zhongyi 86 22 27406454
dworld@mail.zlnet.com.cn
Li, Shiyu 86 22 27404177
shyli@tju.edu.cn
Li, Shufen 86 22 27408578
shfli@tju.edu.cn
Li, Xianggao 86 22 27406893
Lixianggao@hotmail.com
Li, Xingang 86 22 27404473
xingli@tju.edu.cn
Li, Yong-dan 86 22 27405613
ydli@tju.edu.cn
Liu, Changjun 86 22 27406498
skcct@tju.edu.cn
Liu, Dongzhi 86 22 27473134
tpd@tju.edu.cn
Liu, Jiaqi 86 22 27404190
gwliu@tju.edu.cn
Ma, Pei-sheng 86 22 27406081
mapeisheng@tju.edu.cn
Ma, Pingsheng 86 22 27401654
pingshengm@yahoo.com
Ma, Youguang 86 22 27404772
Ma-youguang@sina.com
Mi, Zhentao 86 22 27403714
zhtmi@tju.edu.cn
Qin, Yong-ning 86 22 27401455
ynqin@tju.edu.cn
Song, Haihua 86 22 27403684
song_hh@990.net.cn
Sun, Duoxian 86 22-27407213
tdxsun@eyou.com
Sun, Yan 86 22 27406590
ysun@tju.edu.cn
Tang, Zhiyuan 86 22 27404786
zytang@public.tpt.tj.cn
Wang, Baoguo 86 22 27405642
bgwang@tju.edu.cn

Wang, Chengyang 86 22 27472656
cywang@tju.edu.cn
Wang, Jing-kang 86 22 27405754
jkwang@tju.edu.cn
Wang, Shichang 86 22 27404347
jiswang@tju.edu.cn
Wang, Shuying 86 22 27405445
sywang@public.tpt.tj.cn
Wang, Wei 86 22 27404243
wwangg@eyou.com
Wang, Yaquan 86 22 27406498
wangyaquan@eyou.com
Wang, Yiping 86 22 27404674
xinxing@tju.edu.cn
Wang, Yuxin 86 22 27404347
wyx@shell.tjvan.net.cn
Wang, Zhi 86 22 27404347
wangzhi@tju.edu.cn
Wei, Hongyuan 86 22 27405754
hongyuanwei@hotmail.com
Wu, Jinchuan 86 22 27408154
jchwu@tju.edu.cn
Xin, Feng 86 22 27403215
xinf@mail.zlnet.com.cn
Xu, Chen 86 22 27403496
xchen@tju.edu.cn
Xu, Gen-hui 86 22-27405267
ghxu@tju.edu.cn
Xu, Wen 86 22 27405643
yangpaul@eyou.com
Xu, Yongshen 86 22 27407139
yongsheng@public.tpt.tj.cn
Yang, Zhicai 86 22 27404493
zcyang@public.tpt.tj.cn
Yao, Suwei 86 22 27415680
gongzl@public.tpt.tj.cn
Yin, Qiuxiang 86 22 27405754
srcict@tju.edu.cn
Yu, K.T.(Guocong) 86 22 27404332
ktyu@tju.edu.cn
Yuan, Xigang 86 22 27405679
yuanxg@tju.edu.cn
Yuan, Ying-jin (Associate Dean) 86 22 27403888
yjyuan@tju.edu.cn
Zhang, Fengbao (Associate Dean) 86 22 87890013
zhangfengbao@eyou.com
Zhang, Guoliang 86 22 27403515
zhangguoliang@eyou.com
Zhang, Ji-yan 86 22 27406469
jyzhang@tju.edu.cn
Zhang, Weijiang 86 22 27402028
wizh@tju.edu.cn
Zhao, Xue-ming (Dean) 86 22 27404770
xmzhao@tju.edu.cn
Zhong, Shun-he 86 22 27401453
shzhong@publicl.tpt.tj.cn
Zhou, Li 86 22 27406301
Zhouli@public.tpt.tj.cn
Zhou, Ming 86 22 27403945
mxzhou@tju.edu.cn

Associate Professors

Ban, Rui 86 22 27401654
banrui@eyou.com
Chang, Heying 86 22 27400859
Chen, Jiuling 86 22 27401446
jlchen@public.tpt.tj.cn
Dong, Anjie 86 22 27403171
ajdong@tju.edu.cn
Feng, Huisheng 86 22 27401915
Gao, Ruichang 86 22 27404347
Guan, Yi 86 22 27401455
guanyi36@eyou.com
Guo, Cuili 86 22 27401476
Guo, Hongyu 86 22 27404307
hyguo@eyou.cn
Guo, Jintang 86 22 27408829
jintang_guo@sina.com
Guo, Ruiwei 86 22 27405773
rwguo@public.tpt.tj.cn
Han, Zhenwei 86 22 27404705
zwhan@tju.edu.cn
He, Fei 86 22 27404472
Jiang, Bin 86 22 27404473
Kang, Shifang 86 22 27405824
Kang, Yong 86 22 27408813
falSsp@263.net
Li, Pingli 86 22 27404347
Li, Shizhong 86 22 27406876
lishizhong@hotmail.com
Li, Wei 86 22 23370279
Li, Yonghong 86 22 27406498
skcct@tju.edu.cn
Li, Zengxi 86 22 27402975
Zengxi_Li@263.net
Li, Zhenhua 86 22 27406498
skcct@tju.edu.cn
Li, Zhenya 86 22 27477479
zhy_li@eyou.com
Liu, Jianhua 86 22 27402344
lihtju@263.net
Liu, Mingyan 86 22 27404614
Ma, Hongqin 86 22 27404481
jhtian@eyou.com
Ma, Xinbin 86 22 27406498
xbma@tju.edu.cn
Ma, Zhi 86 22 27401451
mazhi@eyou.com
Na, Ping 86 22 27401999
naping@tju.edu.cn
Niu, Wentai 86 22 27404249
Ren, Yan 86 22 27404347
Shan, Zhongqiang 86 22 27403237
Sun, Jinsheng 86 22 27404473
jssun@tju.edu.cn
Tan, Wei 86 22 27408728
wtan@eyou.com
Tan, Xin 86 22 27403337
Tian, Jianhua 86 22 3127509
jhtian@eyou.com
Wang, Fumin 86 22 27890041
Wang, Jun 86 22 27890773
Wang, Rijie 86 22 27401018
rjwang@tju,edu,cn
Wang, Shirong 86 22 27406893
Wang, Shiwen 86 22 27406054
Wang, Shue 86 22 27405125
Wang, Xiaojing 86 22 27473115
Wang, Yanjun 86 22 23304452
tdxjwang@eyou.com
Wang, Yongli 86 22 27405754
srcict@tju.edu.cn
Wen, Jianping 86 22 27404070
Wu, Peng 86 22 27404061
Xia, Qing 86 22 27404447
Xu, Chunjian 86 22 27404495
cjxu@tju.edu.cn

Xu, Li 86 22 27890515
Xu, Shichang 86 22 27402168
xshchang@sina.com
Xu, Shimin 86 22 27890628
Xu, Songlin 86 22 27404701
slxu@tju.edu.cn
Yao, Fanglian 86 22 27491146
Zeng, Aiwu 86 22 27404705
awzeng@tju.edu.cn
Zhang, Baoquan 86 22 27405080
bqzhang@tju.edu.cn
Zhang, Dongming 86 22 27404701
nercdt@tju.edu.cn
Zhang, Jinli 86 22 27401476
xinxing@tju.edu.cn
Zhang, Lvhong 86 22 27401476
jiangjia@public.tpt.tj.cn
Zhang, Meijing 86 22 27400004
srcict@tju.edu.cn
Zhang, Pengfei 86 22 27407950
Zhang, Tianyong 86 22 27406610
tyzhang@tju.edu.cn
Zhang, Xiangwen 86 22 27402604
zliuying@public.tpt.tj.cn
Zhang, Yimin 86 22 27401944
zhangym001@263.net
Zhang, Ying 86 22 27403195
Zhao, Guangrong 86 22 27401149
grzhao@tju.edu.cn
Zheng, Jiaming 86 22 27473523
Zhou, Shaogang 86 22 37832522
Zhu, Hongji 86 22 27890645
zhuhongji163@163.com
Zhu, Huiming 86 22 27408779
hzhu@mail.tju.edu.cn
Zong, Runkuan 86 22 27404676

Accredited by: State Education Commission of China

Degrees granted 2001-2002:

B.S.: 376 M.S.: 82 Ph.D.: 35

Tsinghua University

Department of Chemical Engineering
Beijing, 100084, China

University phone .... (86) 10 62782035
Department phone .... (86) 10 62784523
Department FAX .... (86) 10 62770304

Professors

Cao, Zhuan 62784120
cza-dce@tsinghua.edu.cn
Chen, Bingzhen 62784129
dcecbz@tsinghua.edu.cn
Chen, Cuixian 62783747
chencx@tsinghua.edu.cn
Chen, Jian 62789195
cj-dce@tsinghua.edu.cn
Dai, Youyuan 62784153
daiyy@tsinghua.edu.cn
Ding, Fuxin 62785410
dingfx@tsinghua.edu.cn
Duan, Zhanting 62784175
duanzt@tsinghua.edu.cn
Fei, Weiyang 62784185
fwy-dce@tsinghua.edu.cn
Gao, Guanghua 62786448
gaogh@tsinghua.edu.cn
He, Xiaorong 62783913
hexr@tsinghua.edu.cn
Hu, Ping 62787617
hspinghu@tsinghua.edu.cn
Hu, Xianhua 62784263
hxh-dce@tsinghua.edu.cn
Jin, Yong 62784294
jiny@tsinghua.edu.cn
Li, Jiding 62785929
lijiding@tsinghua.edu.cn
Li, Yourun 62771039
liyr@tsinghua.edu.cn
Lin, Zhanglin
zhanglinlin@tsinghua.edu.cn
Liu, Dehua 62772825
dhliu@ tsinghua.edu.cn
Liu, Deshan 62784263
Liu, Zheng (Chair) 62788858
liuzheng@tsinghua.edu.cn
Luo, Guangsheng 62788545
gsluo@tsinghua.edu.cn
Shen, Jingzhu 62784479
shenjz@tsinghua.edu.cn
Wang, Dezheng 62794404
wangdezheng@tsinghua.edu.cn
Wang, Jiading 62785145
wjd-dce@tsinghua..edu.cn
Wang, Jinfu 62789233
wangjfu@tsinghua.edu.cn
Wang, Xiaogong (Associate Chair) 62787673
wxg-dce@tsinghua.edu.cn
Wang, Xiaolin 62773732
xl-wang@tsinghua.edu.cn
Wang, Yundong 62782748
wangyd@chemeng.tsinghua.edu.cn
Wang, Zhanwen 62786622
wangzw@tsinghua.edu.cn
Wei, Fei 62783626
wf-dce@tsinghua.edu.cn
Xie, Xuming 62786785
xxm-dce@tsinghua.edu.cn
Xing, Xinhui 62785514
xing@chemeng.tsinghua.edu.cn
Yang, Jichu 62786176
yjc-dce@tsinghua.edu.cn
Yu, Jian 62788201
yuj@tsinghua.edu.cn
Zhou, Rongqi 62784431
zhourq@tsinghua.edu.cn
Zhou, Xiao 62783483
zhoux@tsinghua.edu.cn
Zhu, Shenlin 62787911
zhusl@tsinghua.edu.cn

Associate Professors

Chen, Fuming 62785922
chenfm@tsinghua.edu.cn
Gao, Yanfang 62786448
yfgao@tsinghua.edu.cn
Guo, Baohua 62786645
bhguo@tsinghua.edu.cn
Guo, Qingfeng 62786303
zhp-dms@tsinghua.edu.cn
Guo, Zhigang 62785930
guozhig@tsinghua.edu.cn
Han, Minghan 62773374
hanmh@tsinghua.edu.cn

Hao, Jihua 62773082
haojh@tsinghua.edu.cn
Hu, Shanying 62771249
hxr-dce@tsinghua.edu.cn
Kan, Chengyou 62794191
kancy@tsinghua.edu.cn
Li, Qiang 62787870
liqiang@tsinghua.edu.cn
Lian, Yanqing 82865079
yqlian@tsinghua.edu.cn
Lin, Aiguang 62786090
linaig@tsinghua.edu.cn
Liu, Fuzhen 62786096
lfz-dce@tsinghua.edu.cn
Liu, Ruizhi 62784470
Luo, Guohua 62773733
luoguoh@tsinghua.edu.cn
Qin, Wei 62789197
qinw@chemeng.tsinghua.edu.cn
Shen, Jinyu 62787604
shenjy@tsinghua.edu.cn
Shi, Dianwen (Associate Chair) 62786483
Sun, Dengwen 62785105
Tang, Liming 62771539
tanglm@tsinghua.edu.cn
Tang, Zhigang 62773572
zhg-tang@tsinghua.edu.cn
Wang, Baoguo 62783869
bgwang@tsinghua.edu.cn
Wang, Tao 62789196
taowang@tsinghua.edu.cn
Wang, Tingjie 62789232
wangtj@tsinghua.edu.cn
Xiang, Lan 62329289
lxiang@chemeng.tsinghua.edu.cn
Yu, Lixin 62789238
yulixin@tsinghua.edu.cn
Yu, Yangxin 62773533
yuyx@tsinghua.edu.cn
Zhang, Liping 62754895
Zhao, Hong (Associate Chair) 62789236
zhh@chemeng.tsinghua.edu.cn

Department reports to: Zheng Liu, Chairman

Zhejiang University

Department of Chemical and Biochemical Engineering
38 Zheda Road, Hangzhou 310027
Zhejiang Province

University phone....................... 86-571-8517-2242
Department phone 86-571-8795-1237
Department FAX 86-571-8795-1015

Professors

Cen, Peilin (Director) 86-571-87951220
cenpl@che.zju.edu.cn
Chen, Dezhao 86-571-87951217
chendz@che.zju.edu.cn
Chen, Fengqiu 86-571-87951227
fqchen@mail.hz.zj.cn
Chen, Huanlin 86-571-87951982
chenhl@che.zju.edu.cn
Chen, Jizhong (Associate Chair) 86-571-87951833
jzchen@mail.hz.zj.cn
Chen, Xinzhi 86-571-87951232
chenxz@che.zju.edu.cn
Chen, Zhongxiu 86-571-87951232
chenzx@che.zju.edu.cn
Cheng, Yiyu 86-571-87951138
chengyy@che.zju.edu.cn
Feng, Lianfang 86-571-87951307
fenglf@che.zju.edu.cn
He, Chaohong 86-571-87951231
hech@che.zju.edu.cn
Huang, Zhiming 86-571-87951334
huangzm@che.zju.edu.cn
Jiang, Jialin 86-571-87951223
jiangjl@che.zju.che.cn
Li, Bo-Geng (Dean) 86-571-87951832
bgli@cmsce.zju.edu.cn
Li, Xi 86-571-87952210
lix@che.zju.edu.cn
Lin, Xinhua 86-571-87951216
linxh@che.zju.edu.cn
Lu, Junyi 86-571-87951223
lujy@che.zju.edu.cn
Lu, Xiuyang 86-571-87951224
luxy@che.zju.edu.cn
Mei, Lehe 86-571-87951982
meilh@che.zju.edu.cn
Pan, Qinmin 86-571-87951612
panqm@che.zju.edu.cn
Ren, Qilong 86-571-87951224
renql@che.zju.edu.cn
Rong, Shunxi 86-571-87951227
shxrong@mail.hz.zj.cn
Sun, Jianzhong 86-571-87951612
sunjz@che.zju.edu.cn
Sun, Junquan 86-571-87951833
sunjq@che.zju.edu.cn
Tong, Shuiguang (Director) 86-571-87951929
tongsg@che.zju.edu.cn
Wang, Kai 86-571-87951307
wangk@che.zju.edu.cn
Wang, Leqing 86-571-87951216
wanglq@che.zju.edu.cn
Wang, Xixuan 86-571-87951929
wangxx@che.zju.edu.cn
Wang, Zhangmao 86-571-87951833
wangzm@che.zju.edu.cn
Weng, Zhixue (Director) 86-571-87951334
wengzx@che.zju.edu.cn
Wu, Pingdong 86-571-87951225
wupd@che.zju.edu.cn
Wu, Rongren 86-571-87951223
wurr@che.zju.edu.cn
Xia, Liming 86-571-87951220
xialm@che.zju.edu.cn
Yang, Yongrong (Director) 86-571-87951227
yryang@mail.hz.zj.cn
Yang, Zheng 86-571-86624063
yangz@che.zju.edu.cn
Yao, Shanjing (Chair) 86-571-87951982
yaosj@che.zju.edu.cn
Zheng, Jingyang (Associate Chair) 86-571-87952110
zhenjy@che.zju.edu.cn

Associate Professors

Bao, Yongzhong 86-571-87951334
baoyz@che.zju.edu.cn
Cai, Hong 86-571-87951982
caih@che.zju.edu.cn
Cai, Jin 86-571-87951982
caij@che.zju.edu.cn

Cai, Zhenyun 86-571-87951232
caizy@che.zju.edu.cn
Cao, Kun 86-571-87951832
caok@che.zju.edu.cn
Chen, Yaqiu 86-571-87951138
chenyq@che.zju.edu.cn
Chen, Yingqi 86-571-87951226
chenyg@che.zju.edu.cn
Chen, Zhirong 86-571-87951620
chenzr@che.zju.edu.cn
Dai, Junyi 86-571-87951226
daijy@che.zju.edu.cn
Dou, Mei 86-571-85172244
Fang, Shijiang 86-571-87951334
fangsj@che.zju.edu.cn
Gu, Chaohua 86-571-87951216
guch@che.zju.edu.cn
Guan, Yixin 86-571-87951982
guanyx@che.zju.edu.cn
He, Shizheng 86-571-87951216
hesz@che.zju.edu.cn
Hong, Weirong 86-571-87951241
hongwr@che.zju.edu.cn
Hu, Wangming 86-571-87951232
huwm@che.zju.edu.cn
Jiang, Chunyue 86-571-87951612
jiangcy@che.zju.edu.cn
Jiang, Xin 86-571-87952828
Jiangx@mail.hz.zj.cn
Jin, Guanghai 86-571-87951502
jingh@che.zju.edu.cn
Kuang, Jiyong 86-571-87951641
Cekuangjy@zju.edu.cn
Li, Baofang 86-571-87951832
libf@che.zju.edu.cn
Li, Yong 86-571-87951232
liy@che.zju.edu.cn
Lin, Jiangping 86-571-87951220
Linjp@che.zju.edu.cn
Liu, Longxiao 86-571-87951334
liulx@che.zju.edu.cn
Liu, Ping 86-571-87952509
liuping@zju.edu.cn
Lu, Bo 86-571-87951610
lub@che.zju.edu.cn
Lu, Miaohua 86-571-87951236
lumh@che.zju.edu.cn
Luo, Yingwu 86-571-87951832
luoyw@che.zju.edu.cn
Ma, Hongzhao 86-571-87951233
mahz@che.zju.edu.cn
Meng, Qin 86-571-87951833
mengq@che.zju.edu.cn
Nan, Cuifei 86-571-87951231
nancf@che.zju.edu.cn
Ren, Xiaohong 86-571-87951227
renxh@che.zju.edu.cn
Shan, Guorong 86-571-87951334
shangr@che.zju.edu.cn
Shen, Qinyang 86-571-87951833
shenqy@che.zju.edu.cn
Shen, Zufeng 86-571-87951216
shenzf@che.zju.edu.cn
Shi, Zuyun 86-571-87951977
shizy@che.zju.edu.cn
Song, Xijin 86-571-87951224
songxj@che.zju.edu.cn
Sun, Guoyou 86-571-87952110
sungy@che.zju.edu.cn
Sun, Hong 86-571-87951237
Sunh@rsc.zju.edu.cn
Tan, Shanguang 86-571-87951216
tansg@che.zju.edu.cn
Tang, Jianhua 86-571-87951232
tangjh@che.zju.edu.cn
Wu, Jia 86-571-87951230
Wuj@che.zju.edu.cn
Wu, Sufang 86-571-87951833
wusf@che.zju.edu.cn
Wu, Xinyi 86-571-87951236
wuxy@che.zju.edu.cn
Wu, Yongjiang 86-571-87952509
yjwu@cps.zju.edu.cn
Xie, Rongjing 86-571-87951232
xierj@che.zju.edu.cn
Xu, Wensong 86-571-87951226
xuws@che.zju.edu.cn
Xu, Yuemin 86-571-87951929
xuym@che.zju.edu.cn
Xu, Zhinan 86-571-87951220
xuzn@mail.hz.zj.cn
Yang, Lirong 86-571-87951982
yanglr@che.zju.edu.cn
Yang, Nengwei 86-571-87951226
yangnw@che.zju.edu.cn
Yang, Yiwen 86-571-87951224
yangyw@che.zju.edu.cn
Yin, Hong 86-571-87951620
Yhsyinghong@263.net
Yu, Guozhen 86-571-87951872
yugz@che.zju.edu.cn
Zeng, Sheng 86-571-85982401
Shengzeng@hotmail.com
Zhan, Xiaoli 86-571-87951227
zhanxl@che.zju.edu.cn
Zhang, Weixin 86-571-87951232
zhangwx@che.zju.edu.cn
Zhang, Yunhua 86-571-87951217
zhangy@che.zju.edu.cn
Zhen, Chuanxiang 86-571-87951240
zhencq@che.zju.edu.cn
Zheng, Lianying 86-571-87951982
zhenly@che.zju.edu.cn
Zheng, Shuiying 86-571-87951216
zhensy@che.zju.edu.cn
Zhou, Zhijun 86-571-87952121
Zjz638@hzcnc.com
Zhu, Mingqiao 86-571-87952554

Accredited by: Zhejiang University

Graduate advisor Shanjing Yao

Undergraduate advisor: Jizhong CHEN

Student organization: ChE Students
Advisor: Qiang Fu

Placement service: Yihui Chen

Colombia

Universidad Industrial de Santander

Escuela de Ingenieria Quimica
P.O. Box 678
Bucaramanga

University phone .... (976) 456141
Department phone .... (976) 456141x434
Department FAX .... (976) 352554

Professors

Acevedo, Leonardo (Chair) .... x451 †
Alvarez, Mario x434
Barrera, Alvaro (Emeritus) x451
Centeno, Aristobulo x434
Correa, Rodrigo x434
del Ferrada, Pedro x434
Gonzalez, Cesar (Emeritus) x434
Guerra, Carlos F. (Emeritus) x246
Hernandez, Herzen x434
Kafarov, Vjacheslav x434
Lavente, Dionisio x434
Pulido, Jorge E. x434
Ramirez, Alvaro x451
Retamoso, Clemente (Emeritus) x434
Salazar, Ramiro x434
Sanchez, Luis D. x434
Santaelle, Jose F. x434

Associate Professors

Barajas, Crisostomo x434
Castillo, Edgar x434
Escalante, Humberto x434
del Giraldo, Sonia x434
Idarraga, Luis M. x434
Martinez, Ramiro x434
Valencia, Hugo A. x433

† Dial (976) 459647 and ask for extension.

Department reports to: Orlando Aguirre, Dean, Ph-Ch. Sci. Div.

Placement service: Clara Helena Gomez (x121)

Croatia

University of Zagreb

Faculty of Chemical Engineering and Technology
Marulicev trg 19
10 000 Zagreb

University phone 385 1 45 64 111
Department phone 385 1 45 97 281
Department FAX 385 1 45 97 260

Professors

Bozicevic, Juraj 45 97 131
jbozic@marie.fkit.hr
Budin, Rajka 45 97 138
rbudin@marie.fkit.hr
Cerjan-Stefanovic, Stefica 45 97 210
scerjan@pierre.fkit.hr
Duic, Ljerka 45 97 141
lduic@marie.fkit.hr
Glasnovic, Antun (Associate Dean) 45 97 221
aglasnov@pierre.fkit.hr
Gomzi, Zoran 45 97 105
zgomzi@marie.fkit.hr
Hraste, Marin 45 97 220
mhraste@pierre.fkit.hr
Janovic, Zvonimir 45 97 125
zjanov@marie.fkit.hr
Jelencic, Jasenka (Dean) 45 97 281
jjelen@marie.fkit.hr
Karminski Zamola, Grace 45 97 215
gzamola@pierre.fkit.hr
Kastelan-Macan, Marija 45 97 211
mkastela@jagor.srce.hr
Koprivanac, Natalija 45 97 124
nkopri@marie.fkit.hr
Kovacevic, Vera 48 46 378
vkovac@marie.fkit.hr
Kunst, Branko 45 97 232
kunstb@pierre.fkit.hr
Lopac, Vjera 45 97 106
vlopac@marie.fkit.hr
Matusinovic, Tomislav 45 97 218
tmatusin@pierre.fkit.hr
Metikos-Hukovic, Mirjana 45 97 140
mmetik@marie.fkit.hr
Mintas, Mladen 45 97 214
mmintas@pierre.fkit.hr
Rek, Vesna 48 28 476
vrek@marie.fkit.hr
Sindler, Marija 45 97 246
msindler@pierre.fkit.hr
Soljic, Zvonimir 45 97 204
zsoljic@pierre.fkit.hr
Vasic-Racki, ura 45 97 104
dvracki@marie.fkit.hr
Zrncevic, Stanka 45 97 102
szrnce@marie.fkit.hr

Associate Professors

Ivankovic, Hrvoje 45 97 228
hivan@pierre.fkit.hr
Ivankovic, Marica 45 97 230
mivan@pierre.fkit.hr
Sertic-Bionda, Katica 45 97 129
kserti@marie.fkit.hr
Sipos, Laszlo 45 97 290
laszlo.sipos@zg.tel.hr
Stupnisek-Lisac, Ema (Associate Dean) 45 97 281
elisac@marie.fkit.hr

Assistant Professors

Agic, Ante 48 33 850
aagic@marie.fkit.hr
Bajza, Zeljko 48 33 850
zeljko@zg.tel.hr
Briski, Felicita 45 97 269
fbriski@pierre.fkit.hr
Dananic, Vladimir 45 97 107
vdanan@marie.fkit.hr
Gusic, Ivica 45 97 266
igusic@pierre.fkit.hr
Hrnjak-Murgic, Zlata 45 97 122
zhrnjak@marie.fkit.hr
Kurajica, Stanislav 45 97 226
stankok@pierre.fkit.hr

Mance, Ana D. 45 97 243
admance@pierre.fkit.hr
Matijasevic, Ljubica 45 97 101
ljmatij@marie.fkit.hr
Mestrovic-Markovinovic, Antonija 45 97 113
amarko@marie.fkit.hr
Metes, Azra 45 97 122
ametes@marie.fkit.hr
Milardovic, Stjepan 45 97 286
smilard@pierre.fkit.hr
Papic, Sanja 45 97 122
spapic@marie.fkit.hr
Rogosic, Marko 45 97 299
mrogosic@pierre.fkit.hr
Tomasic, Vesna 45 97 103
vtomas@marie.fkit.hr
Volovsek, Vesna 45 97 135
volovsek@marie.fkit.hr

Czech Republic

Prague Institute of Chemical Technology

Department of Chemical Engineering
Technická 5
166 28 Praha 6 - Dejvice

University phone 420 2 311 7288
Department phone 420 2 2435 3165
Department FAX 420 2 311 7335

Professors

Marek, Milos 2435 3104
Milos.Marek@vscht.cz

Associate Professors

Eckert, Egon 2435 3105
Egon.Eckert@vscht.cz
Hasal, Pavel 2435 3167
Pavel.Hasal@vscht.cz
Linek, Vaclav 2435 3298
Vaclav.Linek@vscht.cz
Machon, Václav (Head) 2435 3165
Vaclav.Machon@vscht.cz
Nekovar, Prokop 2435 3236
Prokop.Nekovar@vscht.cz
Schreiber, Igor 2435 3167
Igor.Schreiber@vscht.cz
Snita, Dalimil 2435 3108
Dalimil.Snita@vscht.cz
Vaclavek, Vladimir 2435 3172
Vladimir.Vaclavek@vscht.cz

Assistant Professors

Basarova, Pavlina 2435 3235
Pavlina.Basarova@vscht.cz
Holecek, Oldrich 2435 3171
Oldrich.Holecek@vscht.cz
Hovorka, Frantisek 2435 3170
Frantisek.Hovorka@vscht.cz
Jahn, Robert 2435 3296
Robert.Jahn@vscht.cz
Jahoda, Milan 2435 3223
Milan.Jahoda@vscht.cz
Kosek, Juraj 2435 3296
Juraj.Kosek@vscht.cz
Moucha, Tomas 2435 3299
Tomas.Moucha@vscht.cz
Nevoral, Vladislav 2435 3293
Vladislav.Nevoral@vscht.cz
Novotna, Alexandra 2435 3172
Alexandra.Novotna@vscht.cz
Schreiberova, Lenka 2435 3171
Lenka.Schreiberova@vscht.cz
Sinevic, Vaclav 2435 3235
Vaclav.Sinevic@vscht.cz
Vanek, Tomas 2435 3106
Tomas.Vanek@vscht.cz

Accredited by: FEANI, Ministry of Education

Graduate advisor: Milos Marek

Undergraduate advisor: Egon Eckert

Student organization: SUS sus@vscht.cz
Advisor: Dalimil Snita

Department reports to: Faculty of Chemical Engineering

Placement service: Svatava.Poupetova@vscht.cz

University of Pardubice

Chemical Engineering Department
nam. Cs. Legii 565
532 10 Pardubice

Department phone 420 46-6037140
Department FAX 420 46-6037068

Professors

Lecjaks, Zdenek (Emeritus)
Machac, Ivan 420 466 037 131
Ivan.Machac@upce.cz
Mikulasek, Petr 420 466 037 130
Petr.Mikulasek@upce.cz

Associate Professors

Cakl, Jiri 420 466 037 128
Jiri.Cakl@upce.cz
Palaty, Zdenek (Head) 420 466 037 503
Zdenek.Palaty@upce.cz

Assistant Professors

Chocholous, Jaroslav 420 466 037 127
Jaroslav.Chocholous@upce.cz
Dolecek, Petr 420 466 037 129
Petr.Dolecek@upce.cz
Jirankova, Hana 420 466 037 134
Hana.Jirankova@upce.cz
Siska, Bedrich 420 466 037 126
Bedrich.Siska@upce.cz
Zakova, Alena 420 466 037 135
Alena.Zakova@upce.cz

Department reports to: Miroslav Ludwig, President of University

Finland

Abo Akademi University

Faculty of Chemical Engineering
Biskopsgatan 8
FIN-20500
Abo

University phone 358 2 21531
Department FAX 358 2 2154967

Professors

Back, Ralph-Johan
Blomquist, Bjorn
Eklund, Dan
Fagervik, Kaj (Dean)
Holmbom, Bjarne
Hupa, Mikko
Ivaska, Ari
Lewenstam, Andrzej
Lilius, Johan
Lonnberg, Bruno
Murzin, Dmitry
Ohman, Goran
Salmi, Tapio
Saxén, Henrik
Sere, Kaisa
Stenlund, Bengt
Toivonen, Hannu
Waller, Kurt
Westerholm, Jan
Westerlund, Tapio
Wilen, Carl-Eric

Department reports to: Kaj Fagervik

Helsinki University of Technology

Department of Chemical Technology
P.O.Box 6100
FIN-02015 HUT
Deliveries: Kemistintie 1 M

University phone 358 9 4511
Department phone 358 9 4511
Department FAX 358 9 462373

Professors

Aittamaa, Juhani 4512630
juhani.aittamaa@hut.fi
Hurme, Markku 4512 632
markku.hurme@hut.fi
Jamsa-Jounela, Sirkka-Liisa 4512 631
sirkka-liisa.jamsa-jounela@hut.fi
Jokela, Reija 4512 531
reija.jokela@hut.fi
Kontturi, Kyösti 4512 575
kyosti.kontturi@hut.fi
Koskinen, Ari 4512 526
ari.koskinen@hut.fi
Krause, Outi 4512 613
outi.krause@hut.fi
Kulmala, Sakari 4512 601
sakari.kulmala@hut.fi
Laakso, Simo 4512 550
simo.laakso@hut.fi
Leisola, Matti (Head) 4512 546
matti.leisola@hut.fi
Liukkonen, Simo 4512 571
simo.liukkonen@hut.fi
Niinistö, Lauri 4512 600
lauri.niinisto@hut.fi
Nordstrom, Katrina 4512 549
katrina.nordstrom@hut.fi
Seppala, Jukka 4512 614
jukka.seppala@hut.fi

Department reports to: Markku Hurme

Lappeenranta University of Technology

Department of Chemical Technology
P.O. Box 20
FIN-53851 Lappeenranta

University phone 358 5 62111
Department phone 358 5 6212101
Department FAX 358 5 6212199

Professors

Kallas, Juha 358 5 621 2132
Juha.Kallas@lut.fi
Lindstrom, Matti 358 5 621 2107
Manner, Hannu (Associate Chair) 358 5 621 2162
Hannu.Manner@lut.fi
Minkkinen, Pentti 358 5 621 2102
Pentti.Minkkinen@lut.fi
Nystrom, Lars 358 5 621 2125
Lars.Nystrom@lut.fi
Nystrom, Marianne 358 5 621 2160
Marianne.Nystrom@lut.fi
Paatero, Erkki 358 5 621 2163
Erkki Paatero@lut.fi
Turunen, Ilkka (Chair) 358 5 621 2165
Ilkka.Turunen@lut.fi

Assistant Professors

Kraslawski, Andrzej 358 5 621 2139
Kraslawski@ltkka.lut.fi
Mennola, Riitta 358 5 621 2110
Riitta.Mennola@lut.fi
Miettinen, Tapio 358 5 621 2109
Tapio.Miettinen@lut.fi
Niemi, Harri 358 5 621 2116
Harri.Niemi@lut.fi
Partanen, Jaakko 358 5 621 2108
Jaakko.Partanen@lut.fi
Ryösö, Kati 358 5 621 2175
Kati.Ryoso@lut.fi

Student organization: http://www.ltky.lut.fi/english.html

Department reports to: The Board of the University, and Markku Lukka, Rector

University of Oulu

Department of Process and Environmental Engineering
P.O. Box 4300
FIN-90014

University phone 358-8-5531011
Department phone 358-8-5532301
Department FAX 358-8-5532304

Professors

Harkki, Jouko
Keiski, Riitta
Kortela, Urpo
Lakso, Esko
Leiviska, Kauko (Chair)
Neubauer, Peter
Niinimaki, Jouko
Pohjola, Veikko
Sillanpää, Mika
Ylinen, Raimo

France

Ecole Nationale Superieure des Ingenieurs en Arts Chimiques et Technologiques (ENSIACET)

Institut National Polytechnique de Toulouse (INPT)
18 Chemin de la Loge
31078 Toulouse Cedex 4

University phone .... (33) 5 62 25 23 00
Department phone .... (33) 5 62 25 23 11
Department FAX .... (33) 5 62 25 23 18

Professors

Andrieu, E. 33 5 62 88 56 62
eandrieu@ensct.fr
Behra, Philippe 33 5 62 88 56 67
phbehra@ensct.fr
Borredon, E. 33 5 62 88 57 26
eborredonensct.fr
Casamatta, G. 33 5 62 25 23 02
Gilbert.Casamatta@ensiacet.fr
Condoret, J.S. 33 5 62 25 23 90
JeanStephane.Condoret@ensiacet.fr
Couderc, J. P. 33 5 62 25 23 06
JeanPierre.Couderc@ensiacet.fr
Delmas, H. 33 5 62 25 23 67
Henri.Delmas@ensiacet.fr
Delmas, M. 33 5 62 88 57 21
mdelmas@ensct.fr
Domenech, S. 33 5 62 25 23 64
Serge.Domenech@ensiacet.fr
Duverneuil, P. 33 5 62 25 23 11
Patrick.Duverneuil@ensiacet.fr
Floquet, Pascal 33 5 62 25 23 72
Pascal.Floquet@ensiacet.fr
Garnier, Patrick (Director) 33 5 62 25 23 01
Patrick.Garnier@ensiacet.fr
Gaset, A. 33 5 62 88 57 50
agaset@ensct.fr
Gleizes, A. 33 5 62 88 56 78
agleizes@ensct.fr
Gorrichon, J.P. 33 5 62 88 57 00
jpgorrichon@ensct.fr
Gourdenne, Albert 33 5 62 88 56 80
agourdenne@ensct.fr
Gourdon, C. 33 5 62 25 23 63
Christophe.Gourdon@ensiacet.fr
Grob, R. 33 5 62 88 56 56
rgrob@ensct.fr
Hemati, M. 33 5 62 25 23 41
Mehrdji.Hemati@ensiacet.fr
Joulia, X. 33 5 62 25 23 55
Xavier.Joulia@ensiacet.fr
Kalck, P. 33 5 62 88 56 90
pkalck@ensct.fr
Koehret, B. 33 5 62 25 23 51
Bernard.Koehret@ensiacet.fr
Lamure, A 33 5 62 88 57 07
alamure@ensct.fr
Le Lann, J.M. 33 5 62 25 23 57
JeanMarc.LeLann@ensiacet.fr
Mijoule, C. 33 5 62 88 57 61
cmijoule@ensct.fr
Molinier, J. 33 5 62 88 56 94
jmolinier@ensct.fr
Najim, K. 33 5 62 25 23 09
Kaddour.Najim@ensiacet.fr
Pibouleau, L. 33 5 62 25 23 66
Luc.Piblouleau@ensiacet.fr
Prevost, M. 33 5 62 25 23 22
Michel.Prevost@ensiacet.fr
Rey, C 33 5 62 88 56 74
crey@cict.fr
Riba, J. P. 33 5 62 25 23 49
JeanPierre.Riba@ensiacet.fr
Ricard, A. 33 5 62 25 23 24
Alain.Ricard@ensiacet.fr
Strehaiano, P. 33 5 62 25 23 47
Pierre.Strehaiano@ensiacet.fr
Terrier, Y.R. 33 5 62 88 56 84
yrterrier@ensct.fr
Wilhelm, A. M. 33 5 62 25 23 71
AnneMarie.Wilhelm@ensigct.fr

Associate Professors

Albet, Joel 33 5 62 88 56 95
jalbet@ensct.fr
Alric, I 33 5 62 88 57 31
ialric@ensct.fr
Barreteau, Denis
Denis.Barreteau@ensiacet.fr
Betbeder, M.C. 33 5 62 25 23 00
MarieClaude.Betbeder@ensiacet.fr
Blanc, C 33 5 62 88 57 08
cblanc@ensct.fr
Canselier, J. P. 33 5 62 25 23 31
JeanPaul.Canselier@ensiacet.fr
Caussat, B. 33 5 62 25 23 32
Brigitte.Caussat@ensiacet.fr
Cecutti, C 33 5 62 88 57 31
ccecutti@ensct.fr
Cognet, P. 33 5 62 25 23 29
Patrick.Cognet@ensiacet.fr
Combes, C 33 5 62 88 56 74
ccombes@cict.fr
Delia, M. L. 33 5 62 25 23 50
MarieLine.Delia@ensiacet.fr
Dufaure, C 33 5 62 88 57 05
cdufaure@ensct.fr
Duquenne, A. M. 33 5 62 25 23 43
AnneMarieDuquenne@ensiacet.fr
Duquenne, Philippe 33 5 62 25 23 60
Philippe.Duquenne@ensiacet.fr
Feurer, R 33 5 62 88 56 68
rfeurer@ensct.fr
Freche, M. 33 5 62 88 56 47
freche@cict.fr
Gabas, N. 33 5 62 25 23 96
Nadine.Gabas@ensiacet.fr
Gaubert, M.

Guiraud, R. 33 5 62 25 23 91
Roland.Guiraud@ensiacet.fr
Hait, Alain 33 5 62 25 24 11
Alain.Hait@ensiacet.fr
Huor, M.H. 33 5 62 25 23 20
MengHeng.Huor@ensiacet.fr
Hétreux, Gilles 33 5 62 25 23 95
Gilles.Hetreux@ensiacet.fr
Labroue, D 33 5 62 88 56 47
dlabroue@ensct.fr
Le Bolay, N. 33 5 62 25 23 44
Nadine.LeBolay@ensiacet.fr
Lebigot, Y 33 5 62 88 57 35
ylebigot@ensct.fr
Liddell, J. 33 5 62 25 23 78
Jean.Liddel@ensiacet.fr
Lucchese, Y. 33 5 62 88 56 44
ylucchese@nsct.fr
Mankowski, G. 33 5 62 88 56 60
gmankoski@ensct.fr
Meyer, M. 33 5 62 25 23 28
Michel.Meyer@ensiacet.fr
Meyer, X. 33 5 62 25 23 54
Xuan.Meyer@ensiacet.fr
Monfort, J. P. 33 5 62 25 23 76
JeanPierre.Monfort@nsiacet.fr
Pantel, C.
Pince, R 33 5 62 88 56 35
rpince@ensct.fr
Prat, Laurent 33 5 62 25 24 33
Laurent.Prat@ensiacet.fr
Quéau, R 33 5 62 88 56 04
rqueau@ensct.fr
Rascol, E. 33 5 62 25 24 03
Eva.Rascol@ensigct.fr
Ratsimba, B. 33 5 62 25 23 25
Berthe.Ratsimba@ensiacet.fr
Riba, M.L. 33 5 62 88 56 46
mriba@ensct.fr
Richard, M 33 5 62 88 56 35
mrichard@ensct.fr
Roques, Y. 33 5 62 56 88 49
yroques@ensct.fr
Serp, P. 33 5 62 88 56 93
pserp@ensct.fr
Silvestre, F. 33 5 62 88 57 44
fsilvestre@ensct.fr
Simon, V 33 5 62 88 57 18
vsimon@ensct.fr
Taillandier, P. 33 5 62 25 23 50
patricia.Taillandier@ensiacet.fr
Thiebaud, S 33 5 62 88 57 19
sthiebaud@ensct.fr
Trompette, P. 33 5 62 25 24 34
JeanLuc.Trompette@ensiacet.fr
Urrutigoity, M 33 5 62 88 56 98
murrutigoity@ensct.fr
Vaca-Garcia, Carlos 33 5 62 88 57 19
cvaca@ensct.fr
Vergnes, H. 33 5 62 25 23 21
Hugues.Vergnes@ensiacet.fr
Viguier, B 33 5 62 88 56 64
bviguier@ensiacet.fr
Zarate, Pascale 33 5 62 25 24 28
Pascale.Zarate@ensi.fr

Accredited by: IChemE(UK)

Undergraduate advisor: P.Duverneuil

Department reports to: M. Morancho, President INPT

Placement service: P.Duverneuil

Ecole Nationale Superieure des Industries Chimiques (ENSIC)

Institut National Polytechnique de Lorraine (INPL)
Chemical Engineering Department - ENSIC
1, rue Grandville - B.P. 451 -
54001 Nancy Cedex

Department phone .... (33) 383 17 50 00
Department FAX .... (33) 383 35 08 11

Professors

Bouchy, Michel .... (33) 383 17 51 37
Michel.Bouchy@ensic.inpl-nancy.fr
Choplin, Lionel (Associate Chair) .... (33) 383 17 50 10
Lionel.Choplin@ensic.inpl-nancy.fr
Corriou, Jean-Pierre .... (33) 383 17 52 13
Jean-Pierre.Corriou@ensic.inpl-nancy.fr
Dellacherie, Edith .... (33) 383 17 52 21
Edith.Dellacherie@ensic.inpl-nancy.fr
Dirand, Michel (Chair) .... (33) 383 17 50 13
Michel.Dirand@ensic.inpl-nancy.fr
Favre, Eric .... (33) 383 17 53 90
Eric.Favre@ensic.inpl-nancy.fr
Houzelot, Jean-Léon .... (33) 383 17 52 35
Jean-Leon.Houzelot@ensic.inpl-nancy.fr
Jamart, Brigitte .... (33) 383 17 52 79
Brigitte.Jamart@ensic.inpl-nancy.fr
Latifi, Abderrazak .... (33) 383 17 52 34
Abderrazak.Latifi@ensic.inpl-nancy.fr
Li, Huai-Zhi .... (33) 383 17 51 09
Huai-Zhi.Li@ensic.inpl-nancy.fr
Matlosz, Michael (Associate Chair) .... (33) 383 17 52 57
Michael.Matlosz@ensic.inpl-nancy.fr
Midoux, Noël .... (33) 383 17 52 50
Noel.Midoux@ensic.inpl-nancy.fr
Molleyre, François .... (33) 383 17 50 71
François.Molleyre@ensic.inpl-nancy.fr
Pla, Fernand .... (33) 383 17 50 49
Fernand.Pla@ensic.inpl-nancy.fr
Plasari, Edouard .... (33) 383 17 50 99
Edouard.Plasari@ensic.inpl-nancy.fr
Roizard, Christine (Associate Chair) .... (33) 383 17 53 06
Christine.Roizard@ensic.inpl-nancy.fr
Saatdjian, Esteban .... (33) 383 17 52 53
Esteban.Saatdjian@ensic.inpl-nancy.fr
Sardin, Michel .... (33) 383 17 52 54
Michel.Sardin@ensic.inpl-nancy.fr
Scacchi, Gérard .... (33) 383 17 50 83
Gerard.Scacchi@ensic.inpl-nancy.fr
Schuffenecker, Louis .... (33) 383 59 59 90
Louis.Schuffenecker@inpl-nancy.fr
Solimando, Roland .... (33) 383 17 53 06
Roland.Solimando@ensic.inpl-nancy.fr

Associate Professors

Arrault, Axelle .... (33) 383 17 52 14
Axelle.Arrault@ensic.inpl-nancy.fr
Bourdet, Jean-Bernard .... (33) 383 17 50 40
Jean-Bernard.Bourdet@ensic.inpl-nancy.fr
Bouroukba, Mohammed .... (33) 383 17 51 32
Mohammed.Bouroukba@ensic.inpl-nancy.fr
Castel, Christophe .... (33) 383 17 51 40
Christophe.Castel@ensic.inpl-nancy.fr

Clément, Robert .... (33) 383 17 50 29
Robert.Clement@ensic.inpl-nancy.fr
Commenge, Jean-Marc .... (33) 383 17 51 80
Jean-marc.Commenge@ensic.inpl-nancy.fr
Durand, Alain .... (33) 383 17 52 92
Alain.Durand@ensic.inpl-nancy.fr
Falk, Véronique .... (33) 383 17 52 75
Veronique.Falk@ensic.inpl-nancy.fr
Ferrer, Monique .... (33) 383 17 50 58
Monique.Ferrer@ensic.inpl-nancy.fr
Foucaut, Jean-françois .... (33) 383 17 50 60
Jean-Francois.Foucaut@ensic.inpl-nancy.fr
Fournet, René .... (33) 383 17 51 01
Rene.Fournet@ensic.inpl-nancy.fr
Gentric, Caroline .... (33) 383 17 53 38
Caroline.Gentric@ensic.inpl-nancy.fr
Gigante, Alexandra .... (33) 383 17 51 11
Alexandra.Gigante@ensic.inpl-nancy.fr
Gorner, Tatiana .... (33) 383 17 51 46
Tatiana.Gorner@ensic.inpl-nancy.fr
Hubert, Nathalie .... (33) 383 17 50 64
Nathalie.Hubert@ensic.inpl-nancy.fr
Jaubert, Jean-Noël .... (33) 383 17 50 81
Jean-Noel.Jaubert@ensic.inpl-nancy.fr
Jaubert, Lucie .... (33) 383 17 50 25
Lucie.Jaubert@ensic.inpl-nancy.fr
Lesage, François .... (33) 383 17 50 00
Francois.Lesage@ensic.inpl-nancy.fr
Muhr, Laurence .... (33) 383 17 53 11
Laurence.Muhr@ensic.inpl-nancy.fr
Mutelet, Fabrice .... (33) 383 17 51 31
Fabrice.Mutelet@ensic.inpl-nancy.fr
Nouvel, Cécile .... (33) 383 17 52 29
Cecile.Nouvel@ensic.inpl-nancy.fr
Perrin, Laurent .... (33) 383 17 50 96
Laurent.Perrin@ensic.inpl-nancy.fr
Petit, Alain .... (33) 383 17 50 95
Alain.Petit@ensic.inpl-nancy.fr
Petitjean, Dominique .... (33) 383 17 50 31
Dominique.Petitjean@ensic.inpl-nancy.fr
Poncin, Souhila .... (33) 383 17 52 23
Souhila.poncin@ensic.inpl-nancy.fr
Richomme, Katia .... (33) 383 17 50 00
Katia.Richomme@ensic.inpl-nancy.fr
Rouat-Rode, Sabine .... (33) 383 17 51 56
Sabine.Rode@ensic.inpl-nancy.fr
Sadtler, Véronique .... (33) 383 17 50 79
Veronique.Sadtler@ensic.inpl-nancy.fr
Schaer, Eric .... (33) 383 17 53 04
Eric.Schaer@ensic.inpl-nancy.fr
Schrauwen, Cornélius .... (33) 383 17 50 98
Cornelius.Schrauwen@ensic.inpl-nancy.fr
Simon, Yves .... (33) 383 17 51 22
Yves.simon@ensic.inpl-nancy.fr
Zahraa, Orfan .... (33) 383 17 51 18
Orfan.Zahraa@ensic.inpl-nancy.fr

Accredited by: IChemE (UK)

Graduate advisor: M. Matlosz

Placement service: G. Scacchi

Germany

Aachen Technical University

Dept. of Chemical Engineering
Turmstrasse 46
D-52056 Aachen

University phone 49-241-80-1
Department phone 49-241-8095470
Department FAX 49-241-8092252

Professors

Buechs, Joachim (Chair) 49-241-80-5546
buechs@bio-vt.rwth-aachen.de
Hartmann, Hugo (Emeritus) 49-241-80-95490
Marquardt, Wolfgang (Chair) 49-241-80-6712
marquardt@lpt.rwth-aachen.de
Melin, Thomas (Chair) 49-241-80-5470
melin@ivt.rwth-aachen.de
Modigell, Michael (Associate Chair) 49-241-80-5159
modigell@ivt.rwth-aachen.de
Pfennig, Andreas (Chair) 49-241-80-95490
andreas.pfennig@tvt.rwth-aachen.de
Schummer, Paul (Emeritus) 49-241-80-5476

Accredited by:

Degrees granted 2003-2004:

M.S.: 25 Ph.D.: 7

Graduate advisor: betreuung-vt@lpt.rwth-aachen.de

Technical University of Berlin

Department of Chemical Engineering
Strasse des 17. Juni 135, TK7
D-10623 Berlin

University phone 0049 30 314 0
Department phone 0049 30 314 22755
Department FAX 0049 30 314 22406

Professors

Arlt, Wolfgang (Head) 314 22755
Knapp, Helmut (Emeritus) 314 25063

Technische Universitaet Braunschweig

Department of Mechanical Engineering (Maschinenbau)
Division of Chemical Engineering (Energie und Verfahrenstechnik)
Box 3329
D-38023 Braunschweig
Deliveries: Schleinitzstrasse 20, D-38106 Braunschweig

University phone 49 531-391-0
Department phone 49 531-391-7683
Department FAX 49 531-391-5947

Professors

Bohnet, Matthias (Emeritus) 49 531-391-2790
m.bohnet@tu-bs.de
Hempel, Dietmar-Christian (Dean) 49 531-391-7650
d.hempel@tu-bs.de
Koehler, Juergen (Head) 49 531-391-2627
juergen.koehler@tu-bs.de

Kosyna, Gunter (Head) 49 531-391-2918
g.kosyna@tu-bs.de
Leithner, Reinhard (Head) 49 531-391-3030
r.leithner@tu-bs.de
Scholl, Stephan (Head) 49 531-391-2780
s.scholl@tu-bs.de
Schwedes, Joerg (Head) 49 531-391-9610
j.schwedes@tu-bs.de

Technical University of Clausthal

Institut fuer Chemische Verfahrenstechnik
Leibnizstrasse 17
D-38678 Clausthal-Zellerfeld

University phone 49 5323 72-0
Department phone 49 5323 72-2187
Department FAX 49 5323 72-2182

Professors

Hoffmann, Ulrich (Chair) .... 72-2184 †
hoffmann@icvt.tu-clausthal.de

Associate Professors

Kunz, Ulrich 72-2181
kunz@icvt.tu-clausthal.de

Assistant Professors

Horst, Christian 72-2561
horst@icvt.tu-clausthal.de

† Faculty numbers have prefix 49 5323.

Graduate advisor: U. Hoffmann

Undergraduate advisor: Zentrale Studienberatung, Studienberatung@tu-clausthal.de

Student organization: Fachschaft MVC

Darmstadt University of Technology

Fachbereich Maschinenbau
Fachgebiet Thermische Verfahrenstechnik
Petersenstr. 30
64287 Darmstadt

University phone 49 (6151) 16-0
Department phone 49 (6151) 162164
Department FAX 49 (6151) 164516

Professors

Hampe, Manfred J. (Chair) 49 6151 162164
hampe@tvt.tu-darmstadt.de
Kast, W. (Emeritus) 49 6151 163364

Associate Professors

Jaeschke, Lothar (Emeritus)
Korkhaus, Juergen (Visiting)
Schadler, Norbert (Visiting)
schadler@axiva.com
Schneider, Klaus (Emeritus)

Accredited by: Deutscher Akkreditierungsrat

Graduate advisor: Barbara Seifert

University of Dortmund

Fachbereich Chemietechnik
Universität Dortmund
D-44221 Dortmund
Deliveries: Emil-Figge-Str. 70, D-44227 Dortmund

University phone 231-755-1
Department phone 231-755-2362
Department FAX 231-755-2251

Professors

Agar, D. (Associate Dean) 231-755-2697
agar@bci.uni-dortmund.de
Behr, A. 231-755-2310
behr@bci.uni-dortmund.de
Engell, S. 231-755-5126
s.engell@bci.uni-dortmund.de
Fahlenkamp, H. 231-755-2322
h.fahlenkamp@bci.uni-dortmund.de
Friedrich, C. 231-755-5115
c.friedrich@bci.uni-dortmund.de
Giesekus, H. (Emeritus) 231-755-2301
Górak, A. (Associate Dean) 231-755-2323
a.gorak@bci.uni-dortmund.de
Köster, U. 231-755-2678
koester@bci.uni-dortmund.de
Onken, U. (Emeritus) 231-755-2696
Sadowski, G. 231-755-2635
g.sadowski@bci.uni-dortmund.de
Schecker, H.-G. (Emeritus) 231-755-2014
schecker@bci.uni-dortmund.de
Schmidt-Traub, H. 231-755-2338
schmtr@bci.uni-dortmund.de
Schulz, S. (Emeritus) 231-755-2635
Schwind, H. (Emeritus) 231-755-2577
Simmrock, K.H. (Emeritus) 231-755-2310
Strauss, K. 231-755-3214
strauss@bci.uni-dortmund.de
Walzel, P. (Dean) 231-755-6088
p.walzel@bci.uni-dortmund.de
Weinspach, P.-M. (Emeritus) 231-755-2323
Weiss, E. 231-755-2523
weiss@bci.uni-dortmund.de
Werner, U. (Emeritus) 231-755-2326
Wichmann, R. 231-755-3205
wichmann@bci.uni-dortmund.de

Associate Professors

Steiff, A. 231-755-2356

Assistant Professors

Joerissen, J. 231-755-2317
joerissen@bci.uni-dortmund.de
Kenig, E. 231-755-2357
kenig@bci.uni-dortmund.de

Universitat Erlangen-Nurnberg

Department of Chemical and Bioengineering
University of Erlangen-Nuremberg
Cauerstrasse 4
D-91058 Erlangen
Germany

University phone 49 9131 85-0
Department phone 49 9131 85-29597
Department FAX 49 9131 85-29503

Professors

Arlt, Wolfgang .... 85-27440 †
wolfgang.arlt@cbi.uni-erlangen.de
Buchholz, Rainer 85-23000
rainer.buchholz@bvt.cbi.uni-erlangen.de
Durst, Franz 85-29500
durst@lstm.uni-erlangen.de
Emig, Gerhard (Emeritus) 85-27424
emig@tc.uni-erlangen.de
Hofmann, Hanns (Emeritus) 85-27424
hanns.hofmann@tc.uni-erlangen.de
Leipertz, Alfred 85-29900
sek@ltt.uni-erlangen.de
Molerus, Otto (Emeritus) 85-29401
o.molerus@lfg.uni-erlangen.de
Neesse, Thomas (Director) 85-23200
thomas.neesse@uvt.cbi.uni-erlangen.de
Peter, Siegfried (Emeritus) 85-27443
siegfried.peter@rzmail.uni-erlangen.de
Peukert, Wolfgang 85-29400
w.peukert@lfg.uni-erlangen.de
Schlücker, Eberhard 85-29450
sl@ipat.uni-erlangen.de
Steiner, Rudolf (Emeritus) 85-27957
rudolf.steiner@rzmail.uni-erlangen.de
Vetter, Gerhard (Emeritus) 85-29457
vetter@ipat.uni-erlangen.de
Wasserscheid, Peter 85-27420
wasserscheid@crt.cbi.uni-erlangen.de

Associate Professors

Brunn, Peter O. 85-29566
pbrunn@lstm.uni-erlangen.de
Dörnenburg, Heike 85-23005
heike.doernenburg@bvt.cbi.uni-erlangen.de
König, Axel 85-27446
axel.koenig@rzmail.uni-erlangen.de
Schwieger, Wilhelm 85-28910
schwieger@rzmail.uni-erlangen.de
Wirth, Karl E. 85-29403
k.e.wirth@lfg.uni-erlangen.de

Assistant Professors

Raszillier, Hans 85-29505
raszillier@lstm.uni-erlangen.de

† Numbers above have prefix 49 9131.

Accredited by:

Degrees granted 2003-2004:

B.S.: 6 M.S.: 68 Ph.D.: 16

Graduate advisor: Lüder Depmeier, Christian Schneider

Undergraduate advisor: Lüder Depmeier, Christian Schneider

Department reports to: Technische Fakultaet, Erwin-Rommel-Str. 60, D-91058 Erlangen

Technical University Hamburg-Harburg

Studiendekanat Verfahrenstechnik
Kasernenstr. 12
21073 Hamburg

University phone .... (40) 42878-0
Department phone .... (40) 42878 3117
Department FAX .... (40) 42878 2582

Professors

Brunner, G. .... (40) 42878-3040
Forstner, U. .... (40) 42878-3008
Friedel, L. .... (40) 42878-3052
Hapke, J. .... (40) 42878-3048
Kasche, V. .... (40) 42878-3018
kasche@tu-harburg.de
Keil, F. .... (40) 42878-3042
Maerkl, H. .... (40) 42878-3017
maerkl@tu-harburg.de
Werther, J. .... (40) 42878-3239

Associate Professors

Antranikian, G. (Dean) .... (40) 42878-3117
Calmano, W. .... (40) 42878-3108
Eggers, R. .... (40) 42878-3191
Gruhn, G. .... (40) 42878-3041
Muller, R. .... (40) 42878-3118

Department reports to: The President

University of Hannover

Institut fuer Verfahrenstechnik
Chemical and Process Engineering
Callinstr. 36
30167 Hannover
Germany

University phone 49 511 762-0
Department phone 49 511 762-3828
Department FAX 49 511 762-3031

Professors

Hallensleben, M.L (Emeritus) 49 761 430452
Luke, Andrea 49 511 762 2877
luke@ift.uni-hannover.de
Merker, Guenter F. 49 511 762 2418
merker@itv.uni-hannover.der
Mewes, Dieter 49 511 762 3638
dms@ifv.uni-hannover.de
Seume, Joerg 49 511 762 2733
seume@ifs.uni-hannover.de

Associate Professors

Alshuth, T. 49 511 842010
Bederna, Ch. 49 511 97645993
christoph.bederna@contitech.de
Bode, H. 49 6196 750018
Freund, B. 49 2233 964890
Giese, U. 49 511 8420143
Hempel, J. 49 6201 804880
Herrmann, W. 49 511 93859501
wolfram.herrmann@contitech.de
Härtel, V. 49 511 9763346
volker.haertel@conti.de

Kammann, A. 49 6201 806562
andreas.kammann@freudenberg.de
Lechtenböhmer, A. 352 81993626
annette.lechtenboehmer@goodyear.com
Obrecht, W. 49 214 3072496
w.obrecht.wo@bayer-ag.de
Peinemann, K.-V. 49 4152 872420
klaus-viktor.peinemann@gkss.de
Schuster, R.H. 49 511 842010
Stark, R. 49 5131 51130
starkgarbsen@aol.com
Treffenfeldt, W. 49 6196 566242
wtreffenfeldt@dow.com
von Brook, U. 49 40 7667 1416
ulrich.vonbroock@vibracoustic.de
Wahl, G. 49 511 9763046
guenter.wahl@conti.de
Weiss, R. 49 6201 8046331
Wrana, C. 49 214 3081048
claus.wrana.cw@bayer-ag.de
Zetzmann, K. 49 2208 6999

Accredited by:

Degrees granted 2003-2004:

B.S.: 220 M.S.: 120 Ph.D.: 60

Graduate advisor: Professor L. Overmeyer

Undergraduate advisor: Professor L. Overmeyer

Student organization: International Office

Placement service: internationaloffice@uni-hannover.de

University of Hannover, ITC

Institut fur Technische Chemie
Callinstr. 3
30167 Hannover

University phone.......................... 49 511 462 0
Department phone 49 511 762 2269
Department FAX 49 511 762 3004

Professors

Bellgardt, Karl-Heinz 49 511 762 3167
Bellgardt@IFTC.Uni-Hannover.de
Hesse, Diethard 49 511 762 2354
Hesse@IFTC.Uni-Hannover.de
Scheper, Thomas (Head) 49 511 762 2509
Scheper@IFTC.Uni-Hannover.de
Schügerl, Karl (Emeritus) 49 511 762 2253
Schuegerl@IFTC.Uni-Hannover.de

Assistant Professors

Hitzmann, Bernd 49 511 762 2963
Hitzmann@IFTC.Uni-Hannover.de
Kretzmer, Gerlinde 49 511 762 5560
Kretzmer@IFTC.Uni-Hannover.de

Department reports to: Fachbereich Chemie

Universitat Kaiserslautern

Fachbereich Maschinenbau und Verfahrenstechnik
Studienrichtung Vefahrenstechnik
Gottlieb Daimler Strasse
D-67663 Kaiserslautern

University phone.......................... 0049 631 205 0
Department phone 0049 631 205 2417
Department FAX 0049 631 205 3600

Professors

Bart, Hans-Jorg .... 2414 †
bart@mv.uni-kl.de
Garcia, Luis L. 3775
Maurer, Gerd 2410
gmaurer@rhrk.uni-kl.de
Ripperger, Siegfried 2121
ripperger@mv.uni-kl.de
Wuestenberg, Dieter 3031
wuestenberg@mv.uni-kl.de

Associate Professors

Buettner, Helmut 2412
buettner@mv.uni-kl.de

Assistant Professors

Dau, Guenter 2560
dau@mv.uni-kl.de
Kraetz, Lorenz 2765
kraetz@mv.uni-kl.de
Perez-Salado, Alvaro 2761
salado@rhrk.uni-kl.de

† Faculty numbers are extensions of Dept. phone.

Accredited by:

Degrees granted 2003-2004:

M.S.: 14 Ph.D.: 5

Department reports to: Helmut J. Schmidt, President

Placement service: Harald Kuehn, x2215

Universitat Karlsruhe

Fakultat fuer Chemieingenieurwesen und Verfahrenstechnik
Box 6980
D76128 Karlsruhe

Department phone .... (721) 608-3786
Department FAX .... (721) 608-7531

Professors

Bockhorn, Henning .... (721) 608-2570
Braun, Andre M. .... (721) 608-2557
Buggisch, Hans .... (721) 608-2661
Frimmel, Fritz H. .... (721) 608-2580
Kasper, Gerhard .... (721) 608-6561
Kind, Matthias .... (721) 608-2390
Kraushaar-Czarnetzki, B. .... (721) 608-4133
Leuckel, W. (Emeritus) .... (721) 608-2570
Nirschl, Hermann .... (721) 608-2404
Reimert, Rainer .... (721) 608-2560
Schaber, Karlheinz .... (721) 608-2321
Schluender, E. (Emeritus) .... (721) 608-2390
Schubert, Helmar .... (721) 608-2497
Stahl, Werner .... (721) 608-2401
Syldatk, Christoph .... (721) 608-2124

Associate Professors

Lintz, Hans G. (Emeritus) .... (721) 608-3939
Martin, Holger .... (721) 608-2386
Oellrich, Lothar .... (721) 608-2332
Posten, Clemens .... (721) 608-2410
Schaub, Georg .... (721) 608-2572
Zarzalis, Nikolaos .... (721) 608-4231

Accredited by:

Georg-Simon-Ohm Fachhochschule Nuernberg

Fachbereich Angewandte Chemie
Kesslerplatz 12
Postfach
D-90121 Nuernberg

University phone .... (0911) 5880-0
Department phone .... (0911) 5880-1260
Department FAX .... (0911) 5880-5139

Professors

Aust, Eberhard
Bartsch, Stephan
Bauer, Hermann
Dorn, Alfred
Herold, Thomas
Jacob, Karl-Heinz (Head)
Kinkel, Hans-Joachim
Stark, Walter
Stephan, Rainer
Volgnandt, Peter
Wehnert, Gerd

Graduate advisor: Prof. Dr. Hermann Bauer

Department reports to: Prof. Dr. Karl-Heinz Jacob, Dekan

Fachhochschule Mannheim, University of Applied Sciences

Fachbereich Verfahrens- und Chemietechnik
Windeckstr 110
68163 Mannheim

Department phone 49 621 292-6424
Department FAX 49 621 292-6555

Professors

Adrian, Till .... 6305 †
t.adrian@fh-mannheim.de
Diewald, Werner 6489
w.diewald@fh-mannheim.de
Fosshag, Erich 6585
e.fosshag@fh-mannheim.de
Fritz, Wolfgang 6306
w.fritz@fh-mannheim.de
Hagen, Jens 6382
j.hagen@fh-mannheim.de
Hassenpflug, Hans-Uwe 6484
hassenpflug@fh-mannheim.de
Hoyningen-Huene v., Dietmar 6400
hoy@zv.fh-mannheim.de
Kern, Heinz 6305
Kunz, Peter 6304
kunz.p.m.prof@t-online.de
Landwehr, Brigitta 6486
b.landwehr@fh-mannheim.de
Michel, Hartmut 6306
dr.h.michel@t-online.de
Mueller, Herbert 6305
Peschges, Klaus-Juergen 6498
k.peschges@fh-mannheim.de
Raedle, Matthias 6330
m.raedle@fh-mannheim.de
Schinke, Bernd (Head) 6387
b.schinke@fh-mannheim.de
Schmidt, Volkmar 6307
v.m.schmidt@t-online.de
Schmitt, Wolfgang 6489
w.ph.schmitt@fh-mannheim.de
Steinert, Heike 6306
h.steinert@fh-mannheim.de
Traegner, Ulrich 6296
u.traegner@fh-mannheim.de

† Numbers are extensions of Dept. phone.

Accredited by: IChemE(UK) in progress

Department reports to: Hoyningen-Huene v., Dietmar; Rector; ext. 6400/6401

Ghana

Kwame Nkrumah University of Science and Technology

Department of Chemical Engineering
KNUST
Kumasi-Ghana

University phone .... (233) 51-60137
Department phone .... (233) 51-60234
Department FAX .... (233) 51-60234

Professors

Afrane, G.
Ankudey, E. G.
Baah-Ennumh, E. K.
Bediako, P. C. K.
Momade, Zs.
Nyadziehe, T. K.
Twimasi, K. A.
Woode, M. Y. (Head)

Accredited by: National Accreditation Board (NAB)

Graduate advisor: Head of Department

Student organization: CHEESA

Greece

Aristotle University

Chemical Engineering Department
54124 Thessaloniki

Department phone 30310 99-6267
Department FAX 30310 99-6168

Professors

Assael, M. J. .... 6163 †
assael@auth.gr

Karabelas, A. J. 6201
karabaj@cperi.certh.gr

Kiparissides, C. 6211
cypress@alexandros.cperi.forth.gr

Liakopoloulou-Kyriakidou, M. (Head) 6193
markyr@vergina.eng.auth.gr

Nychas, S. G. 6231
nychas@vergina.eng.auth.gr

Panayiotou, C. 6223
cpanayio@mailhost.ccf.auth.gr

Papageorgiou, V. P. 6241
vaspap@vergina.eng.auth.gr

Platakis, N. 6261
platakis@eng.auth.gr

Sakelaropoulos, G. 6271
sakel@vergina.eng.auth.gr

Stoukides, M. 6165

Vasalos, I. A. 6191
vasalos@filippos.techpath.gr

Associate Professors

Kastrinakis, E. G. 6183
kastr@vergina.eng.auth.gr

Stamatoudis, M. 6233
stamatou@vergina.eng.auth.gr

Assistant Professors

Bacola-Christianopoulou, M. N. 6213
bakola@vergina.eng.auth.gr

Kotali, A. 6253
kotali@vergina.eng.auth.gr

Lemonidou, A. 6273
lemonido@vergina.eng.auth.gr

Markopoulos, J. 6203
jonimark@vergina.eng.auth.gr

Mellidis, A. S. 6164
melidis@vergina.eng.auth.gr

Paras, S. P. 6174
paras@vergina.eng.auth.gr

Sikalides, K. 6185
sikalidi@vergina.eng.auth.gr

Stoforos, N. 6450
Stoforos@cheng.auth.gr

Tsitouridou, R. 6194
roxani@vergina.eng.auth.gr

Zlatanos, S. N. 6173
szlatano@vergina.eng.auth.gr

† Direct dial with prefix 3031 99.

University of Patras

University Campus
GR-26504 Patras
Greece

University phone 302610-991822
Department phone 302610-997580
Department FAX 302610-997849

Professors

Dassios, George 30610997373
dassios@iceht.forth.gr

Dondos, Anastasios (Emeritus) 30610997652
dondos@iceht.forth.gr

Koutsoukos, Petros G. 30610997265
pgk@iceht.forth.gr

Kravaris, Costas 30610996339
kravaris@chemeng.upatras.gr

Ladas, Spyridon 30610997631
ladas@iceht.forth.gr

Lyberatos, Gerasimos 30610997573
lyberatos@chemeng.upatras.gr

Nikolopoulos, Panajotis S. (Associate Chair) . 30610997563
nikolop@chemeng.upatras.gr

Pandis, Spyros
pandis@chemeng.upatras.gr

Papatheodorou, George N. 30610997570
gpap@iceht.forth.gr

Pavlou, Stavros 30610997640
sp@chemeng.upatras.gr

Payatakes, Alkiviades C. 30610997574
acp@iceht.forth.gr

Rapakoulias, Dimitrios E. (Chair) 30610993361
rap@chemeng.upatras.gr

Tsahalis, Demosthenes T. 30610997577
tsahalis@lfme.chemeng.upatras.gr

Tsamopoulos, John 30610997203
tsamo@chemeng.upatras.gr

Vayenas, Constantinos G. 30610997576
cat@chemeng.upatras.gr

Verykios, Xenophon E. 30610991527
verykios@iceht.forth.gr

Associate Professors

Kennou, Styliani 30610993255
kennou@iceht.forth.gr

Mavrantzas, Vlasis
vlasis@chemeng.upatras.gr

Staikos, George 30610997501
staikos@iceht.forth.gr

Tsitsilianis, Constantinos 30610997500
tsic@iceht.forth.gr

Assistant Professors

Angelopoulos, George N. 30610997509
angel@chemeng.upatras.gr

Bebelis, Symeon I. 30610997756
simeon@iceht.forth.gr

Boghosian, Soghomon B. 30610997854
bogosian@iceht.forth.gr

Mataras, Dimitrios 30610996340
dim@chemeng.upatras.gr

Accredited by:

Degrees granted 2003-2004:

M.S.: 70 Ph.D.: 12

Graduate advisor: Gerasimos Lyberatos

Undergraduate advisor: Gerasimos Lyberatos

Hong Kong, SAR of China

Hong Kong University of Science and Technology

Department of Chemical Engineering
Clear Water Bay
Kowloon, Hong Kong

Department phone 852 2358-7130
Department FAX 852 2358-0054

Professors

Chan, Chi-Ming .... 2358-7125 †
kecmchan@ust.hk

Ng, Ka M. (Head) 2358-7238
kekmng@ust.hk

Yue, Po-Lock 2358-8370
keplyue@ust.hk

Associate Professors

Barford, John 2358-7237
barford@ust.hk

Chan, Chak K. 2358-7124
keckchan@ust.hk

Chen, Guohua 2358-7138
kechengh@ust.hk

Gao, Furong 2358-7139
kefgao@ust.hk

Gao, Ping 2358-7126
kepgao@ust.hk

Hsing, I-Ming 2358-7131
kehsing@ust.hk

Hu, Xijun 2358-7134
kexhu@ust.hk

Hui, David 2358-7137
kehui@ust.hk

Mckay, Gordon 2358-8412
kemckayg@ust.hk

Mi, Yongli 2358-7127
keymix@ust.hk

Yeung, King L. 2358-7123
kekyeung@ust.hk

Assistant Professors

Porter, John F. 2358-7132
kejep@ust.hk

† Dial counrty code 852 for faculty numbers

Accredited by: Institution of Chemical Engineers, Hong Kong Institution of Chemical Engineers

Graduate advisor: Dr. Xijun Hu

Undergraduate advisor: Dr. C.K. Chan

Department reports to: Prof. Philip Chan, Acting Dean of Engineering (852 2358-6952)

Hungary

Budapest University of Technology and Economics (Budapesti Muszaki es Gazdasagtudomanyi Egyetem)

Faculty of Chemical Engineering
H-1521, Budapest, Muegyetem
rakpart 3.

Department phone .... (36-1) 463-3571
Department FAX .... (36-1) 463-3570

Professors

Bitter, Istvan .... 463-1379 †
ibitter@mail.bme.hu

Borsa, Judit (Associate Dean) 463-1376
jborsa@mail.bme.hu

Faigl, Ferenc (Associate Dean) 463-5889
ffaigl@mail.bme.hu

Fekete, Jeno 463-1596
fekete@mail.bme.hu

Fogassy, Elemer 463-1883
efogassy@mail.bme.hu

Fonyo, Zsolt (Chair) 463-3196
fonyo@mail.bme.hu

Gal, Sandor 463-1216
gal@mail.bme.hu

Grofcsik, Andras 463-1484
agrofcsik@mail.bme.hu

Hargittai, Istvan 463-1286
hargittai@mail.bme.hu

Hencsei, Pal 463-2294
hencsei@mail.bme.hu

Horvai, Gyorgy (Chair) 463-1480
ghorvai@mail.bme.hu

Huszthy, Peter (Chair) 463-1071
huszthy@mail.bme.hu

Kalaus, Gyorgy 463-1285
kalaus@mail.bme.hu

Keglevich, Gyorgy (Chair) 463-5883
gkeglevich@mail.bme.hu

Kemeny, Sandor 463-2209
kemeny@mail.bme.hu

Kubinyi, Miklos 463-2137
kubinyi@mail.bme.hu

Lasztity, Radomir (Emeritus) 463-1627
lasztity@mail.bme.hu

Mihaltz, Pal 463-2574
mihaltz@mail.bme.hu

Novak, Bela 463-1364
bnovak@mail.bme.hu

Novak, Lajos 463-2207
l-novak@mail.bme.hu

Nyitrai, Jozsef 463-2205
nyitrai@mail.bme.hu

Nyulaszi, Laszlo (Chair) 463-1281
nyulaszi@mail.bme.hu

Orsi, Ferenc 463-2283
orsi@mail.bme.hu

Pokol, Gyorgy (Dean) 463-1593
pokol@mail.bme.hu

Pukanszky, Bela (Chair) 463-2015
bpukanszky@mail.bme.hu

Pungor, Erno (Emeritus) 463-4054
pungor@mail.bme.hu

Reffy, Jozsef 463-3281
jreffy@mail.bme.hu

Salgo, Andras (Chair) 463-3854
salgo@mail.bme.hu

Sevella, Bela (Chair) 463-2595
bsevella@mail.bme.hu

Szantay, Csaba (Emeritus) 463-1195
szantay@mail.bme.hu

Toke, Laszlo 463-3653
ltoke@mail.bme.hu

Toth, Klara 463-2273
ktoth@mail.bme.hu

Tungler, Antal (Associate Dean) 463-1203
atungler@mail.bme.hu

Vesztpremi, Tamas 463-1793
tveszpremi@mail.bme.hu

Zrinyi, Miklos (Chair) 463-3229
zrinyi@mail.bme.hu

† Faculty numbers have prefix 36-1.

Graduate advisor: Dr. Mariann Lengyelne

Undergraduate advisor: Marta Takacs

Department reports to: Gyorgy Pokol, Dean of Faculty

Placement service: Dr. Mariann Lengyelne

University of Veszprem

Institute of Chemical Engineering
Veszprem, Egyetem u. 10 H-8201

University phone .... (36) 88-422022
Department phone .... (36) 88-421905
Department FAX .... (36) 88-421905

Professors

Arva, Peter .... x4510 †
arva@fmt.vein.hu

Bartha, Laszlo (Head) x4217
bartha@almos.vein.hu

Deak, Gyula x4411
deak@almos.vein.hu

Hancsok, Jeno x4413
hancsokj@almos.vein.hu

Horvath, Geza (Dean) .... (36) 88-421905
horvathg@almos.vein.hu

Kotsis, Levente x4301
kotsis@almos.vein.hu

Lakatos, Bela x4210
lakatos@fmt.vein.hu

Marton, Gyula x4513
martongy@almos.vein.hu

Szanya, Tibor x4268
szanyat@almos.vein.hu

Szeifert, Ferenc (Head) x4290
szeifert@fmt.vein.hu

Associate Professors

Abonyi, Janos x4209
abonyj@fmt.vein.hu

Argyelan, Janos x4154
gyela@almos.vein.hu

Chovan, Tibor x4209
chovan@fmt.vein.hu

Dencs, Belane x4177
jutka@almos.vein.hu

Marton, Zsuzsa x4187
martonzs@almos.vein.hu

Moser, Karoly x4105
moser@fmt.vein.hu

Nagy, Lajos x4276
nagyl@fmt.vein.hu

Nemeth, Sandor x4276
nemeth@fmt.vein.hu

Proder, Jozsef x4381
proder@almos.vein.hu

Szokonya, Laszlo x4270
szokonya@almos.vein.hu

Assistant Professors

Varga, Zoltan x4414
vargaz@almos.vein.hu

† Extensions have prefix (36) 88-422022.

Accredited by: MAB (Hungarian Accreditation Board), IChemE, FEANI

Graduate advisor: Nagyne Stribl Tunde

Undergraduate advisor: Nagyne Stribl Tunde

Student organization: H.O.K.

Department reports to: Dr. Géza Horváth, Dean of Engineering

India

Anna University

Department of Chemical Engineering
Alagappa College of Technology
Madras 600 025

Professors

Balu, K.
Mohan, V.
Rajendran, M.
Ramamoorthy, V.
Sathiamoorthy, N.

Associate Professors

Murugesan, T.
Velan, M.

Department reports to: Mu.Rajendran

Placement service: N.Nagendra Gandhi

Annamalai University

Department of Technology
Annamalai University, Annamalai Nagar, Tamil Nadu
608 002

University phone 914144-238259
Department phone 914144-239737
Department FAX 914144-238275

Professors

Karunanithi, T. 914144-221235
tkarunanithi@hotmail.com
Viruthagiri, T. (Head) 914144-239737
drtvgiri@rediffmail.com

Assistant Professors

Anand, G. M. F.
mariafanand@yahoo.co.in
Bhaba, P. K. 04144-238691
Pkbhaba@yahoo.com
Dhanasekaran, R. 914144-238022
rdhanasekar28@rediffmail.com
Karthikeyan, C. 914144-224679
drcktech@rediffmail.com
Manickam, N. 914144-224494
Meyyappan, R. M. 221654
vijimani_cdm@yahoo.com
Vaithiyanathan, K. 914144223026
Vijayagopal, V. 914144249236
vevevin@yahoo.co.in

Accredited by: AICTE

Department reports to: Dr.T.Viruthagiri

Placement service: Dr.T.Viruthagiri

Birla Institute of Technology & Science (BITS) - Pilani

Department of Chemical Engineering
Vidya Vihar Campus, B.I.T.S.,
PILANI - 333 031 Rajasthan

University phone 91-1596-245073
University FAX 91-1596-244183
Department phone 91-1596-245073x224
Department FAX 91-1596-244183

Professors

Mathur, T.N.S. (Visiting) .... x8404 †
tns@bits-pilani.ac.in
Natarajan, B.R. x239
brnt@bits-pilani.ac.in
Vaid, R.P. x201
rpvaid@bits-pilani.ac.in

Associate Professors

Babu, B.V. (Head) x205/224
bvbabu@bits-pilani.ac.in

† Faculty numbers are extensions of 91-1596-245073.

Accredited by: NAAC

Degrees granted 2003-2004:

B.S.: 90 M.S.: 20 Ph.D.: 2

Graduate advisor: Prof. Ravi Prakash

Undergraduate advisor: Prof. S. Venkateswaran

Student organization: Chemical Engineering Association
Advisor: Mr. S.D.Manjare

Department reports to: Prof. A.K. Sarkar, Dean, Faculty Division-I

Placement service: Prof. G. Raghurama

Calcutta University

Department of Chemical Engineering
University College of Tech.
92 Acharyya Prafulla Chandra Rd.
Kolkata, 700 009

University phone 91-33-2241-0071
Department phone .... 91-33-2350-8386 ext 201
Department FAX 91-33-2351-9755

Professors

Banik, Ajit K.
Bhattacharjee, Sekhar .... 91-33-23508386 x 203
sekharbhatta@rediffmail.com
Das, Manas .... 91-33-23508386 x 296
De, Parameswar. (Head) .... 91-33-23508386 x 217
parameswar_de@rediffmail.com
Dutta, Benay. K. 91-33-23508386 x208
dutta_asc@vsnl.net
Pal, Sitangshu S. .... 91-33-23508386 x 290
Ray, P. .... 91-33-23508386 x 218
praycuce@vsnl.com
Saha, Ahin C. .... 91-33-23508386 x 203

Associate Professors

Ash, Soumendra N. .... 91-33-23508386 x 289
Basu, Ranjan K. .... 91-33-23508386 x 289
Chaudhury, Basab .... 91-33-23508386 x 203
basabc@vsnl.net
Das, Sudip K. .... 91-33-23508386 x 296
drsudipkdas@vsnl.net
Saha, Bibhuti R. .... 91-33-23508386 x 206

Assistant Professors

Das, Bhaskar C. .... 91-33-23508386 x 203
De, Asim K. .... 91-33-23508386 x 203
Ganguly, Kausik. .... 91-33-23508386 x 203
kausik_ganguly_cal@yahoo.com

Accredited by: Indian Institute of Chemical Engineering, Institute of Engineers, All India Council for Technical Education

Degrees granted 2003-2004:

B.S.: 35 M.S.: 5 Ph.D.: 2

Graduate advisor: Head of Department

Department reports to: Head of Department

Government Engineering College

Chemical Engineering Department
Raipur (Chattisgarh) 492010

Department phone 91 771 254299
Department FAX 91 771 254600

Professors

Kumar, H.

Associate Professors

Chandrakar, S. K.
Pande, A. L.
Poonia, A. K.

Singh, V. K.
Soni, A. B.

Accredited by: All India Council of Technical Education

Department reports to: R. P. Sukhija, Principal
91-771-254200

Placement service: A. K. Poonia

Indian Institute of Science

Department of Chemical Engineering
Bangalore 560 012

University phone 91 80 23600411
University FAX 91 80 3600085
Department phone 91 80 22932318
Department FAX 91 80 23608121

Professors

Gandhi, K. S. 91 80 22932320
gandhi@chemeng.iisc.ernet.in
Govindarao, V.M.H. 91 80 22932768
govind@chemeng.iisc.ernet.in
Modak, J. M. (Chair) 91 80 22933108
modak@chemeng.iisc.ernet.in
Rao, K. K. 91 80 22932341
kesava@chemeng.iisc.ernet.in

Associate Professors

Ayappa, K. G. 91 80 22932769
ayappa@chemeng.iisc.ernet.in
Gupta, S. K. 91 80 22933110
sanjeev@chemeng.iisc.ernet.in
Kumaran, V. 91 80 22933112
kumaran@chemeng.iisc.ernet.in
Madras, Giridhar 91 80 22932321
giridhar@chemeng.iisc.ernet.in
Nott, Prabhu R. 91 80 22932317
prnott@chemeng.iisc.ernet.in

Accredited by: AICTE

Degrees granted 2003-2004:

M.S.: 22 Ph.D.: 4

Graduate advisor: Prof. Giridhar Madras

Student organization: Chemical Engineering Association
Advisor: Prof. Giridhar Madras

Department reports to: Prof. M. L. Munjal, Chairman, Division of Mechanical Sciences

Placement service: Prof. Vikram Jairam, Advisor, Placement Ctr.

Indian Institute of Technology, Bombay

Department of Chemical Engineering
Powai, Mumbai 400076, India

University phone 91 22 572 2545
Department phone 91 22 576 7200
Department FAX 91 22 572 6895

Professors

Bellare, Jayesh .... 5767207 †
jb@che.iitb.ac.in
Dwivedi, M. C. 5767229
dwivedi@che.iitb.ac.in
Gurjar, V. G. 5767234
vgg@che.iitb.ac.in
Juvekar, V.A. 5767236
vaj@che.iitb.ac.in
Khakhar, D. V. 5767212
khakhar@che.iitb.ac.in
Khilar, K. C. 5767230
kartic@che.iitb.ac.in
Kudchadker, A. P. (Emeritus)
Madhavan, K. P. 5767237
kpm@che.iitb.ac.in
Mahajan, S. P. 5767225
mahajan@che.iitb.ac.in
Malik, Ranjan K. (Head) 5767796
rkmalik@che.iitb.ac.in
Mehra, Anurag 5767217
mehra@che.iitb.ac.in
Misra, Ashok (Director) 5768000
director@cc.iitb.ac.in
Moharir, A. S. 5767795
amoharir@che.iitb.ac.in
Moudgalya, K.M. (Associate Dean) 5767213
kannan@che.iitb.ac.in
Mukhopadhyay, Ms M. 5767248
mm@che.iitb.ac.in
Narayana Murthy, S. L. (Dean) 5767216
slnm@che.iitb.ac.in
Patwardhan, S. R. 5767214
srp@che.iitb.ac.in
Raghunathan, T. S. 5767235
raghu@che.iitb.ac.in
Rao, V. G. 5767220
vgr@che.iitb.ac.in
Shankar, H. S. 5767239
hss@che.iitb.ac.in
Shenoy, U. V. (Adjunct) 5767222
uvs@che.iitb.ac.in
Suresh, A. K. 5767240
aksuresh@che.iitb.ac.in
Trivedi, M. K. 5767241
trivedi@che.iitb.ac.in

Associate Professors

Gudi, Ravindra D. 5767204
ravindra@che.iitb.ac.in
Patwardhan, Sachin C. 5767211
sachinp@che.iitb.ac.in
Roy, S. 5767249
sr@che.iitb.ac.in
Sureshkumar, G. K. 5767208
gksuresh@che.iitb.ac.in
Venkatesh, K. V. 5767223
venks@che.iitb.ac.in

Assistant Professors

Mahajani, Sanjay M. 5767246
sanjaym@che.iitb.ac.in
Nanavati, H. 5767215
hnanavati@che.iitb.ac.in
Noronha, S.B. 5767238
noronha@che.iitb.ac.in
Wangikar, P.W. 5767232
pramodw@che.iitb.ac.in

† All numbers have prefix 91-22.

Accredited by: IIT Act, Indian Parliament

Graduate advisor: Prof. A. K. Suresh/Prof. D. V. Khakhar

Student organization: ChEA
Advisor: Head, Department of Chemical Engineering

Department reports to: Prof. Ashok Misra, Director

Placement service: N. S. Rathi

Indian Institute of Technology, Kanpur

Chemical Engineering Department
Kanpur, U.P., 208 016, India

University phone 91-512-2597629
University FAX 91-512-2590104
Department phone 91-512-2597406
Department FAX 91-512-2590104

Professors

Bhattacharya, P. K. 2597093
pkbhatta@iitk.ac.in
Chhabra, R. P. 2597393
chhabra@iitk.ac.in
Gupta, J. P. 2597175
jpg@iitk.ac.in
Gupta, S. K. 2597031
skg@iitk.ac.in
Khanna, Ashok 2597117
akhanna@iitk.ac.in
Kumar, Anil 2597195
anilk@iitk.ac.in
Kunzru, D.K. (Dean) 2597193
dkunzru@iitk.ac.in
Rao, D. P. 2597873
dprao@iitk.ac.in
Sharma, Ashutosh (Head) 2597026
ashutos@iitk.ac.in

Associate Professors

Deo, Goutom 2597363
goutam@iitk.ac.in
Verma, Nishith 2597704
nishith@iitk.ac.in

Assistant Professors

Bandyopadhyay, Rajdip 2597697
rajdip@iitk.ac.in
Garg, Sanjeev 2597736
sgarg@iitk.ac.in
Ghatak, Animangsu 2597146
aghatak@iitk.ac.in
Joshi, Yogesh M 2597629
ymjoshi@iitk.ac.in
Kaistha, Nitin 2597513
nkaistha@iitk.ac.in
Shankar, V 2597377
vshankar@iitk.ac.in

Accredited by:

Degrees granted 2003-2004:

B.S.: 45 M.S.: 36 Ph.D.: 4

Graduate advisor: Dr. Sanjeev Garg

Undergraduate advisor: Dr. Nitin Kaistha

Student organization: Student Gymkhana
Advisor: Dean of Students Affairs

Placement service: Dr. Ashok Khanna

Indian Institute of Technology, Kharagpur

Department of Chemical Engineering
Kharagpur, 721 302 West Bengal

University phone 91 3222 755221
Department phone 91 3222 82250
Department FAX 91 3222 755303

Professors

Acharjee, D. K. .... 83910 †
dka@che.iitkgp.ernet.im
Biswas, A. K. (Head) 83918
akb@che.iitkgp.ernet.in
Biswas, M. N. 83920
mani@che.iitkgp.ernet.in
De, D. S. 83924
dsde@che.iitkgp.ernet.in
Ghar, R. N. 83930
rng@che.iitkgp.ernet.in
Saha, R. K. 83946
rks@che.iitkgp.ernet.in

Associate Professors

Basu, J. K. 83914
jkb@che.iitkgp.ernet.in
Dasgupta, S. 83922
sunando@che.iitkgp.ernet.in
De, S. 83926
sde@che.iitkgp.ernet.in
Ganguly, S. 83928
sganguly@che.iitkgp.ernet.in
Mukherjee, D. 83938
dibyendu@che.iitkgp.ernet.in
Samanta, A. N. 83948
amar@che.iitkgp.ernet.in

Assistant Professors

Bhattacharya, D. 83916
db@che.iitkgp.ernet.in
Das, Gargi 83952
gargi@che.iitkgp.ernet.in
Kar, D. D. 83932
dkar@che.iitkgp.ernet.in
Kundu, G. 83934
gk@che.iitkgp.ernet.in
Pradhan, N. C. 83940
ncp@che.iitkgp.ernet.in
Ray, S. 83944
sray@che.iitkgp.ernet.in

† Faculty numbers are extensions of 91 3222 55221.

Accredited by: Indian Parliament

Graduate advisor: R. K. Saha

Undergraduate advisor: S. DasGupta

Student organization: Chemical Engineering Association
Advisor: J. K. Basu

Department reports to: The Director, IIT Kharagpur

Placement service: G. Kundu

Indian Institute of Technology, Madras

Chemical Engineering Department
Chennai 600 036

University phone .... (044) 2257 1365
Department phone .... (044) 2257 8207
Department FAX .... (044) 2257 0509

Professors

Ananth, M.S. (Director) 257 8208
ananth@iitm.ac.in
Balakrishnan, A. R. 257 8209
arbala@iitm.ac.in
Chidambaram, M. (Head) 257 8210
chidam@iitm.ac.in
Krishnaiah, K. 257 8211
krishnak@iitm.ac.in
Murthy, D.V.S. 257 8212
dvs@iitm.ac.in
Neelakantan, N.R. 257 8213
nrn@iitm.ac.in
Panda, T. 257 8217
tpanda@che.iitm.ac.in
Pushpavanam, P. 257 8218
spush@iitm.ac.in
Rao, V. S. R. 257 8214
vsrr@iitm.ac.in
Shankar, Narasimhan 257 8221
naras@iitm.ac.in
Swaminathan, T. 257 8222
tswami@iitm.ac.in

Associate Professors

Jayaraman, Guhan 257 8225
guhanj@iitm.ac.in
Ravi, R. 257 8229
rravi@iitm.ac.in
Sai, P. S. T. 257 8220
psts@iitm.ac.in
Sankaranarayanan, S. 257 8219
ssnarayan@iitm.ac.in

Assistant Professors

Basak, Tanmay
Chandraraj, K.
Deshpande, Abhijit P. 257 8223
abhijit@iitm.ac.in
Gummadi, Satyanarayana N.
Jayanti, Sreenivas 445 8227
sjayanti@iitm.ac.in
Kannan, A. 257 8224
kannan@iitm.ac.in
Varughese, Susy 257 8228
susy@iitm.ac.in

Placement service: Deputy Registrar, Placement Office

Indian Institute of Technology, Delhi

Department of Chemical Engineering,
Hauz Khas, New Delhi, India -110016

University phone 91 11 2686 1977
Department phone 91 11 2659 1021
Department FAX 91 11 2658 1120

Professors

Guha, B. K. 91 11 659 1038
bkguha@iitd.ernet.in
Gupta, A. K. 91 11 659 1019
ashokkg@chemical.iitd.ernet.in
Gupta, S. K. 91 11 659 1023
sgupta@chemical.iitd.ernet.in
Iyer, P. V. R. 91 11 659 1024
pvr@chemical.iitd.ernet.in
Khanna, Rajesh (Visiting) 91 11 659 1031
rajesh@chemical.iitd.ernet.in
Nigam, K. D. P. (Head) 91 11 659 1021
nigamkdp@hotmail.com
Pitchumani, B. 91 11 6591022
bpmani@chemical.iitd.ernet.in
Rao, D. P. 91 11 6591016
dprao@chemical.iitd.ernet.in
Rao, D. S. 91 11 6591018
subbarao@chemical.iitd.ernet.in
Rao, T. R. 91 11 6591026
trrao@chemical.iitd.ernet.in
Srivastava, V. K. 91 11 6591025
vks@chemical.iitd.ernet.in
Walia, D. S. 91 11 6591017
dswalia@chemical.iitd.ernet.in

Associate Professors

Bhaskarwar, A. N. 91 11 659 1028
ashoknb@chemical.iitd.ernet.in
Dhingra, S. C. 91 11 659 1027
dhingra@chemical.iitd.ernet.in

Assistant Professors

Amar, O. P. 91 11 659 1030
Basu, S. 91 11 659 1036
sbasu@chemical.iitd.ernet.in
Mohan, Ratan 91 11 6591033
ratan@chemical.iitd.ernet.in
Pant, K. K 91 11 6596172
kkpant@chemical.iitd.ernet.in

Student organization: CHESS
Advisor: Dr. Rajesh Khanna

Department reports to: Director, IITD

Placement service: Y. S. Goel, Placement Cell

Karnataka Regional Engineering College, Surathkal

Chemical Engineering Department
Mangalore, P. O. Srinivasnagar
PIN 574157
Karnataka

University phone 91-824-475984
Department phone 91-824-475984 ext.255

Professors

Mugeraya, Gopal 475984
master@krec.ernet.in
Murthy, D. V. R. (Head) 475984
dvrm@krec.ernet.in
Singh, P. N. (Director) 475984
Sriniketan, G. 475984
srini@krec.ernet.in

Associate Professors

Saidutta, M. B. 475984
mbs@krec.ernet.in

Assistant Professors

Gangamma, S. 475984
Rajmohan, B. 475984
Shetty Vidya, K. 475984

Accredited by: AIChE, ISTE, Institute of Engineers, Indian Institute of Chemical Engineers

Department reports to: P. N. Singh, Principal

Placement service: G. Premkumar, Professor of Training, Placement and Student's Welfare

Laxminarayan Institute of Technology, Nagpur University

Chemical Engineering Department
Nagpur, 440 010

Department phone 531659 (0712)

Professors

Nageshwar, G. D. (Director)

Associate Professors

Chandak, B. S.
Sonolikar, R. L.
Thorat, R. T.

Assistant Professors

Bhagade, S. S.
Dawande, S. D.
Koranne, K. V.
Pandharipande, S. L.
Thorat, R. T.
Vyas, R. P.

Accredited by: Indian Institute of Chemical Engineers

Department reports to: G. D. Nageshwar, Director

Placement service: S. P. Ghisad

Manipal Institute of Technology

Department of Chemical Engineering
University of MAHE
Manipal (S.K.)
Postal Pin Code 576 119

Department phone .... (08252)71061 ext 24311
Department FAX .... (08252) 71071

Professors

Bhat, Jayadev (Associate Dean) .... x24311 †
soodajbhat@eudoramail.com
Muniswaran, P. K. A. (Head) x24311
munishpka@yahoo.com
Murthy, P. S. x24316
parimimurthy@rediffmail.com

Associate Professors

Prabhu, Balakrishna x24312
balakrishnaprabhu@yahoo.com
Reddy, B. N. x24316
reddy_botta@hotmail.com
Sivasankaran, S. x24314
sivasankaraniyer@yahoo.com

Assistant Professors

Guru, Bharath R. x24312
bharathrajag@yahoo.co.in
Kini, Srinivasa x24315
srinivas_kini@rediffmail.com
Kumar, S. H. x24314
harishmanipal@rediffmail.com
Murthy, V. R. x24314
vytlarama@yahoo.com

† Faculty phone numbers are extensions of 2571060.

Accredited by: AICTE, National Board of Accreditation (Technical Education)

Department reports to: Dr. B.S. Prabhu, Director Office : 71072

Placement service: K. J. Kamath

Malaviya National Institute of Technology, Jaipur

Chemical Engineering Department
Malaviya National Institute of Technology
Jaipur 302 017
Rajasthan

University phone 91 141 2702954
Department phone 91 141 2702591

Professors

Gupta, Alok 91 141 2702591
agch1@rediffmail.com

Associate Professors

Chaurasia, S. P. (Head) 91-141-2702591
chch3@mnit.ac.in
Vyas, R.K. 91-141-2702591

Assistant Professors

Agarwal, Madhu
George, Suja
Jana, Susanta K.
Pandit, Prabhat
Singh, Kailash 91-141-2702591
ksch@mnit.ac.in
Vasistha, Manish

Accredited by: AICTE

Undergraduate advisor: Dr. R. K. Vyas

Student organization: Chemical Engineering Students' Society (ChESS)
Advisor: Kailash Singh

Department reports to: Director

Placement service: Professor, Training and Placement.

University of Mumbai

Chemical Engineering Division
Institute of Chemical Technology
Matunga, Mumbai
400 019

University phone 91-22-2703248
Department phone 91-22-4145616
Department FAX 91-22-4145614

Professors

Gaikar, V. G. (Chair) . . . . x286 †
vgg@udct.ernet.in

Joshi, J. B. (Director) x282
jbj@udct.ernet.in

Mahajani, V. V. x287
vvm@udct.ernet.in

Mann, Reginald (Visiting)
Misra, D.P. (Visiting)
Modak, Jayant (Visiting)
Niemantsverdriet, J.W. (Visiting)

Pandit, A. B. x278
abp@udct.ernet.in

Pangarkar, V. G. (Head) x284
vgp@udct.ernet.in

Venkataramanan (Visiting)

Yadav, G. D. (Chair) x291
gdy@udct.ernet.in

Associate Professors

Bhagwat, S. S. x285
sunil@udct.ernet.in

Lali, A. M. x292
arvind@udct.ernet.in

Marathe, K. V. x288
kvm@udct.ernet.in

Sawant, S. B. x281
sbs@udct.ernet.in

Assistant Professors

Patwardhan, A. W. x283
awp@udct.ernet.in

Thorat, B. N. x289
bnt@udct.ernet.in

† Faculty numbers are extensions of 91-22-4145616.

Accredited by: All India Council for Technical Education

Graduate advisor: S.S. Bhagwat

Undergraduate advisor: V. G. Gaikar

Student organization: Technological Association
Advisor: V.G. Gaikar

Department reports to: J.B.Joshi, Director, Institute of Chemical Technology, 91-22-4140865

Placement service: S.S. Bhagwat

National Institute of Technology Warangal

Department of Chemical Engineering
National Institute of Technology
Warangal - 506 004 (A.P.)INDIA

University phone 091-870-2459191
Department phone 091-870-2459191 x461
Department FAX 091-870-2459547

Professors

Sastry, R.C. (Head) . . . . 2459482 †
sastry@nitw.ernet.in

Venkat Reddy, G. 2459105
vreddy@nitw.ernet.in

Associate Professors

Pramanik, K. (Mrs.) 2430506
kpr@nitw.ernet.in

Pydisetty, Y. 2459292
psetty@nitw.ernet.in

Sarat Babu, A. 2459615
sarat@nitw.ernet.in

Assistant Professors

Anand Kishore, K. 2430574
kola@nitw.ernet.in

Prasanna Lakshmi, L. (Mrs.) 2430196
prasan@nitw.ernet.in

Srinath, S. 2432290
srinath@nitw.ernet.in

Srivani, K. (Mrs.) 2459165
vani@nitw.ernet.in

Venu Vinod, A. 2457731
avv@nitw.ernet.in

† Faculty numbers are extensions of 091-870.

Accredited by: NBA

Student organization: Chemical Engineering Association
Advisor: Dr.(Mrs.).K.Pramanik

Placement service: Training & Placement Section

Pravara Rural Education Society's Pravara Rural Engineering College, Loni

Chemical Engineering Department
Loni, 413 736
Tal. Rahata
Dist. Ahmednagar (M.S.)

Department phone 09-02422-73202
Department FAX 09-02422-73537

Professors

Dhirendra, S. (Chair) 09-02422-73303
dhirendra_shukla@hotmail.com

Assistant Professors

Dange, P. N.

Gaikwad, R. W. 09-02422-73240
gaikwad_ravindra@usa.net

Kadam, S. R.
Misal, S. A.
Shirsath, S. R.
Singh, S. K. P. 09-02422-74699

Student organization: Chemical Engineering Students Association

Department reports to: Principal, Pravara Rural, Engineering College

Placement service: S.U.Raut

R. V. College of Engineering, Affiliated with Visveswaraiah Technical University

Department of Chemical Engineering
Bangalore 560 059

University phone 0831 472500
Department phone 080-860-1701x246
Department FAX 080-860-2148

Professors

Rao, Y. J. (Head)

Associate Professors

Suresh, R.

Assistant Professors

Archna, Kumar
Bindu, G.
Kallur, Vinod
Ranganath, D.

Accredited by: Indian Institute of Chemical Engineers

Department reports to: K A Ranganatha Setty, Principal

Placement service: Prof. C.S. Chandrasekhara Murthy, Placement Officer

Regional Engineering College, Affiliated with Bharathidasan University

Chemical Engineering Department
Regional Engineering College
Tiruchirappalli-620 015
(Tamil Nadu)

University phone 91-431-500181
Department phone 91-431-500281
Department FAX 91-431-500133

Professors

Pandey, S. K.
Sundaram, S. (Head) 91 431 500 295
Venkataramani, V.

Assistant Professors

Anantharaman, N
Prabhu, H. J.
Radhakrishnan, T. K.
Shanmugasundaram, P.
Sivashanmugam, P.

Accredited by: NBA

Graduate advisor: S. Sundaram

Undergraduate advisor: S. Sundaram

Department reports to: Dr. S. Sundaram

Placement service: Dr. (Mrs.) S. K. Pandey, Head-in-Charge, Training and Placement Cell

Indian Institute of Technology

Chemical Engineering Department
Roorkee (Uttrancal)
PIN-247667

University phone 91-1332-272349
Department phone 91-1332-276534
Department FAX 91-1332-276535

Professors

Agarwal, C. P. .... 285712 †
cpaurfch@iitr.ernet.in
Bhattacharya, S. D. 285708
sdbiofch@iitr.ernet.in
Gupta, S. C. 285709
satisfch@iitr.ernet.in
Kumar, Surendra 285714
skumar@iitr.ernet.in
Mishra, I. M. 285715
imishfch@iitr.ernet.in
Mohanty, B. (Head) 285710
bmohanty@iitr.ernet.in

Associate Professors

Chand, Shri 285383
schanfch@iitr.ernet.in
Mall, I. D. 285319
invigfch@iitr.ernet.in

Assistant Professors

Agarwal, V. K. 285718
vijayfch@iitr.ernet.in
Bhargava, R. 285382
ravibfch@mail.iitr.ernet.in
Majumder, C. 285321
chandfch@iitr.ernet.in
Prasad, B. 285323
bashefch@iitr.ernet.in
Shashi 285672
shashifch@iitr.ernet.in
Sinha, S. N. 285384
sudhifch@iitr.ernet.in

† All faculty numbers are preceded by 91-1332

Accredited by: AICTE, Indian Institute of Chemical Engineers, Institution of Engineers (India)

Graduate advisor: Dr. B. Prasad

Undergraduate advisor: Dr. Shri Chand

Student organization: Institution of Engineers(India)
Advisor: Dr. V. K Agarwal

Department reports to: D. V. Singh (72742), Vice Chancellor

Placement service: H. K. Verma (85245), Chief, Employment Bureau

Siddaganga Institute of Technology, Affiliated with Viveswaraiah University

Chemical Engineering Department
Siddaganga Institute of Technolgy
Tumkur (Karnataka)
Pin 572103

Department phone 91- 816-2092070
Department FAX 91-816-2282994

Professors

Nirgunababu, P. 91 9880031145
Ramakrishnan, K. (Head) 91 9844052242
ramkey5@rediffmail.com

Associate Professors

K.R.S., Murthy 91 9844327238
Rajaguru, S. 91 8162274167
Shivabasappa, K. L. 91 9845768784
shivukl@yahoo.com

Assistant Professors

B.S., Gowrishankar 91 9444190002
bsgowri@rediffmail.com
Hegde, Vinayak. M. 91 8162290818
vinumh@rediffmail.com
Nagaraj, Naveen 91 9880385009
navee74@yahoo.co.in
Shyam Prasad, K.B. 91 9448531582
kbs_prasad@yahoo.com

Accredited by: NBA, New Delhi

Degrees granted 2003-2004:

B.S.: 48 M.S.: 10

Undergraduate advisor: P Nirguna Babu

Student organization: IIChE
Advisor: VIkram M Pai

Department reports to: Dr. M. N. Channabasappa, Principal

Placement service: Vinayak M Hegde, Placement Officer

S. V. National Institute of Technology

Department of Chemical Engineering
Ichchhanath
Surat 395 007

University phone..................... 91 261 2223371 - 74
University FAX...................91 261 2227334, 2229394
Department phone .... 91 261 2223371 Ext. 332
Department FAX 91 261 2227334

Associate Professors

Murthy, Z. V. P. .... 334 †
zvpm2000@yahoo.com
Parikh, P.A (Head) 331
parimal_svr@yahoo.co.uk

Assistant Professors

Chakraborty, M. 332
mousumi_chakra@yahoo.com
Mukhopadhyay, M. 333
mausumi_mukhopadhyay@yahoo.com

† Faculty numbers are extensions of university main number.

Accredited by: AICTE

Degrees granted 2003-2004:

B.S.: 60

Undergraduate advisor: Dr. Z.V.P. Murthy

Placement service: Prof. D. B. Naik

Thapar Institute of Engineering and Technology

Department of Chemical Engineering
Thapar Technology Campus
Patiala-147004

University phone......................... 91 175 2393001
Department phone 91 175 2393063
Department FAX 91 175 2364498

Professors

Bajpai, Pramod K. (Head) 91 175 393551
pkbajpai@mail.tiet.ac.in
Bansal, T.K. 91 175 393306
tkbansal@mail.tiet.ac.in

Associate Professors

Gangacharyulu, D. 91 175 393305
dgangacharyulu@mail.tiet.ac.in
Kumar, Vineet 91 175 393309
vkumar@mail.tiet.ac.in
Mehta, Rajeev 91 175 393310
rmehta@mail.tiet.ac.in
Sinha, Shishir 91 175 2393308
ssinha@mail.tiet.ac.in

Assistant Professors

Ahuja, Sanjeev 91 175 393119
skahuja@mail.tiet.ac.in
Gupta, Raj K. 91 175 393310
rkgupta@mail.tiet.ac.in
Gupta, Ritu (Mrs.) 91 175 393120
rgupta@mail.tiet.ac.in
Toor, A.P. (Mrs.) 91 175 393120
aptoor@mail.tiet.ac.in

Accredited by: AICTE

Graduate advisor: Dr. Vineet Kumar

Undergraduate advisor: Dr. D. Gangacharyulu

Student organization: I.I.Ch.E.
Advisor: Dr. Shishir Sinha

Department reports to: Director

Placement service: Mr. H.S. Bawa

Vellore Institute of Technology

Room 114, Hexagon Building
Katpadi-Thiruvalam Road
Deemed University
Vellore, Tamil Nadu, 632014

University phone 91-416-2243091
Department phone 91-416-2243091x523
Department FAX 91-416-2243092

Professors

Pullabhotla, S. R. (Head) .... x144 †
raop47@hotmail.com

Assistant Professors

Dr. Kumar, P. x557
kumar_perumal@hotmail.com
Dr. Murthy Shekhar, S. x537
smshekhar@hotmail.com

† Faculty number are extensions of 91-416-243091

Accredited by: UGC/AICTE

Graduate advisor: Prof. S. R. Pullabhotla

Undergraduate advisor: Prof. S. R. Pullabhotla

Student organization: SChE
Advisor: David. K. Daniel

Department reports to: Anand A. Samuel

Placement service: Prof. S. R. Pullabhotla

Iran

Amirkabir University of Technology(Tehran Polytechnic)

Chemical Engineering Department
No. 424, Hafez Avenue
Tehran, Iran
PO.Box 15875-4413

University phone ....
University FAX ....
Department phone .... (9821) 6499066
Department FAX .... (9821) 6405847

Professors

Dabir, Bahram
dabir@aut.ac.ir

Edrisi, Mohammad
edrisi@aut.ac.ir

Modarress, Hamid M.
hmodares@aut.ac.ir

Rashidi, Fariborz
rashidi@aut.ac.ir

Sohrabi, Morteza
sohrabi@aut.ac.ir

Jamshidi, Esmail
jamshidi@aut.ac.ir

Kaghazchi, Tahereh
kaghazch@aut.ac.ir

Associate Professors

Kalbasi, Mansour
mkabasi@aut.ac.ir

Accredited by:

Graduate advisor:

Undergraduate advisor: Eng. Mohammadhossein Namazi Ghadim

Department reports to:
http://www.aut.ac.ir/departments/chem/1024/index.htm

Placement service: Chem.Eng@aut.ac.ir

Isfahan University of Technology

Department of Chemical Engineering
Isfahan, Iran
Post Code: 8415683111

University phone .... (98) (311) 3912737
University FAX .... (98) (311) 3912862
Department phone .... (98) 311 3915618
Department FAX .... (98) 311 3912677

Associate Professors

Shams, Kayghobad
k_shams@cc.iut.ac.ir
Ehsani, Mohammad R. .... (98) 311 3915607
ehsanimr@cc.iut.ac.ir
Etemad, S.Gholamreza
etemad@cc.iut.ac.ir
Ghoreishi, Seyyed M. 98 311 3915604
ghoreshi@cc.iut.ac.ir
Mehrabani Z., Arjomand
Arjomand@cc.iut.ac.ir

Assistant Professors

Roodpeyma, Shapoor
shapoor@cc.iut.ac.ir

Accredited by:

Graduate advisor: Ahmad Moheb

Sharif University of Technology

Chemical and Petroleum Engineering Department
P. O. Box 11365-9465
Azadi Avenue, Tehran

University phone 98-21-600 5210
Department phone 98-21-600 5819
Department FAX 98-21-602 2853

Professors

Alemzadeh, Iran .... Ext. 5486
alemzadeh@sharif.edu
Badakhshan, Amir (Visiting) Ext.5461
abadakhs@sharif.edu
Goodarznia, Iraj .... Ext. 5456
goodarz@sharif.edu
Kazemeini, Mohammad .... Ext. 5425
kazemeini@sharif.edu
Razavi, Jalil Ext.5414
razavi@sharif.edu

Safe Kordi, Ali A. .... Ext. 5418
safekordi@sharif.edu

Shahrokhi, Mohammad 98-21-600 5819
shahrokhi@sharif.edu

Soltanieh, Mohammad .... Ext. 5417
soltani@sharif.edu

Vosoughi, Manouchehr .... Ext. 5478
vosoughi@sharif.edu

Associate Professors

Abdekhodaie, Mohammad J. .... Ext. 5426
abdmj@sharif.edu

Borghei, Mehdi .... Ext. 5483
mborghei@sharif.edu

Farhadi, Fathollah (Associate Chair) .... Ext. 5423
farhadi@ sharif.edu

Frounchi, Masoud .... Ext. 5422
Frounchi@sharif.edu

Ghotbi, Cyrus (Chair) 98-21-600 5819
ghotbi@sharif.edu

Kheirolomoom, Azadeh Ext.5412
akheir@sharif.edu

Khorasheh, Farhad Ext.5411
khorashe@sharif.edu

Rashtchian, Davood (Associate Chair) 98 21 600 5819
rashtchian@sharif.edu

Shayegan, Jaleleddin .... Ext. 5420
shayegan@sharif.edu

Assistant Professors

Aghajan, Abdol H. .... Ext. 5408
aghajan@sharif.edu

baghalha, Morteza .... Ext. 5469

Bastani, Daryoush .... Ext. 5409
bastani@sharif.edu

Bozorgmehry, Ramin (Associate Chair) 98-21-600 5819
brbozorg@sharif.edu

Kazemi, Akhtarelmolouk Ext.5484
a_kazemi@sharif.edu

Molaee dehkordi, Asghar .... Ext. 5411

Pishvaie, Mohammad R. Ext.5429
Pishvaie@sharif.edu

Ramazani, Ahmad Ext.5431
ramazani@sharif.edu

Roosta Azad, Reza (Associate Chair) 98-21-600 5417
roosta@sharif.edu

Shojaee, Akbar .... Ext. 5429
shojaee@sharif.edu

Taghikhani, Vahid Ext.5458
Taghikhani@sharif.edu

Yaghmaei, Soheila (Associate Chair) Ext.5428
yaghmaei@sharif.edu

Accredited by:

Degrees granted 2003-2004:

B.S.: 100 M.S.: 65 Ph.D.: 4

Graduate advisor: Davood Rashtchian

Undergraduate advisor: R. Bozorgmehri

Student organization: Soheila Yaghmaei
Advisor: Soheila Yaghmaei

Placement service: Cyrus Ghotbi

Ireland

Cork Institute of Technology

Department of Chemical Engineering
Rossa Avenue
Cork, Ireland
Deliveries: Mrs. P. Copley

University phone 353-21-4326100
Department phone 353-21-4326885
Department FAX 353-21-4326851

Professors

Cunningham, J. D. 353 21 344864
cleantechnology@cit.ie

Duffy, N. B. 353 21 324407
nduffy@cit.ie

O'Shea, J. T. (Head) 353 21 326404
joshea@cit.ie

Associate Professors

Moroney, P. J.

O'Gorman, A. 353 21 326411
aogorman@cit.ie

Petrie, G. R. 353 21 326430
gpetrie@cit.ie

Power, P. C.

Raissian, S. M. 353 21 326430
mraissian@cit.ie

Ryan, L. B. 353 21 326411
bryan@cit.ie

Assistant Professors

Doyle, B.

Kennedy, P 353 21 326416
pkennedy@cit.ie

O'Suilleabhain, C 353 21 326416
cosuilleabhain@cit.ie

Accredited by: National Council for Educational Awards, Institution of Chemical Engineers (UK), Institution of Engineers of Ireland

Degrees granted 2003-2004:

B.S.: 29 M.S.: 2 Ph.D.: 1

Graduate advisor: A.O'Gorman

Undergraduate advisor: J. T. O'Shea

Student organization: A.O'Gorman
Advisor: A.O'Gorman

Department reports to: Dr B. J. Murphy, Director

Placement service: J. T. O'Shea

University College Dublin

Chemical and Biochemical Engineering Department
Faculty of Engineering and Architecture
University College Dublin
Belfield, Dublin 4

University phone 353-1-716 7777
Department phone 353-1-716 1825
Department FAX 353-1-716 1177

Professors

MacElroy, J. M. D. (Head) 353-1-716-1824
don.macelroy@ucd.ie

Associate Professors

Carroll, D. 353-1-716-1897
dan.carroll@ucd.ie

Kieran, Patricia M. 353-1-716-1956
patricia.kieran@ucd.ie

MacLoughlin, P. F. 353-1-716-1894
frank.macloughlin@ucd.ie

Malone, Dermot M. 353-1-716-1895
dermot.malone@ucd.ie

Assistant Professors

Casey, Eoin 353-1-716 1877
eoin.casey@ucd.ie

Glennon, Brian A. 353-1-716-1954
brian.glennon@ucd.ie

McDonnell, Susan 353-1-716 1825
susan.mcdonnell@ucd.ie

Mooney, Damian 353-1-716 1827
damian.mooney@ucd.ie

Accredited by: IChemE (UK), IEI (Ireland)

Degrees granted 2003-2004:

B.S.: 34 M.S.: 1 Ph.D.: 3

Undergraduate advisor: Colleen Blaney

Student organization: Students' Union

Department reports to: The President

Placement service: Colm Tobin

Israel

Ben-Gurion University of the Negev

Chemical Engineering Department
Beer Sheva 84105
P.O.B. 653

University phone 972-8-6461-111
Department phone 972-8-6461-480
Department FAX 972-8-6472-916

Professors

Apelblat, Alexander .... 6461-487 †
apelblat@bgumail.bgu.ac.il

Gottlieb, Moshe 6461-486
mosheg@inca.bgu.ac.il

Herskowitz, Mordechai (Dean) 6461-482
herskow@bgumail.bgu.ac.il

Kost, Joseph 6461-766
kost@bgumail.bgu.ac.il

Lang, Sidney B. 6461-490
lang@bgumail.bgu.ac.il

Loeb, Sidney (Emeritus)
sidloeb@bgumail.bgu.ac.il

Merchuk, Jose C. 6461-768
jcm@bgumail.bgu.ac.il

Shacham, Mordechai 6461-481
shacham@bgumail.bgu.ac.il

Tamir, Abraham 6461-765
atamir@bgumail.bgu.ac.il

Wisniak, Jaime 6461-479
wisniak@bgumail.bgu.ac.il

Wolf, David (Emeritus) 6461-489
dwolf@bgumail.bgu.ac.il

Associate Professors

Korin, Eli (Head) 6461-820
ekorin@bgumail.bgu.ac.il

Landau, Miron 6472-141
mlandau@bgumail.bgu.ac.il

Regev, Oren 6472-145
oregev@bgumail.bgu.ac.il

Roy, Ahron (Emeritus) 6461-484
royaaron@bgumail.bgu.ac.il

Assistant Professors

Barenheim-Groswasser, Anne 6472-129
bernheim@bgumail.bgu.ac.il

Yerushalmi-Rosen, Rachel 6472-140
rachely@bgumail.bgu.ac.il

† Direct dial with prefix 972-7.

Accredited by: Israel Council for Higher Education

Graduate advisor: J. Kost

Undergraduate advisor: R. Yerushalmi-Rozen

Department reports to: Yigal Ronen, Dean of Engineering

Technion-Israel Institute of Technology

Chemical Engineering Department
Technion City
Haifa 32000

University phone 972-4-829-2111
Department phone 972-4-829-2820
Department FAX 972-4-8295672

Professors

Aharoni, Chaim (Emeritus) 972-4-8292118
Cohen, Yachin 972-4-8292010
Grader, Gideon 972-4-8292008
Hasson, David (Emeritus) 972-4-8292936
Kehat, Ephraim (Emeritus) 972-4-8292935
Lavie, Ram (Emeritus) 972-4-8292934
Lewin, Daniel 972-4-8292006
Marmur, Abraham 972-4-8293088
Narkis, Moshe (Emeritus) 972-4-8292937
Nir, Avinoam 972-4-8292119
Pismen, Leonid 972-4-8293086
Ram, Arie (Emeritus) 972-4-8292933
Rigbi, Zvi (Emeritus) 972-4-8293089
Semiat, Raphael 972-4-8292009
Sheintuch, Moshe 972-4-8292823
Tadmor, Zehev 972-4-8293087
Talmon, Yeshayahu (Chair) 972-4-8292007

Associate Professors

Brandon, Simon 972-4-8292822
Orell, Aluf (Emeritus) 972-4-8292020
Paz, Yaron 972-4-8292486

Assistant Professors

Bianco-Peled, Havazelet 972-4-8293588
Brenner, Naama 972-4-8293578
Srebnik, Simcha 972-4-8293584
Tsur, Yoed 972-4-8293586

Accredited by: Israel Council for Higher Education

Degrees granted 2003-2004:

B.S.: 63 M.S.: 6 Ph.D.: 6

Undergraduate advisor: Daniel R. Lewin

Department reports to: Prof. Aviv Rosen, Senior Vice President

Italy

Universita di Bologna

DICMA - Dipartimento di Ingegneria Chimica, Mineraria e delle Tecnologie Ambientali
viale Risorgimento, 2
I-40136 Bologna

University phone .... (39) 051-2099349
Department phone .... (39) 051-2093135
Department FAX .... (39) 051-581200

Professors

Doghieri, Ferruccio .... (39) 051-2090426
ferruccio.doghieri@mail.ing.unibo.it
Gostoli, Carlo .... (39) 051-2093144
carlo.gostoli@mail.ing.unibo.it
Magelli, Franco (Chair) .... (39) 051-2093147
franco.magelli@mail.ing.unibo.it
Santarelli, Francesco .... (39) 051-2093148
francesco.santarelli@mail.ing.unibo.it
Sarti, Giulio C. .... (39) 051-2093142
giulio.sarti@mail.ing.unibo.it
Spadoni, Gigliola .... (39) 051-2093146
gigliola.spadoni@mail.ing.unibo.it

Associate Professors

Bandini, Serena .... (39) 051-2093138
serena.bandini@mail.ing.unibo.it
Camera-Roda, Giovanni .... (39) 051-2093137
giovanni.camera.roda@mail.ing.unibo.it
Cozzani, Valerio .... (39) 051-2093140
valerio.cozzani@mail.ing.unibo.it
Gatta, Alceo .... (39) 051-2093154
alceo.gatta@mail.ing.unibo.it
Nocentini, Massimo .... (39) 051-2090424
massimo.nocentini@mail.ing.unibo.it
Pasquali, Gabriele .... (39) 051-2093139
gabriele.pasquali@mail.ing.unibo.it
Pinelli, Davide .... (39) 051-2090427
davide.pinelli@mail.ing.unibo.it
Stramigioli, Carlo .... (39) 051-2093153
carlo.stramigioli@mail.ing.unibo.it

Assistant Professors

Bonvicini, Sarah .... (39) 051-2093141
sarah.bonvicini@mail.ing.unibo.it
Giacinti Baschetti, Marco .... (39) 051-2090408
marco.giacinti@mail.ing.unibo.it
Leonelli, Paolo .... (39) 051-2093150
paolo.leonelli@mail.ing.unibo.it
Montante, Giusi .... (39) 051-2090406
giusi.montante@mail.ing.unibo.it

Department reports to: F. Magelli, Chairman

Universita di Cagliari

Dipartimento di Ingegneria Chimica e Materiali
Piazza d'Armi
I-09123 Cagliari

University phone 390706751
Department phone 390706755071
Department FAX 390706755067

Professors

Cao, Giacomo 390706755058
cao@dicm.unica.it
Massidda, Luigi 390706755080
massidda@dicm.unica.it
Sanna, Ulrico 390706755063
sanna@dicm.unica.it
Viola, Antonio 390706755055
av@dicm.unica.it

Associate Professors

Amat, Paolo 390706755075
amat@dicm.unica.it
Baratti, Roberto 390706755056
baratti@dicm.unica.it
Carta, Renzo 390706755068
carta@dicm.unica.it
Dernini, Stella 390706755077
dernini@dicm.unica.it
Lallai, Antonio 390706755060
lallai@dicm.unica.it
Mura, Giampaolo 390706755051
mura@dicm.unica.it
Orru', Roberto 390706755076
orru@dicm.unica.it
Polcaro, Anna M. 390706755059
polcaro@dicm.unica.it
Tola, Giuseppe (Chair) 390706755074
tola@dicm.unica.it
Usai, Giorgio 390706755061

Assistant Professors

Cincotti, Alberto 390706755066
cincotti@dicm.unica.it
Delogu, Francesco 390706755073
delogu@dicm.unica.it
Mascia, Michele 390706755066
mmascia@dicm.unica.it
Meloni, Paola 390706755077
meloni@dicm.unica.it
Palmas, Simona 390706755069
spalmas@dicm.unica.it
Pilloni, Giovanni 390706755072
pilloni@dicm.unica.it
Tronci, Stefania 390706755073
tronci@dicm.unica.it

Accredited by:

Degrees granted 2003-2004:

B.S.: 20 M.S.: 15 Ph.D.: 1

Universita de L'Aquila

Dipartimento di Chimica
Ingegneria Chimica e Materiali
I-67040 Monteluco Di Roio - L'Aquila

University phone ....(39) 0862-43-4002
Department phone ....(39) 862-43-4214
Department FAX ....(39) 862-43-4203

Professors

Andruzzi, Romano .... (39) 0862-434216 nike@ing.univaq.it
Barba, Diego .... (39) 0862-434238 moscalia@ing.univaq.it
Brandani, Vincenzo .... (39) 0862-434229 brandani@ing.univaq.it
Cantarella, Maria .... (39) 0862-434215 cantarel@ing.univaq.it
Del Re, Giovanni .... (39) 0862-434219 delre@ing.univaq.it
Di Giacomo, Gabriele .... (39) 0862-434225 digiacom@ing.univaq.it
Foscolo, Pier U. (Associate Dean) .... (39) 0862-434214 foscolo@ing.univaq.it
Gibilaro, Giovanni L. .... (39) 0862-434214 gibilaro@ing.univaq.it
Inesi, Achille .... (39) 0862-434246 inesi@ing.univaq.it
Pelino, Mario .... (39) 0862-434224 pelino@ing.univaq.it
Scoccia, Giancarlo .... (39) 0862-434222 scoccia@ing.univaq.it
Villa, Pierluigi .... (39) 0862-434245 villa@ing.univaq.it
Volpe, Roberto (Head) .... (39) 0862-434222 volpe@ing.univaq.it

Associate Professors

Cantalini, Carlo .... (39) 0862-434233 canta@ing.univaq.it
Evangelista, Franco .... (39) 0862-434243 frev@ing.univaq.it
Fantauzzi, Felice .... (39) 0862-434216 felice@ing.univaq.it
Fumarola, Giuseppe .... (39) 0862-434242 fumarola@ing.univaq.it
Gallifuoco, Alberto .... (39) 0862-434232 gallifuo@ing.univaq.it
Germanà, Antonino .... (39) 0862-434214 germana@ing.univaq.it
Pajewsky, Leonardo .... (39) 0862-434227 pajewsky@ing.univaq.it
Quaresima, Raimondo .... (39) 0862-434226 raimondo@ing.univaq.it
Rossi, Leucio .... (39) 0862-434246 rossil@ing.univaq.it
Vegliò, Francesco .... (39) 0862-434223 veglio@ing.univaq.it

Assistant Professors

Beolchini, Francesca .... (39) 0862-434236 beolffra@ing.univaq.it
Jand, Nader .... (39) 0862-434237 nader@ing.univaq.it
Mucciante, Vittoria .... (39) 0862-434216 nike@ing.univaq.it
Prisciandaro, Marina .... (39) 0862-434241 maripri@ing.univaq.it
Taglieri, Giuliana .... (39) 0862-434227 taglieri@ing.univaq.it

Graduate advisor: Prof. Giovanni Del Re

Undergraduate advisor: Prof. Giovanni Del Re

Student organization: Dr. Katia Gallucci
Advisor: Dr. Nader Jand

Department reports to: Prof. Pier Ugo Foscolo, Director of Studies

Placement service: Prof. Alberto Gallifuoco

Universita di Napoli

Dipartimento di Ingegneria Chimica
Piazzale V. Tecchio 80
80125 NAPOLI

University phone ....(39) 081 2531111
Department phone ....(39) 081 7682265
Department FAX ....(39) 081 2391800

Professors

Caprio, Vincenzo (Head) .... (39) 0817682250 caprio@unina.it
Crescitelli, Silvestro (Chair) .... (39) 0817682257 crescit@unina.it
D'Alessio, Antonio (Chair) .... (39) 0817682256 andaless@unina.it
Gioia, Francesco (Chair) .... (39) 081 7682277 gioia@unina.it
Greco, Guido (Chair) .... (39) 0817682274 grecog@unina.it
Marrucci, Giuseppe (Chair) .... (39) 0817682273 marrucci@unina.it
Russo, Gennaro (Chair) .... (39) 0817682266 genrusso@unina.it
Salatino, Piero (Chair) .... (39) 0817682258 salatino@unina.it
Volpicelli, Gennaro (Chair) .... (39) 0817682267 volpicel@unina.it

Associate Professors

Andreozzi, Roberto .... (39) 081 7682251 randreoz@unina.it
Cavaliere, Antonio .... (39) 0817682264 antcaval@unina.it
D'Anna, Andrea .... (39) 0817682241 danna@irc.na.cnr.it
Di Blasi, Colomba .... (39) 0817682232 diblasi@unina.it
Diana, Massimo .... (39) 0817682241 diana@unina.it
Grizzuti, Nino .... (39) 0817682285 grizzuti@unina.it
Guido, Stefano .... (39) 0817682271 steguido@unina.it
Lancia, Amedeo .... (39) 0817682243 lancia@unina.it
Maffettone, Pier L. (Visiting) .... (39) 0817682282 maffetto@unina.it
Marzocchella, Antonio .... (39) 0817682541 marzoc@hp3.irc.na.cnr.it
Murena, Fabio .... (39) 0817682272 murena@unina.it

Turco, Maria ..... (39) 0817682259
turco@unina.it

Assistant Professors

Bagnasco, Giovanni ..... (39) 0817682259
bagnasco@unina.it

Grosso, Massimiliano ..... (39) 0817682282
grosso@unina.it

Ianniruberto, Giovanni ..... (39) 0817682270
iannirub@unina.it

Marotta, Raffaele ..... (39) 0817682253
rmarotta@unina.it

Nigro, Roberto ..... (39) 081 7682249
rnigro@unina.it

Pirozzi, Domenico ..... (39) 0817682284
dpirozzi@unina.it

Simeone, Marino ..... (39) 0817682539
simeone@unina.it

Toscano, Giuseppe ..... (39) 0817682278
giutosca@unina.it

Student organization: Ingrid Florenzano
Advisor: Antonio Brasiello

Department reports to: Vincenzo Caprio

Universita di Pisa

Dipartimento di Ingegneria Chimica,
Chimica Industriale e Scienza dei Materiali
Via Diotisalvi, 2
56126 Pisa
Italy

University phone 39-050-2212-111
Department phone 39-050-511-111
Department FAX 39-050-511-266

Professors

Brambilla, Alessandro ..... 39-050-511-243 †
a.brambilla@ing.unipi.it

Butta, Enzo (Emeritus) 39-050-511-203
Giusti, Paolo 39-050-511-287
p.giusti@ing.unipi.it

Levita, Giovanni (Associate Chair) 39-050-511-201
levita@ing.unipi.it

Magagnini, Pierluigi (Chair) 39-050-511-222
Magagnini@ing.unipi.it

Marconi, Pierfilippo 39-050-511-211
marconi@ing.unipi.it

Nencetti, Gianfranco 39-050-511-215
g.nencetti@ing.unipi.it

Scali, Claudio 39-050-511-241
scali@ing.unipi.it

Tartarelli, Roberto 39-050-511-256
r.tartarelli@ing.unipi.it

Tognotti, Leonardo 39-050-511-240
tognotti@ing.unipi.it

Zanelli, Severino 39-050-511-212
zanelli@ing.unipi.it

Associate Professors

Bartolozzi, Mauro 39-050-511209
m.bartolozzi@ing.unipi.it

Davini, Paolo 39-050-511-216
davini@ing.unipi.it

De Sanctis, Massimo 39-050-511-227
desanctis@ing.unipi.it

Di Maina, Marcello 39-050-511-111
Lazzeri, Andrea 39-050-511-207
a.lazzeri@ing.unipi.it

Lazzeri, Luigi 39-050-511-277
lu.lazzeri@ing.unipi.it

Lupinacci, Domenico 39-050-511-231
d.lupinacci@ing.unipi.it

Marchetti, Augusto 39-050-511-207
marchetti@ing.unipi.it

Mauri, Roberto 39-050-511-248
r.mauri@ing.unipi.it

Nicolella, Cristiano 39-050-511294
c.nicolella@ing.unipi.it

Paci, Massimo 39-050-511-230
m.paci@ing.unipi.it

Petarca, Luigi 39-050-511-224
petarca@ing.unipi.it

Solina, Adriano 39-050-511-257
solina@ing.unipi.it

Vitolo, Sandra 39-050-511-278
vitolo@ing.unipi.it

Assistant Professors

Bresci, Bruno 39-050-511-278
Brunazzi, Elisabetta 39-050-511-213
e.brunazzi@ing.unipi.it

Cascone, Maria G. 39-050-511-221
mg.cascone@ing.unipi.it

Polacco, Giovanni 39-050-511-220
g.polacco@ing.unipi.it

Rizzo, Cosimo 39-050-511-204
c.rizzo@ing.unipi.it

Seggiani, Maurizia 39-050-511-281
m.seggiani@ing.unipi.it

Tricoli, Vincenzo 39-050-511-246
v.tricoli@ing.unipi.it

Valentini, Renzo 39-050-511-259
r.valentini@ing.unipi.it

Vatistas, Nicolaos 39-050-511-239
vatistas@ing.unipi.it

† Direct dial with prefix 39-50.

Accredited by:

Degrees granted 2003-2004:

M.S.: 112 Ph.D.: 2

Graduate advisor: Prof. L. Tognotti

Undergraduate advisor: Prof. R. Mauri

University of Salerno

Dipartimento di Ingegneria Chimica e Alimentare
Via Ponte Don Melillo
I-84084 Fisciano (SA)
Deliveries: Prof. Ernesto Reverchon

University phone 39 089 964185
Department phone 39 089 964185
Department FAX 39 089 964 057

Professors

Ciambelli, Paolo 39 089964151
pciambelli@unisa.it

D'Amore, Matteo 39 089 964 136
damore@dica.unisa.it

Di Matteo, Marisa 39 089 964 131
mdimatteo@unisa.it

Donsi', Giorgio 39 089 694 150
donsi@unisa.it
Ferrari, Giovanna 39 089 964 134
gferrari@unisa.it
Reverchon, Ernesto (Head) 39 089 964 116
ereverchon@unisa.it
Titomanlio, Giuseppe 39 089 964 152
gtitomanlio@unisa.it
Vittoria, Vittoria 39 089 964 114
vittoria@dica.unisa.it

Associate Professors

Dovinola, Vincenzo 39 089 964 146
Iannelli, Pio 39 089 964 103
piannelli@unisa.it
Incarnato, Loredana 39 089 964 144
lincarnato@unisa.it
Miccio, Michele 39 089 964 148
mmiccio@unisa.it
Nobile, Maria R. 39 089 964 143
mrnobile@unisa.it
Poletto, Massimo 39 089 964 132
mpoletto@unisa.it
Romano, Vittorio 39 089 964 079
vromano@unisa.it
Sesti Osseo, Libero 39 089 964 133
lsestiosseo@unisa.it
Vaccaro, Salvatore 39 089 964 153
svaccaro@unisa.it

Assistant Professors

Di Maio, Luciano 39 089 964 102
ldimaio@unisa.it
Gorrasi, Giuliana 39 089 964115
ggorrasi@unisa.it
Guadagno, Liberata 39 089 964 115
lguadagno@unisa.it
Palma, Vincenzo 39 089 964 147
vpalma@unisa.it
Pantani, Roberto
rpantani@unisa.it
Parascandola, Palma 39 089 964 078
pparascandola@unisa.it
Russo, Paola 39 089 964 027
parusso@unisa.it
Sannino, Diana 39 089 964 027
sannino@dica.unisa.it

Undergraduate advisor: Prof. G. Ferrari

Politecnico di Milano

Dipartimento di Chimica, Materiali ed Ingegneria Chimica
Giulio Natta
Piazza Leonardo da Vinci, 32
20133 Milano

University phone..........................(39) 02-23 991
Department phone .... (39) 02-23 99 3201
Department FAX .... (39) 02-70 63 8173

Professors

Allegra, Giuseppe .... (39) 02-23 99 3023
giuseppe.allegra@polimi.it
Biardi, Giuseppe .... (39) 02-23 99 3263
giuseppe.biardi@polimi.it
Buzzi Ferraris, Guido .... (39) 02-23 99 3257
guido.buzziferraris@polimi.it
Centola, Paolo .... (39) 02-23 99 3251
paolo.centola@polimi.it
Cigada, Alberto .... (39) 02-23 99 3164
alberto.cigada@polimi.it
Dente, Mario (D rector) .... (39) 02-23 99 3295
mario.dente@polimi.it
Forzatti, Pio .... (39) 02-23 99 3238
pio.forzatti@polimi.it
Lietti, Luca .... (39) 02-23 99 3272
luca.lietti@polimi.it
Pasquon, Italo .... (39) 02-23 99 3221
italo.pasquon@polimi.it
Pavan, Andrea .... (39) 02-23 99 3217
andrea.pavan@polimi.it
Pegoraro, Mario .... (39) 02-23 99 3243
mario.pegoraro@polimi.it
Pierucci, Sauro .... (39) 02-23 99 3256
sauro.pierucci@polimi.it
Porri, Lido .... (39) 02-23 99 3236
lido.porri@polimi.it
Ranzi, Eliseo .... (39) 02-23 99 3250
eliseo.ranzi@polimi.it
Rink, Marta .... (39) 02-23 99 3217
marta.rink@polimi.it
Tronconi, Enrico .... (39) 02-23 99 3264
enrico.tronconi@polimi.it
Zerbi, Giuseppe .... (39) 02-23 99 3235
giuseppe.zerbi@polimi.it

Associate Professors

Beretta, Alessandra .... (39) 02-23 99 3284
alessandra.beretta@polimi.it
Botta, Donatella .... (39) 02-23 99 3234
donatella.botta@polimi.it
Castiglioni, Chiara .... (39) 02-23 99 3230
chiara.castiglioni@polimi.it
Del Rosso, Renato .... (39) 02-23 99 3239
renato.delrosso@polimi.it
Di Landro, Luca .... (39) 02-23 99 3253
luca.dilandro@polimi.it
Faravelli, Tiziano .... (39) 02-23 99 3273
tiziano.faravelli@polimi.it
Gallazzi, Maria C .... (39) 02-23 99 3224
maria.gallazzi@polimi.it
Gronchi, Paolo .... (39) 02-23 99 3274
paolo.gronchi@polimi.it
Guerreri, Gianfranco .... (39) 02-23 99 3259
gianfranco.guerreri@polimi.it
Mazzocchia, Carlo .... (39) 02-23 99 3223
carlo.mazzocchia@polimi.it
Morpurgo, Armando .... (39) 02-23 99 3276
armando.morpurgo@polimi.it
Pellegrini, Laura .... (39) 02-23 99 3237
laura.pellegrini@polimi.it
Rovaglio, Maurizio .... (39) 02-23 99 3273
maurizio.rovaglio@polimi.it
Severini, Febo .... (39) 02-23 99 3227
febo.severini@polimi.it
Tieghi, Giuseppe .... (39) 02-23 99 3218
giuseppe.tieghi@polimi.it

Assistant Professors

Agrigento, Valeria .... (39) 02-23 99 3135
valeria.agrigento@polimi.ti
Bozzano, Giulia .... (39) 02-23 99 3295
giulia.bozzano@polimi.it
Cristiani, Cinzia .... (39) 02-23 99 3248
cinzia.cristiani@polimi.it

Dotelli, Giovanni .... (39) 02-23 99 3232
giovanni.dotelli@polimi.it
Giarrusso, Antonino .... (39) 02-23 99 3226
antonino.giarrusso@polimi.it
Manca, Davide .... (39) 02-23 99 3281
davide.manca@polimi.it

Department reports to: A. DeMaio, Chair of Politecmico

Japan

Fukuoka University

Department of Chemical Engineering
8-19-1 Nanakuma, Jonan-ku
Fukuoka 814-0180

University phone 81-92-871-6631
Department phone 81-92-871-6631
Department FAX 81-92-865-6031

Professors

Ide, Mitsuharu
Ishikura, Toshifumi
Kariyasaki, Akira
Morooka, Shigeharu
Nakahara, Shunsuke
Nakano, Katsuyuki
Tanaka, Ryuichi

Associate Professors

Kato, Takafumi
Mishima, Kenji
Shibata, Hiromichi

Accredited by: JABEE

Department reports to: Toshifumi Ishikura

Placement service: Katsuyuki Nakano

Himeji Institute of Technology

Graduate School of Engineering
2167 Shosha
Himeji Hyogo 671-2201

University phone 81 792 66 1661
Department phone 81 792 67 4830
Department FAX 81 792 67 4830

Professors

Fukui, Keisuke (Chair) 81 792 67 4848
fukui@cheng.eng.himeji-tech.ac.jp
Hirota, Mitsuaki 81 792 67 4842
hirota@cheng.eng.himeji-tech.ac.jp
Kanki, Tatsuo 81 792 67 4849
kanki@cheng.eng.himeji-tech.ac.jp
Kawashima, Yousuke 81 792 67 4852
kawasima@cheng.eng.himeji-tech.ac.jp
Murakami, Atsushi 81 792 67 4846
murakami@cheng.eng.himeji-tech.ac.jp

Associate Professors

Asano, Tsuyoshi 81 792 67 4850
asano@cheng.eng.himeji-tech.ac.jp
Honda, Etsurou 81 792 67 4965
pitot@mech.eng.himeji-tech.ac.jp
Kishi, Hajime 81 792 67 4843
kishi@cheng.eng.himeji-tech.ac.jp
Suzuki, Michitaka 81 792 67 4851
suzuki@cheng.eng.himeji-tech.ac.jp

Assistant Professors

Iimura, Kenji 81 792 67 4960
iimura@cheng.eng.himeji-tech.ac.jp
Maeda, Kouji 81 792 67 4854
maeda@cheng.eng.himeji-tech.ac.jp
Matsuda, Satoshi 81 792 67 4853
matsuda@cheng.eng.himeji-tech.ac.jp
Sano, Noriaki 81 792 67 4845
sano@cheng.eng.himeji-tech.ac.jp

Accredited by: Society of Chemical Engineers Japan

Graduate advisor: Keisuke Fukui, Prof.

Undergraduate advisor: Suzuki Michitaka Dr.

Student organization: Kengo Iwatani
Advisor: Kanki Tatsuo Prof.

Placement service: Kanki Tatsuo, Prof.

Hiroshima University

Department of Chemical Engineering
Graduate School of Engineering
Kagamiyama 1-4-1
Higashi-Hiroshima, 739-8527

Department phone .... (81)82-424-7718
Department FAX .... (81)82-424-5494

Professors

Asaeda, Masashi .... (81)82-424-7719
asaeda@hiroshima-u.ac.jp
Okada, Mitsumasa 24-5526
mokada@hiroshima-u.ac.jp
Okuyama, Kikuo (Chair) .... (81)82-424-7716
okuyama@hiroshima-u.ac.jp
Sakohara, Shuji .... (81)82-424-7720
sakohara@hiroshima-u.ac.jp
Takishima, Shigeki .... (81)82-424-7713
r736735@hiroshima-u.ac.jp
Yoshida, Hideto .... (81)82-424-7853
r736619@hiroshima-u.ac.jp

Associate Professors

Iizawa, Takashi .... (81)82-424-7711
tiizawa@hiroshima-u.ac.jp
Isomoto, Yoshinori .... (81)82-424-7845
iyoshi@hiroshima-u.ac.jp
Kitamura, Mitsutaka .... (81)82-424-7715
mkitamu@hiroshima-u.ac.jp
Nishijima, Wataru 24-6199
wataru@hiroshima-u.ac.jp
Shimada, Manabu .... (81)82-424-7717
smd@hiroshima-u.ac.jp
Takimoto, Kazuto .... (81)82-424-7621
takimoto@hiroshima-u.ac.jp
Tsuru, Toshinori .... (81)82-424-7714
tsuru@hiroshima-u.ac.jp

Assistant Professors

Fukui, Kunihiro .... (81)82-424-7853
kfukui@hiroshima-u.ac.jp
Gotoh, Takehiko .... (81)82-424-7720
tgoto@hiroshima-u.ac.jp
Lenggoro, Wuled .... (81)82-424-7850
wld@hiroshima-u.ac.jp
Mukai, Tetsuo .... (81)82-424-7626
tmukai@hiroshima-u.ac.jp
Nakano, Yoichi 24-6195
ynakano@hiroshima-u.ac.jp
Okuda, Tetsuji 24-6197
aqua@hiroshima-u.ac.jp
Tokuyama, Hideaki .... (81)82-424-7720
tokuyama@hiroshima-u.ac.jp
Tsumura, Toshinori .... (81)82-424-7723
ttsumura@hiroshima-u.ac.jp
Yabuki, Akihiro .... (81)82-424-7852
ayabuki@hiroshima-u.ac.jp
Yang, Kwan-sik .... (81)82-424-7853
yangks@hiroshima-u.ac.jp
Yoshioka, Tomohisa .... (81)82-424-7719
tom@hiroshima-u.ac.jp

Accredited by:

Degrees granted 2003-2004:

B.S.: 34 M.S.: 31 Ph.D.: 11

Kanazawa University

Engineering Campus
Chemistry and Chemical Engineering Department
40-20, Kodatsuno 2-chome
Kanazawa, 920-8667

University phone 076-264-5111
Department phone 076-234-4830
Department FAX 076-234-4829

Professors

Emi, Hitoshi (Emeritus) 234-4834
emi@t.kanazawa-u.ac.jp
Hayashi, Yoshishige 234-4806
yohayasi@t.kanazawa-u.ac.jp
Komura, Teruhisa 234-4770
komura@t.kanazawa-u.ac.jp
Kunimoto, Kouki 264-5805
lee@t.kanazawa-u.ac.jp
Miyagishi, Shigeyoshi 234-4764
miyagisi@t.kanazawa-u.ac.jp
Mori, Shigeru 234-4825
smori@t.kanazawa-u.ac.jp
Motoi, Masatoshi 234-4781
motoi@t.kanazawa-u.ac.jp
Nakamoto, Yoshiaki 234-4774
nakamoto@t.kanazawa-u.ac.jp
Ohta, Tatsuhiko 234-4803
tohta@t.kanazawa-u.ac.jp
Ohtani, Yoshio 234-4813
otani@t.kanazawa-u.ac.jp
Senda, Hitoshi 264-5802
senda@t.kanazawa-u.ac.jp
Shimizu, Nobuaki 234-4807
nshimizu@t.kanazawa-u.ac.jp
Tamura, Kazuhiro (Head) 234-4804
tamura@t.kanazawa-u.ac.jp
Ueda, Kazumasa 234-4791
kueda@t.kanazawa-u.ac.jp
Yamada, Toshiro 234-4802
tyamada@t.kanazawa-u.ac.jp

Associate Professors

Asakawa, Tsuyoshi 234-4765
asakawa@t.kanazawa-u.ac.jp
Hasegawa, Hiroshi 234-4792
hhiroshi@t.kanazawa-u.ac.jp
Kanoh, Shigeyoshi 234-4782
kanoh@t.kanazawa-u.ac.jp
Kawanishi, Takuya 234-4809
kawanisi@t.kanazawa-u.ac.jp
Nakamura, Yoshitoshi 234-4819
ynakamu@t.kanazawa-u.ac.jp
Namiki, Norikazu 234-4815
namiki@t.kanazawa-u.ac.jp
Nitta, Koh-hei 234-4818
nitta@t.kanazawa-u.ac.jp
Segi, Masahito 234-4787
segi@t.kanazawa-u.ac.jp
Takahashi, Kenji 234-4828
ktkenji@t.kanazawa-u.ac.jp
Takahashi, Koushin 234-4771
ktakaha@t.kanazawa-u.ac-jp
Yamagishi, Tadaaki 234-4776
polymer@kenroku.kanazawa-u.ac.jp

Department reports to: Chairman of Department

Placement service: Chairman of Department

Kansai University

Faculty of Engineering
Chemical Engineering Department
Suita Osaka 564-8680

University phone 81-6-6368-1121
Department phone 81-6-6368-1121
Department FAX 81-6-6388-8869

Professors

Miyake, Takanori (Chair) .... 81-6-6368-0918†
tmiyake@ipcku.kansai-u.ac.jp
Miyake, Yoshikazu 81-6-6368-0950
ymiyake@kansai-u.ac.jp
Muroyama, Katsuhiko 81-6-6368-0945
muroyama@kansai-u.ac.jp
Oda, Hirokazu 81-6-6368-0808
oda@ipcku.kansai-u.ac.jp
Shibata, Junji 81-6-6368-0856
shibata@kansai-u.ac.jp
Suzuki, Toshimitsu 81-6-6368-0865
tsuzuki@ipcku.kansai-u.ac.jp
Yamamoto, Hideki 81-6-6368-0972
yhideki@kansai-u.ac.jp

Associate Professors

Hayashi, Junichi 81-6-6368-0913
hayashi7@ipcku.kansai-u.ac.jp
Ikenaga, Naiki 81-6-6368-0792
ikenaga@ipcku.kansai-u.ac.jp
Iyoki, Shigeki 81-6-6368-0786
iyoki@ipcku.kansai-u.ac.jp

† Dial 81-6-368-1121 and ask for extension.

Accredited by:

Degrees granted 2003-2004:

B.S.: 104 M.S.: 34 Ph.D.: 4

Graduate advisor: Takanori MIyake

Undergraduate advisor: Takanori MIyake

Department reports to: Takanori MIyake

Kyoto University

Department of Chemical Engineering
Faculty of Engineering
Kyoto University Katsura, Nishikyo-ku
Kyoto, 615-8510

University phone 81-75-753-7531
Department phone 81-75-383-2071
Department FAX 81-75-383-2078

Professors

Adachi, Motonari (Joint) 81-774-38-3518
adachi@iae.kyoto-u.ac.jp
Harada, Makoto (Emeritus)
Hasebe, Shinji 81-75-383-2667
hasebe@cheme.kyoto-u.ac.jp
Hashimoto, Iori (Emeritus)
iori@cheme.kyoto-u.ac.jp
Hashimoto, Kenji (Emeritus)
Higashitani, Ko (Head) 81-75-383-2662
k_higa@cheme.kyoto-u.ac.jp
Mae, Kazuhiro 81-75-383-2668
kaz@cheme.kyoto-u.ac.jp
Masuda, Hiroaki 81-75-383-2665
masuda@cheme.kyoto-u.ac.jp
Miura, Kouichi 81-75-383-2663
miura@cheme.kyoto-u.ac.jp
Ogino, Fumimaru (Emeritus)
Ohshima, Masahiro 81-75-383-2666
oshima@cheme.kyoto-u.ac.jp
Okazaki, Morio (Emeritus)
Sada, Eizo (Emeritus)
Takamatsu, Takeichiro (Emeritus)
Tamon, Hajime 81-75-383-2664
tamon@cheme.kyoto-u.ac.jp
Tanigaki, Masataka (Joint) 81-75-753-5566
tanigaki@cheme.kyoto-u.ac.jp
Yoshida, Fumitake (Emeritus)

Associate Professors

Kano, Manabu 81-75-383-2687
kano@cheme.kyoto-u.ac.jp
Kawase, Motoaki 81-75-383-2683
kawase@cheme.kyoto-u.ac.jp
Kinoshita, Masahiro (Joint) 81-774-38-3503
kinoshit@iae.kyoto-u.ac.jp
Maruyama, Toshiro 81-75-383-2631
maruyama@cheme.kyoto-u.ac.jp
Matsusaka, Shuji 81-75-383-2685
matsu@cheme.kyoto-u.ac.jp
Miyahara, Minoru 81-75-383-2682
miyahara@cheme.kyoto-u.ac.jp
Mukai, Shin 81-75-383-2684
mukai@cheme.kyoto-u.ac.jp

Accredited by:

Degrees granted 2003-2004:

B.S.: 51 M.S.: 25 Ph.D.: 4

Kyushu University

Department of Chemical Engineering
Faculty of Engineering
Hakozaki, Higashiku
Fukuoka, 812-8581

University phone 81-92-642-2111
Department phone 81-92-642-3540
Department FAX 81-92-642-3539

Professors

Arai, Yasuhiko 81-92-642-3495
arai@chem-eng.kyushu-u.ac.jp
Fukai, Jun 81-92-642-3515
jfukai@chem-eng.kyushu-u.ac.jp
Funatsu, Kazumori 81-92-642-3508
funatsu@chem-eng.kyushu-u.ac.jp
Goto, Masahiro 81-92-642-3575
mgototcm@mbox.nc.kyushu-u.ac.jp
Kawakami, Koei 81-92-642-4127
kawakami@chem-eng.kyushu-u.ac.jp
Kusunoki, Koichiro (Emeritus)
Matsuyama, Hisayoshi (Emeritus) 81-92-642-3532
Minemoto, Masaki 81-92-642-3521
minemoto@chem-eng.kyushu-u.ac.jp
Miyatake, Osamu (Emeritus)
Miyazaki, Noriyuki 81-92-642-3526
miyazaki@chem-eng.kyushu-u.ac.jp
Morooka, Shigeharu (Emeritus)
Nakashio, Fumiyuki (Emeritus)
Shinohara, Hisashi (Emeritus)
Wakabayashi, Katsuhiko (Emeritus)

Associate Professors

Ijima, Hiroyuki 81-92-642-4114
ijima@chem-eng.kyushu-u.ac.jp
Ikeda, Toru 81-92-642-3527
ikeda@chem-eng.kyushu-u.ac.jp
Iwai, Yoshio 81-92-642-3496
iwai@chem-eng.kyushu-u.ac.jp
Kajiwara, Toshihisa 81-92-642-3509
kajiwara@chem-eng.kyushu-u.ac.jp
Kamiya, Noriho 81-92-642-3577
noritcm@mbox.nc.kyushu-u.ac.jp
Kishida, Masahiro 81-92-642-3505
kishida@chem-eng.kyushu-u.ac.jp
Kusakabe, Katsuki 81-92-642-3552
Matsukuma, Yosuke 81-92-642-3520
ymatsu@chem-eng.kyushu-u.ac.jp
Tsuge, Yoshifumi 81-92-642-3533
tsuge@chem-eng.kyushu-u.ac.jp

Nagoya University

Department of Chemical Engineering
Furo-cho, Chikusa-ku
Nagoya, 464-8603

University phone 052-781-5111
Department phone 052-789-4822
Department FAX 052-789-3180

Professors

Goto, Shigeo 052-789-3261
goto@park.nuce.nagoya-u.ac.jp

Hasatani, Masanobu 052-789-3381
hasatani@nuce.nagoya-u.ac.jp
Iritani, Eiji 052-789-3374
iritani@nuce.nagoya-u.ac.jp
Katagiri, Haruo 052-789-3911
h-kata@nuce.nagoya-u.ac.jp
Koda, Shinobu 052-789-3275
koda@nuce.nagoya-u.ac.jp
Matsubara, Masakazu (Emeritus)
Matsuda, Hitoki 052-789-3382
matsuda@nuce.nagoya-u.ac.jp
Miyahara, Yutaka (Emeritus)
Mori, Shigekatsu 052-789-3377
mori@nuce.nagoya-u.ac.jp
Morisue, Toshiya (Emeritus)
Murase, Toshiro (Emeritus)
Nakamura, Masaaki (Chair) 052-789-3618
nakamura@nuce.nagoya-u.ac.jp
Nomura, Hiroyasu (Emeritus)
Onogi, Katsuaki 052-789-3263
onogi@nuce.nagoya-u.ac.jp
Shirato, Mompei (Emeritus)
Sugiyama, Sachio (Emeritus)
Takahashi, Katsuroku 052-789-3391
takaroku@nuce.nagoya-u.ac.jp
Takeuchi, Hiroshi (Emeritus)
Toyama, Shigeki (Emeritus)
Tsubaki, Junichiro 052-789-3096
tsubaki@nuce.nagoya-u.ac.jp

Associate Professors

Bando, Yoshiyuki 052-789-3622
bando@nuce.nagoya-u.ac.jp
Deguchi, Seiichi 052-789-2733
deguchi@nuce.nagoya-u.ac.jp
Hashizume, Susumu 052-789-3594
hashi@nuce.nagoya-u.ac.jp
Itaya, Yoshinori 052-789-3378
yitaya@nuce.nagoya-u.ac.jp
Kawaizumi, Fumio 052-789-3392
kawaizum@nuce.nagoya-u.ac.jp
Kobayashi, Noriyuki 052-789-3383
koba@nuce.nagoya-u.ac.jp
Kurimoto, Hidekazu 052-789-4757
kurimoto@nuce.nagoya-u.ac.jp
Matsuoka, Tatsuro 052-789-3274
matsuoka@nuce.nagoya-u.ac.jp
Mori, Hidetoshi 052-789-2568
hmori@nuce.nagoya-u.ac.jp
Mukai, Yasuhito 052-789-3375
mukai@nuce.nagoya-u.ac.jp
Nii, Susumu 052-789-3390
nii@nuce.nagoya-u.ac.jp
Tagawa, Tomohiko 052-789-3388
tagawa@park.nuce.nagoya-u.ac.jp
Yasuda, Keiji 052-789-3623
yasuda@nuce.nagoya-u.ac.jp

Niigata University

Department of Chemistry and Chemical Engineering
Ikarashi 2-8050
Niigata ZC 950-2181

University phone 81 25 223 6161
Department phone 81 25 262 6781
Department FAX 81 25 262 7010

Professors

Ohkawa, Akira .... x6786 †
biochem@eng.niigata-u.ac.jp
Taguchi, Yoji (Head) x6787
taguchi1@eng.niigata-u.ac.jp
Tanaka, Masato x6791
tanaka@eng.niigata-u.ac.jp
Watanabe, Atsuo x7958
watanabe@eng.niigata-u.ac.jp
Yamagiwa, Kazuaki x6785
yamagiwa@eng.niigata-u.ac.jp

Associate Professors

Hotta, Noriyasu x6782
nhotta@eng.niigata-u.ac.jp
Ito, Akira x6789
aito@eng.niigata-u.ac.jp
Kimura, Isao x7194
ikim@eng.niigata-u.ac.jp
Shimizu, Tadaaki x6783
tshimizu@eng.niigata-u.ac.jp

Assistant Professors

Taguchi, Yoshinari x6784
puchi@eng.niigata-u.ac.jp
Yoshida, Masanori x7343
myoshida@eng.niigata-u.ac.jp

† Dial 025-262-6098 and ask for extension.

Accredited by:

Degrees granted 2003-2004:

B.S.: 42 M.S.: 23 Ph.D.: 1

Osaka Prefecture University

Chemical Engineering Department
1-1, Gakuen-cho, Sakai
Osaka, 599-8531

University phone 81-722-52-1161
Department phone 81-722-54-9297
Department FAX 81-722-54-9911

Professors

Asai, Satoru (Emeritus)
Ishikawa, Haruo 0722-54-9296
ishikawa@chemeng.osakafu-u.ac.jp
Kataoka, Takeshi (Emeritus)
Konishi, Yasuhiro (Chair) 0722-54-9297
yasuhiro@chemeng.osakafu-u.ac.jp
Kousaka, Yasuo (Emeritus)
Miyanami, Kei (Emeritus)
Nishiki, Tadaaki 0722-54-9302
nishiki@chemeng.osakafu-u.ac.jp
Satoh, Munetake 0722-54-9304
musatoh@chemeng.osakafu-u.ac.jp
Yano, Takeo (Emeritus)
Yoshida, Hiroyuki 0722-54-9298
yoshida@chemeng.osakafu-u.ac.jp

Associate Professors

Akiyama, Tomohiro 0722-54-9303
takiyama@chemeng.osakafu-u.ac.jp
Ishimi, Kousaku 0722-54-9299
ishimi@chemeng.osakafu-u.ac.jp

Niida, Tohru 0722-54-9300
niida@chemeng.osakafu-u.ac.jp
Terashima, Masaaki 0722-54-9301
terashima@chemeng.osakafu-u.ac.jp
Watano, Satoru 0722-54-9305
watano@chemeng.osakafu-u.ac.jp

Assistant Professors

Nakamura, Hidemi 0722-54-9306
nakamura@chemeng.osakafu-u.ac.jp
Ogino, Hiroyasu 0722-54-9299
ogino@chemeng.osakafu-u.ac.jp
Terashita, Keijiro 0722-54-9305
terasita@chemeng.osakafu-u.ac.jp

Accredited by: Ministry of Education Japan

Graduate advisor: Konishi, Yasahiro

Placement service: Konishi, Yasahiro

Osaka University

Division of Chemical Engineering
Department of Materials Engineering Science
Graduate School of Engineering Science
Toyonaka
Osaka 560-8531

University phone........................... 06-6850-6111
Department phone 06-6850-6295
Department FAX 06-6850-6296

Professors

Fueno, Takayuki (Emeritus)
Hirata, Yushi 81-6-6850-6275
hirata@cheng.es.osaka-u.ac.jp
Imanaka, Toshinobu (Emeritus)
Inoue, Yoshiro 81-6-6850-6276
inoue@cheng.es.osaka-u.ac.jp
Ito, Ryuzo (Emeritus)
Kaneda, Kiyotomi 81-6-6850-6260
kaneda@cheng.es.osaka-u.ac.jp
Katayama, Takashi (Emeritus)
Komasawa, Isao (Emeritus)
Kuboi, Ryoichi 81-6-6850-6285
kuboi@cheng.es.osaka-u.ac.jp
Kunugita, Eiichi (Emeritus)
Nitta, Tomoshige 81-6-6850-6265
nitta@cheng.es.osaka-u.ac.jp
Ohgaki, Kazunari (Chair) 81-6-6850-6290
ohgaki@cheng.es.osaka-u.ac.jp
Otake, Tsutao (Emeritus)
Taya, Masahito 81-6-6850-6251
taya@cheng.es.osaka-u.ac.jp
Tone, Setsuji (Emeritus)
Ueyama, Korekazu 81-6-6850-6255
ueyama@cheng.es.osaka-u.ac.jp

Associate Professors

Ebitani, Kohki 81-6-6850-6262
ebitani@cheng.es.osaka-u.ac.jp
Egashira, Yasuyuki 81-6-6850-6259
egashira@cheng.es.osaka-u.ac.jp
Kino-oka, Masahiro 81-6-6850-6253
kino-oka@cheng.es.osaka-u.ac.jp
Nakano, Masayoshi 81-6-6850-6268
mnaka@cheng.es.osaka-u.ac.jp
Nishiyama, Norikazu 81-6-6850-6256
nisiyama@cheng.es.osaka-u.ac.jp
Sato, Hiroshi 81-6-6850-6291
hsato@cheng.es.osaka-u.ac.jp
Takahashi, Hideaki 81-6-6850-6267
takahasi@cheng.es.osaka-u.ac.jp
Umakoshi, Hiroshi 81-6-6850-6287
umakoshi@cheng.es.osaka-u.ac.jp

Graduate advisor: Ohgaki, Kazunari

Undergraduate advisor: Ohgaki, Kazunari

Placement service: Chairman, Division of Chemical Engineering

Seikei University

Department of Applied Chemistry
Seikei University
Musashino, Tokyo 180-8633
Deliveries: Prof. T. Kojima

University phone......................... 81-422-37-3705
Department phone 81-422-37-3750
Department FAX 81-422-37-3871

Professors

Higuchi, Akon 0422-37-3748
higuchi@ch.seikei.ac.jp
Katoh, Akira 0422-37-3747
katoh@ch.seikei.ac.jp
Kojima, Toshinori 0422-37-3750
kojima@ch.seikei.ac.jp
Kurita, Keisuke 0422-37-3745
kurita@ch.seikei.ac.jp
Morita, Makoto 0422-37-3749
morita@ch.seikei.ac.jp
Ozaki, Yoshiharu 0422-37-3751
ozaki@ch.seikei.ac.jp
Totani, Yoichiro 0422-37-3744
totani@ch.seikei.ac.jp
Tsubomura, Taro (Head) 0422-37-3752
tsubo@ch.seikei.ac.jp

Associate Professors

Hara, Setsuko 0422-37-3753
shara@ch.seikei.ac.jp

Accredited by: Ministry of Education Science and Culture

Degrees granted 2003-2004:

B.S.: 92 M.S.: 21 Ph.D.: 1

Graduate advisor: Prof. A. Higuchi

Undergraduate advisor: Prof. A. Katoh

Tohoku University

Department of Chemical Engineering
Aoba-yama 07,Aoba-ku Sendai 980-8579
Japan

University phone......................... 81-22-217-4844
Department phone 81-22-217-7255
Department FAX 81-22-217-7293

Professors

Arai, Kunio .... 217-7245 †
karai@arai.scw.che.tohoku.ac.jp
Hozawa, Mitsunori 217-5649
hozawa@icrs.tohoku.ac.jp
Inomata, Hiroshi 217-7283
inomata@scf.che.tohoku.ac.jp
Konno, Mikio 217-7239
konno@mickey.che.tohoku.ac.jp
Matsumoto, Shigeru (Chair) 217-7259
smatsu@pse.che.tohoku.ac.jp
Miura, Takatoshi 217-7250
miura@tranpo.che.tohoku.ac.jp
Saito, Shozaburo (Emeritus)
Suzuki, Mutsumi 217-7265
msuzuki@pse.che.tohoku.ac.jp
Tadaki, Teiriki (Emeritus)
Takahashi, Shinji (Emeritus)
Yokoyama, Chiaki 217-5647
chiaki@icrs.tohoku.ac.jp
Yonemoto, Toshikuni 217-7255
toshiy@rpel.che.tohoku.ac.jp

Associate Professors

Adschiri, Tadafumi 217-7246
ajiri@arai.scw.che.tohoku.ac.jp
Aoki, Hideyuki 217-7251
aoki@tranpo.che.tohoku.ac.jp
Smith, Richard L. 217-7247
smith@scf.che.tohoku.ac.jp
Tsukada, Takao 217-5650
tsukada@icrs.tohoku.ac.jp
Yamashita, Yoshiyuki 217-7266
yyama@pse.che.tohoku.ac.jp

† Faculty numbers have prefix 81-22.

Department reports to: Toshikuni Yonemoto

Placement service: Toshikuni Yonemoto

University of Tokyo

Department of Chemical System Engineering
Hongo, Bunkyo-ku, Tokyo 113-8656

University phone 03-3812-2111
Department FAX 03-5841-7321

Professors

Hirano, Toshisuke (Emeritus)
Iizuka, Yoshinori 5841-7298
Inoue, Hakuai (Emeritus)
Kawazoe, Kunitaro (Emeritus)
Kimura, Shoji (Emeritus)
Koda, Seiichiro (Emeritus)
Komiyama, Hiroshi 5841-7323
Koshi, Mitsuo 5841-7295
Kume, Hitoshi (Emeritus)
Kunii, Daizo (Emeritus)
Nakao, Shin-ichi (Chair) 5841-7326
Nishimura, Hajime (Emeritus)
Oshima, Yoshito 5841-3027
Sadakata, Masayoshi 5841-7344
Sakoda, Akiyoshi 5452-6350
Sekizawa, Ai 5841-7287
Takahashi, Hiroshi 5841-7751
Tamura, Masamitsu 5841-7291
Yamaguchi, Yukio 5841-7303
Yamamoto, Kenji 5841-7303
Yamashita, Koichi 5841-7228
Yanagisawa, Yukio 5841-7324
Yoshida, Kunio (Emeritus)

Associate Professors

Akutsu, Yoshiaki 5841-7384
Arai, Mitsuru 5841-7292
Asai, Yosihiro 029-861-5373
Dobashi, Ritsu 5841-7304
Hirao, Masahiko 5841-7343
Hori, Keiichi 0427-59-8283
Miyoshi, Akira 5841-7296
Okada, Fumio 5841-7352
Okubo, Tatsuya 5841-7348
Sakai, Yasuyuki 5452-6352
Tsutsumi, Atsushi 5841-7336
Yamaguchi, Takeo 5841-7345
Yoshinaga, Jun 5841-8859

Assistant Professors

Kraines, Steven 5841-8837

Tokyo Institute of Technology

Department of Chemical Engineering
2-12-1, O-okayama, Meguro-ku,
Tokyo 152-8552

University phone 81-3-3726-1111
Department phone 81-3-5734-2475
Department FAX 81-3-5734-2475

Professors

Kawasaki, Junjiro .... 5734-3285 †
jkawasak@chemeng.titech.ac.jp
Kuroda, Chiaki 5734-2115
ckuroda@chemeng.titech.ac.jp
Masuko, Masabumi 5734-3036
mmasuko@chemeng.titech.ac.jp
Niiyama, Hiroo 5734-2121
hniiyama@chemeng.titech.ac.jp
Ogawa, Kohei 5734-2117
kogawa@chemeng.titech.ac.jp
Ohtaguchi, Kazuhisa 5734-2113
kotaguch@chemeng.titech.ac.jp
Suzuki, Masaaki (Head) 5734-2112
masaaki@chemeng.titech.ac.jp
Tsuda, Ken 5734-2116
ktsuda@chemeng.titech.ac.jp

Associate Professors

Aida, Takashi 5734-2883
taida@chemeng.titech.ac.jp
Fuchino, Tetsuo 5734-2474
fuchino@chemeng.titech.ac.jp
Kosuge, Hitoshi 5734-2151
hkosuge@chemeng.titech.ac.jp
Kubouchi, Masatoshi 5734-2119
mkubouch@chemeng.titech.ac.jp
Sekiguchi, Hidetoshi 5734-2110
hsekiguc@chemeng.titech.ac.jp
Taniguchi, Izumi 5734-2155
itaniguc@chemeng.titech.ac.jp
Yoshikawa, Shiro 5734-3278
syoshika@chemeng.titech.ac.jp

Assistant Professors

Asami, Kazuhiro 5734-3034
kasami@chemeng.titech.ac.jp

Iwakabe, Koichi 5734-3287
kiwakabe@chemeng.titech.ac.jp

Matsumoto, Hideyuki 5734-2115
hmatsumo@ chemeng.titech.ac.jp

Mori, Shinsuke 5734-2629
smori@chemeng.titech.ac.jp

Ookawara, Shinichi 5734-3035
sokawara@chemeng.titech.ac.jp

Sakai, Tetsuya 5734-2124
tsakai@chemeng.titech.ac.jp

Suzuki, Akihito 5734-2628
asuzuki@chemeng.titech.ac.jp

† Faculty numbers have prefix 81-3.

Accredited by:

Degrees granted 2003-2004:

B.S.: 32 M.S.: 33 Ph.D.: 3

Graduate advisor: Suzuki, Masaaki

Undergraduate advisor: Fuchino, Tetsuo

Placement service: Kawasaki, Junjiro

Toyama University

Chemical Process Engineering Department
3190, Gofuku
Toyama, 930-8555

Department FAX .... (81) (76) 445-6697

Professors

Kumazawa, Hidehiro 445-6859
kumazawa@eng.toyama-u.ac.jp

Morohashi, Syoichi (Chair) 445-6856
morohasi@eng.toyama-u.ac.jp

Yamamoto, Ken-ichi 445-6831
kyamamot@eng.toyama-u.ac.jp

Yamazaki, Ryouhei 445-6884
ryamaza@eng.toyama-u.ac.jp

Associate Professors

Hoshino, Kazuhiro 445-6857
khoshino@eng.toyama-u.ac.jp

Kawasaki, Hiroyuki 445-6860
hkawasak@eng.toyama-u.ac.jp

Kurooka, Taketoshi 445-6829
kurooka@eng.toyama-u.ac.jp

Takase, Hitoshi 445-6830
htakase@eng.toyama-u.ac.jp

Yoshida, Masamichi 445-6836
yoshida@parrot.eng.toyama-u.ac.jp

Yamagata University

School of Chemical Engineering
4-3-16 Jonan, Yonezawa
Yamagata, Japan 992-8510

Department phone ....(81) 238-26-3160
Department FAX ....(81) 238-26-3414

Professors

Hasegawa, Masahiro (Chair) ....(81) 238-26-3162
mhase@yz.yamagata-u.ac.jp

Kanda, Yoshiteru ....(81) 238-26-3161
kanda@yz.yamagata-u.ac.jp

Kuriyama, Masafumi ....(81) 238-26-3150
kuri@yz.yamagata-u.ac.jp

Sato, Shimio ....(81) 238-26-3145
shimio@yz.yamagata-u.ac.jp

Takahashi, Koji ....(81) 238-26-3156
koji@yz.yamagata-u.ac.jp

Toda, Masayuki ....(81) 238-26-3165
toda@yz.yamagata-u.ac.jp

Yokota, Toshiyuki ....(81) 238-26-3155
yokota@yz.yamagata-u.ac.jp

Associate Professors

Aita, Tadahiro ....(81) 238-26-3146
aita@yz.yamagata-u.ac.jp

Katsuyama, Tetsuo ....(81) 238-26-3131
katsuma@yz.yamagata-u.ac.jp

Shioi, Akihisa ....(81) 238-26-3164
shioi@yz.yamagata-u.ac.jp

Shishido, Masahiro ....(81) 238-26-3166
sisido@yz.yamagata-u.ac.jp

Department reports to: Prof. M. Haswgawa, Chair of Chemical Engineering

Placement service: Ms. H. Ando

Yamaguchi University

Applied Chemistry and
Chemical Engineering Department
Tokiwadai 2-16-1, Ube 755-8611

University phone 81-836-85-9005
Department phone 81-836-85-9200
Department FAX 81-836-85-9201

Professors

Kimura, Makoto (Emeritus)

Kumada, Makoto 836-85-9235
kumada@po.cc.yamaguchi-u.ac.jp

Nakao, Katsumi 836-85-9271
knakao@po.cc.yamaguchi-u.ac.jp

Nishimura, Tatsuo 836-85-9121
tnishimu@po.cc.yamaguchi-u.ac.jp

Nishizawa, Yoshinori 836-85-9291
nisizaw@po.cc.yamaguchi-u.ac.jp

Okamoto, Kenichi 836-85-9660
okamotok@po.cc.yamaguchi-u.ac.jp

Osasa, Kunihisa 836-85-9251
osasa@po.cc.yamaguchi-u.ac.jp

Susumu, Tsuchiya (Emeritus)

Yamamoto, Shuichi 836-85-9241
shu-yama@po.cc.yamaguchi-u.ac.jp

Associate Professors

Akada, Rinji 836-85-9292
rinji@po.cc.yamaguchi-u.ac.jp

Fukunaga, Kimitoshi 836-85-9272
fukun@po.cc.yamaguchi-u.ac.jp

Gotoh, Kuniaki 836-85-9122
kgotoh@po.cc.yamaguchi-u.ac.jp

Kita, Hidetoshi 836-85-9661
kita@po.cc.yamaguchi-u.ac.jp

Kobuchi, Sigetoshi	836-85-9236
	kobuchi@po.cc.yamaguchi-u.ac.jp
Nakakura, Hideo	836-85-9252
	nakakura@po.cc.yamaguchi-u.ac.jp
Tanaka, Kazuhiro	836-85-9662
	tnkkzhr@po.cc.yamaguchi-u.ac.jp

Assistant Professors

Hoshida, Hisashi	836-85-9293
	hoshida@po.cc.yamaguchi-u.ac.jp
Saeki, Takashi	836-85-9242
	saeki@po.cc.yamaguchi-u.ac.jp
Yoshimoto, Makoto	836-85-9273
	yosimoto@po.cc.yamaguchi-u.ac.jp
Yoshino, Makoto	836-85-9662
	yoshinom@po.cc.yamaguchi-u.ac.jp

Department reports to: Dean of Engineering

Yokohama National University

Department of Chemical Engineering
79-5 Tokiwadai, Hodogaya-ku
Yokohama, 240-8501

Department phone	45-339-3991
Department FAX	45-339-4012

Professors

Habuka, Hitoshi	339-3998 †
Hara, Takao	339-4189
Iida, Yoshihiro (Chair)	
Kaminoyama, Meguru	339-3999
	kaminoyama@chemeng.bsk.ynu.ac.jp
Kamiwano, Mitsuo (Emeritus)	339-3988
Koizumi, Jun-ichi	339-4266
	jikoizmi@ynu.ac.jp
Kunieda, Hironobu	339-4190
Matsumoto, Kanji	339-4008
Ohya, Haruhiko (Emeritus)	339-3989
Okuyama, Kunito	339-4009
Tanisho, Shigeharu (Joint)	339-3996
Watanabe, Masatoshi	339-3997

Associate Professors

Aihara, Masahiko	339-4229
Nishi, Kazuhiko	339-3988
Takeda, Minoru	339-4566
Tsunakawa, Hiroshi	339-4006

Assistant Professors

Aramaki, Kenji	339-4190
Hirata, Yoshikuni	339-3976
Misumi, Ryuta	339-3995
Nakamura, Kazuho	339-3980
Suzuki, Ichiro	339-4266
Takeuchi, Takashi	339-3975

† Phone numbers have prefix 81-45.

Accredited by: JABEE

Graduate advisor: Minoru Takeda

Undergraduate advisor: Kazuhiko Nishi

Republic of Korea

Chonnam National University

Chemical Engineering Department
Kwang-ju, 500-757

University phone	(062) 530-0114
Department phone	(062) 530-1850
Department FAX	(062) 530-1869

Professors

Cho, Chong-Hyun (Emeritus)	530-1850 †
Chough, Sung-Hyo	530-1825
Han, Neung-Won (Emeritus)	530-1850
Kang, Choon-Hyoung (Chair)	530-1818
Kang, Sung J.	530-1817
Kim, Jae-Hyung (Emeritus)	530-1850
Kim, Jae-Seung (Emeritus)	530-1850
Kim, Jin-Hwan	530-1815
Lee, Ki-Young	530-1843
Lee, Woo-Tai	530-1813
Park, Don-Hee	530-1841
Park, Nam-Cook	530-1826
Ryu, Hwa-Won	530-1842
Ryu, Min-Su	530-1816
Shin, Jae-Soon	530-1827
Sunwoo, Chang-Shin	530-1814
Yang, Jae-Ho	530-1812
Yu, Eui-Yeon	530-1811

Associate Professors

Kim, Do-Man	530-1844
Kim, Young-Chul	530-1828
Rhee, Jong-Il	530-1847

Assistant Professors

Han, Eun-Mi	530-1829

† Direct dial with prefix 062.

Accredited by: Ministry of Education Korea

Keimyung University

Chemical Engineering Department
1000 Sindangdong Dalseo-gu Taegu
704-701

University phone	81-53-580-5114
Department phone	81-53-580-5042
Department FAX	81-53-581-7686

Professors

Lee, Chun-Soo	81-53-580-5235
Kim, Jong-Sik	81-53-580-5236
Synn, Dong-Su	81-53-580-5238

Associate Professors

Choe, Seok-Bum	81-53-580-5239
Ha, KiRyong (Chief)	81-53-580-5263

Assistant Professors

Rho, Seung-Baik	81-53-580-5237
	rhosb@kmucc.keimyung.ac.kr
Suh, Soong-Hyuck	81-53-580-5545
Byun, Hong-Sik	81-53-580-5569

Korea Advanced Institute of Science and Technology

Department of Chemical and Biomolecular Engineering
373-1 Guseong-dong, Yuseong-gu
Daejeon 305-701, Republic of Korea

University phone 82-42-869-2114
Department phone 82-42-869-3902 4
Department FAX 82-42-869-3910

Professors

Chang, Ho N. 82-42-869-3912
hnchang@kaist.ac.kr
Chang, Yong K. 82-42-869-3927
ychang@kaist.ac.kr
Chung, In J. 82-42-869-3916
chung@kaist.ac.kr
Ham, Mooyoung (Visiting) 82-42-869-3990
mham@kaist.ac.kr
Hong, Won-Hi 82-42-869-3919
whhong@kaist.ac.kr
Ihm, Son-Ki 82-42-869-3915
skihm@kaist.ac.kr
Kim, Do H. 82-42-869-3929
DoHyun.Kim@kaist.ac.kr
Kim, Jong-Duk (Director) 82-42-869-3921
jdkim@kaist.ac.kr
Kim, Sang D. (Director) 82-42-869-3913
kimsd@kaist.ac.kr
Kim, Sung C. (Director) 82-42-869-3914
kimsc@kaist.ac.kr
Lee, Huen 82-42-869-3917
h_lee@kaist.ac.kr
Lee, Sang Y. (Director) 82-42-869-3930
leesy@kaist.ac.kr
Lee, Tai-yong 82-42-869-3926
tylee@kaist.ac.kr
Lee, Won K. (Emeritus) 82-42-869-3911
wklee@kaist.ac.kr
Park, Jung-Ki 82-42-869-3925
jungpark@kaist.ac.kr
Park, O O. (Dean) 82-42-869-3923
ookpark@kaist.ac.kr
Park, Seung-Bin 82-42-869-3928
SeungBinPark@kaist.ac.kr
Park, Sunwon 82-42-869-3920
sunwon@kaist.ac.kr
Woo, Seong I. (Director) 82-42-869-3918
siwoo@kaist.ac.kr
Yang, Ji-Won (Chair) 82-42-869-3924
jwyang@kaist.ac.kr
Yang, Seung-Man 82-42-869-3922
smyang@kaist.ac.kr

Associate Professors

Jung, Hee-Tae 82-42-869-3931
heetae@kaist.ac.kr

Assistant Professors

Park, Hyun G. 82-42-869-3932
hgpark@kaist.ac.kr

Accredited by: Ministry of Education

Degrees granted 2003-2004:

B.S.: 10 M.S.: 40 Ph.D.: 24

Department reports to: Kyung-Yong Chwa, Dean, College of Engineering

Placement service: Ji-Won Yang

Korea University

Department of Chemical and Biological Engineering
1, 5-Ka, Anam-Dong, Sungbuk-Ku
Seoul 136-701

University phone 82-2-3290-1114
Department phone 82-2-3290-3290
Department FAX 82-2- 926-6102

Professors

Chun, Hai S. (Emeritus) 82-2-3290-3400
Doh, Dong S. (Emeritus)
Hong, Suk I. 82-2-3290-3294
sihong@korea.ac.kr
Hyun, Jae C. 82-2-3290-3295
jchyun@grtrkr.korea.ac.kr
Kim, Chongyoup 82-2-3290-3302
cykim@grtrkr.korea.ac.kr
Kim, Seung W. 82-2-3290-3300
kimsw@korea.ac.kr
Kim, Sung H. 82-2-3290-3297
kimsh@korea.ac.kr
Kim, Woo N. 82-2-3290-3296
kimwn@korea.ac.kr
Kim, Yong J. (Emeritus)
Lee, Chul S. 82-2-3290-3293
cslee@korea.ac.kr
Lee, Kwan Y. (Chair) 82-2-3290-3299
kylee@korea.ac.kr
Yang, Dae R. 82-2-3290-3298
dryang@korea.ac.kr
Yoo, Young H. (Emeritus)

Associate Professors

Ahn, Dong J. 82-2-3290-3301
ahn@korea.ac.kr
Ha, Jeong S. 82-2-3290-3303
jeongsha@korea.ac.kr
Lee, Jee W. 82-2-3290-3304
leejw@korea.ac.kr

Assistant Professors

Jung, Hyun W. 82-2-3290-3306
hwjung@grtrkr.korea.ac.kr
Kang, Jung W. 82-2-3290-3305
jwkang@korea.ac.kr

Accredited by: Ministry of Education & Human Resources Development

Graduate advisor: Kwan Young Lee

Undergraduate advisor: Kwan Young Lee

Student organization: Chongyoup Kim
Advisor: Chongyoup Kim

Department reports to: Dean of Engineering

Kunsan National University

Chemical Engineering Department
College of Engineering
Miryong-dong, Kunsan
Chonbuk, 573-701

University phone .... (82) 654-469-4113
Department phone .... (82) 654-469-4771
Department FAX .... (82) 654-469-4779

Professors

Lee, Hee Y. .... (82) 654-469-4772
hylee@knusunl.kunsan.ac.kr

Associate Professors

Song, Byung H. .... (82) 654-469-4773
bhsong@knusunl.kunsan.ac.kr

Assistant Professors

Kim, Young H. .... (82) 654-469-4774
Lee, Sang-Cheol .... (82) 654-469-4775
lee40f@kunsan.ac.kr

Pohang University of Science and Technology

San 31, Hyoja-dong, Nam-gu
Pohang, Korea 790-784
Deliveries: Prof. Kyung Hee Lee

University phone .... (82) 54-275-0900
Department phone .... (82) 54-279-2720
Department FAX .... (82) 54-279-2699

Professors

Chang, Kun S. (Emeritus) .... 279-2262 †
kschang@postech.ac.kr
Cho, Kil W. 279-2270
kwcho@postech.ac.kr
Chung, Jong S. 279-2267
jsc@postech.ac.kr
Kang, In S. 279-2273
iskang@postech.ac.kr
Kim, Jin K. 279-2276
jkkim@postech.ac.kr
Kim, Young G. (Emeritus) 279-2261
ygkim@postech.ac.kr
Lee, In B. 279-2274
iblee@postech.ac.kr
Lee, Jae S. 279-2266
jlee@postech.ac.kr
Lee, Kun H. 279-2271
ce20047@postech.ac.kr
Lee, Kyung H. (Head) 279-2263
kyunglee@postech.ac.kr
Lee, Sun B. 279-2268
sblee@postech.ac.kr
Nam, In-Sik 279-2264
isnam@postech.ac.kr
Park, Chan E. 279-2269
cep@postech.ac.kr
Park, Jong M. 279-2275
jmpark@postech.ac.kr
Rhee, Shi W. 279-2265
srhee@postech.ac.kr

Associate Professors

Cha, Hyung J. 279-2280
hjcha@postech.ac.kr
Yoon, Byung J. 279-2277
bjyoon@postech.ac.kr

Assistant Professors

Yong, Kijung 279-2278
kyong@postech.ac.kr

† Faculty numbers have prefix 82-54.

Accredited by: Ministry of Education

Degrees granted 2003-2004:

B.S.: 18 M.S.: 34 Ph.D.: 14

Graduate advisor: Prof. Kilwon Cho

Undergraduate advisor: Prof. Jong Shik Chung

Seoul National University

School of Chemical Engineering
Seoul 151-744, Korea

University phone 82-2-880-5114
Department phone 82-2-880-7068
Department FAX 82-2-888-1604

Professors

Ahn, Tae-Oan (Emeritus)
Char, Kookheon 82-2-880-7431
khchar@plaza.snu.ac.kr
Choi, Cha Y. 82-2-880-7071
choicy@snu.ac.kr
Choi, Chang K. 82-2-880-7407
ckchoi@snu.ac.kr
Choi, Ung (Emeritus)
Jang, Jyongsik 82-2-880-7069
jsjang@plaza.snu.ac.kr
Kim, Byung-Gee 82-2-880-6774
byungkim@snu.ac.kr
Kim, Hwayong 82-2-880-7406
hwayongk@snu.ac.kr
Kim, Joon Y. (Emeritus)
Lee, Chai-sung (Emeritus)
Lee, Chung-Hak 82-2-880-7075
leech@snu.ac.kr
Lee, Ho-In (Head) 82-2-880-7072
hilee@snu.ac.kr
Lee, Hong H. (Director) 82-2-880-7403
honghlee@plaza.snu.ac.kr
Lee, Jihwa 82-2-880-7076
jiwhalee@plaza.snu.ac.kr
Lee, Ki-Jun 82-2-880-7402
kijunlee@plaza.snu.ac.kr
Lee, Moon-Deuk (Emeritus)
Lee, Seung J. 82-2-880-7410
sjlee@plaza.snu.ac.kr
Lee, Wha Y. 82-2-880-7404
wyl@snu.ac.kr
Lee, Yoon S. 82-2-880-7073
yslee@snu.ac.kr
Moon, Sang H. 82-2-880-7409
shmoon@surf.snu.ac.kr
Oh, Seung M. 82-2-880-7074
seungoh@plaza.snu.ac.kr

Rhee, Hyun-Ku 82-2-880-7405
hkrhee@snu.ac.kr
Shim, Jyong S. (Emeritus)
Shin, Yoon-Kyoung (Emeritus)
Yoo, Young J. (Associate Dean) 82-2-880-7411
yjyoo@plaza.snu.ac.kr
Yoon, En S. 82-2-880-7408
esyoon@pslab.snu.ac.kr

Associate Professors

Hyeon, Taeghwan 82-2-880-7150
thyeon@plaza.snu.ac.kr
Jho, Jae Y. 82-2-880-8346
jyjho@snu.ac.kr
Kim, Young G. 82-2-880-8347
ygkim@plaza.snu.ac.kr
Park, Tai H. 82-2-880-8020
thpark@plaza.snu.ac.kr
Yi, Jongheop 82-2-880-7438
jyiecerl@snu.ac.kr

Assistant Professors

Ahn, Kyung H. 82-2-880-8322
ahnnet@snu.ac.kr
Kim, Jae J. 82-2-880-8863
jjkimm@snu.ac.kr
Lee, Jong-Chan 82-2-880-7070
jongchan@snu.ac.kr
Yoon, Jeyong 82-2-880-8927
jeyong@snu.ac.kr

Accredited by: Ministry of Education Korea

Degrees granted 2003-2004:

B.S.: 109 M.S.: 80 Ph.D.: 32

Department reports to: Jang-Moo Lee, Dean of Engineering

Kuwait

Kuwait University

Chemical Engineering Department
College of Engineering and Petroleum
P.O. Box 5969
13060, Safat

University phone 965 4811188
Department phone 965 4817662
Department FAX 965 4839498

Professors

AL-Haddad, Amir A. .... x5221 †
AL-Sahhaf, Taher x5673
AL-Saif, Walid (Visiting)
Alatiqi, Imad M. x5776
Bouhmra, Walid S. x5613
Ettouney, Hisham x5619
Fahim, M.A x5127
Riazi, Mohammed R. x5772
Shabhan, Habib I. (Head) x5109

Associate Professors

AL-Ameeri, Rashid S. x5148
AL-Enezy, Ghazi 965-4810272
Baker, Christofer G.J x5121
Bishara, Ahmed E. x5617
Lababidi, Haitham M.S. x5778
Mahdi, Khaled A. x5618
k_mahdi@kuc01.kuniv.edu.kw

Assistant Professors

AL-Adwani, Hamad A.H. x7460
Al-Bahri, Tareq A. x7459
Al-Bofersn, Osama A. x5856
Al-Mulla, Adam x5775
Al-Musallam, Abdulwahhab x5612
Al-Ramadan, Hamida A. x5694
AL-Roomi, Yousef M. x7016
Al-Sahali, Mohammad B. x5783
AL-Shayji, Khawla x5108
Ali, Sammi H. G. x5702

† Faculty numbers are extensions of University phone.

Accredited by: ABET

Placement service: Eng. Mubarak Ghanim

Malaysia

University of Malaya

Chemical Engineering Department
Faculty of Engineering
50603 Kuala Lumpur

University phone 60 3 7967-7022
Department phone 60 3 7967-5206
Department FAX 60 3 7967-5319

Professors

Hasan, Masitah .... 7967 5295 †
masitahhasan@um.edu.my
Hashim, Mohd A. 7967 4522
alihashim@um.edu.my
Ramachandran, K. B. 7967 5293
kbram@um.edu.my
Sulaiman, Nik M. 7967 5299
meriam@um.edu.my

Associate Professors

Abdullah, Ezzat C. 7967 5301
ezzat@um.edu.my
Aroua, Mohamed K. 7967 5313
mk_aroua@um.edu.my
Hussain, Mohamed A. (Chair) 7967 5214
mohd_azlan@um.edu.my
Leong, Yub C. 7967 5294
ycleong@um.edu.my
Singh, Harcharan 7967 5290
harcharan@um.edu.my

Assistant Professors

Benamor, Abdelbaki 7967 5292
abdelbaki@um.edu.my
Chua, Adeline S. M.
Ghasem, Nayef M. 7967 5360
nayef@um.edu.my
Hassan, Che R. 7967 5314
rosmani@um.edu.my
Mohamed Nor, Mohamed I. 79675347
misk@um.edu.my

Raman, Abdul A. 7967 5300
azizraman@um.edu.my
Wan Daud, Wan M. A. 7967 5297
ashri@fk.um.edu.my
Yeoh, Hak K.
Yusoff, Rozita

† Direct dial with prefix 03.

Accredited by: IChemE (UK)

Graduate advisor: Ezzat Chan Abdullah

Department reports to: Mohamed Azlan Hussain, Head of Department

Universiti Sains Malaysia

School of Chemical Engineering
Engineering Campus
Seri Ampangan
14300 Nibong Tebal
Seberang Perai Selatan
Pulau Pinang

University phone .... (604) 5937788
Department phone .... (604) 5937788x6400
Department FAX .... (604) 5941013

Professors

Bhatia, Subhash .... x6409 †
chbhatia@eng.usm.my

Associate Professors

Abdullah Aziz, Jalal x6411
chjalal@eng.usm.my
Ahmad, Abdul L. x6418
chlatif@eng.usm.my
Fernando, W.J.N x6428
chnoel@eng.usm.my
Kamaruddin, Azlina H. x6417
chazlina@kimia.eng.usm.my
Mohamed, Abdul R. (Dean) x6400
chrahman@eng.usm.my
Najafpour, Ghasem x6412
chghasem@eng.usm.my

Assistant Professors

Bakar, Mohd Z. A. x6402
chmohdz@eng.usm.my
bin Aziz, Norashid x6457
chnaziz@eng.usm.my
bin Othman, Mohd R. x6426
chroslee@eng.usm.my
Choudhury, J.P. x6424
chjyoti@eng.usm.my
Daud, Mohd Z. M. x6414
chzulk@eng.usm.my
Hameed, Bassim H. x6422
chbassim@eng.usm.my
Hussein, Ye L. M. x6460
chhussin@eng.usm.my
Qassim, Mohd. Q. H. x6423
chqassim@eng.usm.my
Srinivasakannan, C. x6425
chkannan@eng.usm.my
Zakaria, Ridzuan x6403
chduan@eng.usm.my

† Dial (604) 5937788 and ask for extension.

Accredited by: Board of Engineering

Department reports to: Abdul Rahman Mohamed, Dean of the School

Mexico

Instituto Tecnologico y de Estudios Superiores de Monterrey

Chemical Engineering Department
Sucursal de Correos J
Monterrey, N.L. 64849
Deliveries: Dr. Joaquin Acevedo

University phone .... (8) 358-2000
Department phone .... (52) 8158-2034
Department FAX .... (52) 8328-4250

Professors

Garcia, Eugenio (Dean)
jegarcia@itesm.mx
Garcia, Jorge
jorge.garcia@itesm.mx
Mejia, Gerardo
gmejia@itesm.mx
Rito, Marco
mrito@itesm.mx
Romero, Miguel A.
mromero@itesm.mx
Trevino, Belzahet
btrevino@itesm.mx
Yague, Omar
oyague@itesm.mx

Associate Professors

Acevedo, Joaquin (Head)
jacevedo@itesm.mx
Gomez-Maqueo, Eduardo
egomezm@itesm.mx
Lozano, Francisco J.
fjlozano@itesm.mx
Martinez, Jeronimo
jemartin@itesm.mx
Ortiz, Enrique
eortiz@itesm.mx

Assistant Professors

Garza, Vicente J.
vicente.garza@itesm.mx
Mendoza, Alberto
mendoza.alberto@itesm.mx
Patino, Veronica
vpatino@itesm.mx
Ramirez, Darinka
darinka@itesm.mx
Soriano, Ricardo
rsoriano@itesm.mx

Accredited by: SACS, CACEI, ABET

Graduate advisor: Dr. Miguel A. Romero

Undergraduate advisor: Veronica Patiño

Student organization: IMIQ
Advisor: Veronica Patiño

Department reports to: Eugenio García, Dean, Div. of Engineering

Placement service: Eugenio García, Dean, Div. of Engineering

Universidad Iberoamericana

Coordinacion de Ingenieria Quimica
Prolongacion Paseo de la Reforma 880,
Col. Lomas de Santa Fe, Mexico D. F. 01210
Deliveries: Quim. Juan Abud Saint Martin

University phone .... (52) 55 5950 4000
Department phone .... (52) 55 5950 4033
Department FAX .... (52) 55 5950 4279

Professors

Abud Saint Martín, Juan (Chair) x4033
juan.abud@uia.mx

Associate Professors

Chaparro Mercado, Ma. C. x4098
carmen.chaparro@uia.mx
Hernández Esparza, Margarita .... x 4629
margarita.hernandez@uia.mx
Huerta Cevallos, Rene x7088
rene.huerta@uia.mx
López Rubio, Javier F. x7091
javier.lopez@uia.mx
Pedraza Segura, Lorena .... x 4624
lorena.pedraza@uia.mx
Rivera Toledo, Martin x4078
martin.river@uia.mx
Silva Beard, Andrea (Director) x4534
andrea.silva@uia.mx

Accredited by: CACEI

Degrees granted 2003-2004:
B.S.: 40 M.S.: 5

Undergraduate advisor: Quim. Juan Abud

Student organization: SAICQ
Advisor: Eduardo Alvarez F.

Universidad de las Americas, Puebla

Chemical and Food Engineering Department
Ex-hacienda Santa Catarina Martir
San Andres Cholula, Puebla 72820

University phone 52 (222) 229-2000
University FAX 52 (222) 229-2009
Department phone 52 (222) 229-2126
Department FAX 52 (222) 229-2727

Professors

Argaiz-Jamet, Alvaro 52 (222) 229-2000x4092
aargaiz@mail.udlap.mx
Fonseca-Sandoval, Raul G. 52 (222) 229-2026
fonseca@mail.udlap.mx
Lopez-Malo, Aurelio 52 (222) 229-2409
amalo@mail.udlap.mx
Palou-Garcia, Enrique 52 (222) 229-2658
epalou@mail.udlap.mx
Reyes-Mazzoco, Rene 52 (222) 229-2660
rreyes@mail.udlap.mx
Velez-Ruiz, Jorge F. 52 (222) 229-2648
jfvelezr@mail.udlap.mx
Vergara-Balderas, Fidel 52 (222) 229-2146
fvergara@mail.udlap.mx
Welti-Chanes, Jorge S. 52 (222) 229-2728
jwelti@mail.udlap.mx

Associate Professors

Barcenas-Pozos, M. E. 52 (222) 229-2126
mbar@mail.udlap.mx
Lara-Diaz, Rene A. 52 (222) 229-2000x4353
rlara@mail.udlap.mx
Maza-Aranda, J. G. 52 (222) 229-2659
gmaza@mail.udlap.mx
Rios-Casas, Luis G. (Head) 52 (222) 229-2661
lrios@mail.udlap.mx
San Martin-Gonzalez, Fernanda 52 (222) 229-2927
fernanda@mail.udlap.mx

Accredited by: SACS

Degrees granted 2003-2004:
B.S.: 35 M.S.: 19

Graduate advisor: Dr. Aurelio Lopez-Malo (Food Sc.) and Dr. Rene Reyes (Chem. Eng.)

Undergraduate advisor: Ma. Elena Sosa (Food Sc.), René Lara (Chem. Eng.)

Student organization: CEUDLA
Advisor: M. Sc. Luis Rios

Department reports to: Dr. Juan Manuel Ramirez C., Dean of Engineering

Placement service: Silvia Carreño Castillo

Universidad Autonoma Metropolitana-Iztapalapa

Department of Chemical Engineering
San Rafael Atlixco No. 186
Col. Vicentina, Del. Iztapalapa
09340 Mexico, D. F. MEXICO

University phone .... (525) 804-4600
Department phone .... (525) 804-4642/43
Department FAX .... (525) 804-4900

Professors

Alvarez, Jesús .... (525) 804-46-48
jac@xanum.uam.mx
Alvarez-Ramirez, José (Associate Head) .... (525) 804-46-48
jjar@xanum.uam.mx
de los Reyes, José A. (Associate Dean) .... (525) 804-46-48
jarh@xanum.uam.mx
Fuentes, Gustavo A. .... (525) 804-46-48
gfuentes@xanum.uam.mx
Lapidus, Gretchen .... (525) 804-46-48
gtll@xanum.uam.mx
Lobo, Ricardo (Head) .... (525) 804-46-48
lobo@xanum.uam.mx
López, Felipe .... (525) 804-46-48
felipe@xanum.uam.mx

Ochoa, J. A. .... (525) 804-46-48
jaot@xanum.uam.mx
Revah, Sergio .... (525) 804-46-48
srevah@xanum.uam.mx
Ruiz, Richard S. .... (525) 804-46-48
rmr@xanum.uam.mx
Soria, Alberto .... (525) 804-4648
asor@xanum.uam.mx
Vernon, Jaime .... (525) 804-46-48
jvc@xanum.uam.mx
Viveros, Tomás (Dean) .... (525) 804-46-48
tvig@xanum.uam.mx
Vizcarra, Mario .... (525) 804-46-48
mgvm@xanum.uam.mx

Associate Professors

Aréchiga, Uriel .... (525) 804-46-48
uriel@xanum.uam.mx
Escobar, Angel .... (525) 804-46-48
ange@xanum.uam.mx
Gómez, Sergio .... (525) 804-4648
sgomez@xanum.uam.mx
Jarquín, Hugo .... (525) 804-46-48
hja@xanum.uam.mx
Martínez, Carlos .... (525) 804-46-48
cmv@xanum.uam.mx

Accredited by: CACEI Undergraduate Program, CONACyT Graduate Program

Degrees granted 2003-2004:

B.S.: 24 M.S.: 10 Ph.D.: 3

Graduate advisor: Dr. Mario Vizcarra

Undergraduate advisor: Dr. Gretchen Lapidus

Student organization: Marco Antonio Loza
Advisor: Dr. Gretchen Lapidus

Department reports to: Dr. Ricardo Lobo, Head of the Process Engineering and Hydraulics Department

Placement service: gtll@xanum.uam.mx

Universidad Autonoma de San Luis Potosi

Facultad de Ciencias Químicas
Zona Universitaria
Av. Dr. Manuel Nava #6
San Luis Potosi S.L.P. 78210
Deliveries: Dr. Jorge Toro Vazquez

University phone 52 (444) 826 24 50
Department phone 52 (48) 17 53 03
Department FAX 52 (48) 13 07 12

Professors

Cardenas-Galindo, M. G. .... x125 †
cardenas@fcq1.fcq.uaslp.mx
Femat-Flores, Ricardo x108
Handy, Brent E. x125
handy@deimos.tc.uaslp.mx
Moctezuma-Velazques, Edgar x125
edgar@deimos.tc.uaslp.mx
Navarro-Contreras, Jesus x125
ciepfcq@deimos.tc.uaslp.mx
Toro-Vazquez, Fernando (Chair) .... (52-444)8262450
toro@uaslp.mx

Associate Professors

Lopez-Valdivieso, Alejandro x103
valdivieso@deimos.tc.uaslp.mx
Medellin-Rodriguez, Francisco x125
francmr@uaslp.mx
Monroy-Fernandez, Marcos x103
monroyma@deimos.tc.uaslp.mx

Assistant Professors

Alonso-Dovila, Pedro x125
paad@uaslp.mx
Castillo-Huerta, Lilia x128
castilli@uaslp.mx
Flores-Velez, Luisa-Maria x108
lflores@deimos.tc.uaslp.mx
Leyva-Ramos, Roberto x125
rlr@uaslp.mx
Ruiz-Castillo, Ma-Guadalupe x112
gruizcas@deimos.tc.uaslp.mx
Soriano-Perez, Sonia x125
sorianos@deimos.tc.uaslp.mx

† Dial 48 17 53 03 and ask for extension.

Accredited by: CONACYT, CACEI

Graduate advisor: Dr. Francisco Medellín

Undergraduate advisor: Lilia Castillo-Huerta

Student organization: IMIQ-Seccion Estudiantil

Department reports to: Fernando Toro-Vazquez, Chair

Placement service: Ricardo Medina Cerda

Mozambique

Universidade Eduardo Mondlane

Chemical Engineering Department
PO Box 257
Maputo

University phone 258-1428411
Department phone 258-1-475318

Associate Professors

Sapojnikov, Svetslava

Assistant Professors

Amos, Gabriel L. (Dean)
Biquiza, Lucrecio D.
Cumbane, Antonio J.
da Cruz Francisco Jose
da Gama, Vasco J Jr.
de Braganca, Radek
Guiamba, Isabel R.
Khan, Maida A. M. (Head)
Lucas, Carlos
Lucas, Luis H.
Pelembe, Louis A. M.
Tsamba, Alberto J.
Venancio, Joao C. S. M.

Department reports to: Gabriel Luis Amos, Dean

Placement service: Comissao Nac. de, Planificacao

Netherlands

University of Amsterdam

Department of Chemical Engineering
Nieuwe Achtergracht 166
1018 WV Amsterdam

University phone 31 20 525 9111
Department phone 31 20 525 5265
Department FAX 31 20 525 5604

Professors

Bliek, Alfred (Chair) . . . . . 525 6479 †
bliek@science.uva.nl

Das, Hendrik A. (Emeritus)
das@ecn.nl

Fasolino, Annalisa 525 6448
fasolino@science.uva.nl

Fortuin, Jan M. H. (Emeritus) 525 6478
Frenkel, Daan 525 5265
frenkel@amocf3.amolf.nl

Iedema, Piet D. 525 6484
piet@science.uva.nl

Krishna, Rajamani 525 7007
krishna@science.uva.nl

Poppe, Hans (Emeritus) 525 5265
Schoenmakers, Peter J. 525 6642
pjschoen@science.uva.nl

Smilde, Age K. 525 5062
asmilde@science.uva.nl

Smit, Berend 525 5067
B.Smit@science.uva.nl

Tijssen, Robert 525 6550
rtijssen@science.uva.nl

Associate Professors

Dimian, Alexandre C. 525 6034
alexd@science.uva.nl

Hoefsloot, Huub C. J. 525 5867
huubh@science.uva.nl

Jérôme, Blandine 525 6992
jerome@science.uva.nl

Kok, Wim T. 525 6539
wkok@science.uva.nl

van den Heuvel, Johannes C. 525 5078
heuvel@science.uva.nl

Assistant Professors

Bolhuis, Peter G. 525 6447
bolhuis@science.uva.nl

Eiser, Erika 525 6916
eiser@science.uva.nl

Lowe, Christopher P. 525 6485
lowe@science.uva.nl

Meijer, Evert J. 525 6448
ejmeijer@science.uva.nl

Rothenberg, Gadi 525 6963
gadi@science.uva.nl

Westerhuis, Johan A. 525 6546
westerhu@science.uva.nl

† Direct dial with prefix 31 20.

Accredited by: IChemE (UK)

Delft University of Technology

Faculty of Applied Sciences
Department of Chemical Engineering & Bioprocess Technology
Julianalaan 136
2628 BL-Delft

University phone 31 (15) 278-9111
Department phone 31 (15) 278-4841
Department FAX 31 (15) 278-5868

Professors

Akker van den, H.E.A.vanden A. (Director)
Coppens, M.O.Coppens
M.O.Coppens@tnw.tudelft.nl
Dedem van, G.W.K. van D.
G.W.K.vanDedem@tnw.tudelft.nl
Frens, G.Frens (Emeritus)
Grievink, J. . . . . . 278-4351 †
Hagen, W.R 278-5051
W.R.Hagen@tnw.tudelft.nl
Harmsen, G. J. 278-5006
Heijnen, J. J. 278-2341
Jansens, P. P. J. 278-6678
Kapteijn, F 278-4384
Katgerman, L. 278-2249
Kuenen, J. G. 278-5308
Loosdrecht, M.C.M.vanLoosdrecht
M.C.M.vanLoosdrecht@tnw.tudelft.nl
Luyben, K. C. A. M. (Dean) 278-2353
Maschmeijer, Th 278-8826
Moulijn, J.A. 278-5008
Odijk, Th. 278-5590
Ooms, G (Head) 278-2686
g.ooms@tnw.tudelft.nl
Pasman, J 278-5006
Picken, S 278-6946
Pronk, J 278-3214
Schmidt-Ott, A.Schmidt-Ott
A.Schmidt-Ott@tnw.tudelft.nl
Schoonman, J 278-2647
Sheldon, R. A. 278-2675
Swaan Aronds, J.deSwaanArons (Emeritus)
J.deSwaanArons@tnw.tudelft.nl
van den Bleek, C. M. 278-4390
van Loosdrecht, M. 278-1618
van Turnhout, J. 278-2623
Wielen v.d., L.A.M. 278-2361

Associate Professors

de Loos, Th. W. 278-2619
de Loos-Vollebregt, Mrs. M. T. C. 278-6379
Goossens, A 278-4919
Jongejan, J 278-2371
Koper, G.J.M. 2788218
G.J.M.koper@tnw.tudelft.nl
Lameris, G. H. 278-2648
Lemkowitz, S. M. 278-4394
Makkee, M.Makkee
M.Makkee@tnw.tudelft.nl
Nieswaag, H. 278-2234
Nijenhuis, K. te 278-2630
Peters, C. J. 278-2660
Peters, J. A. 278-5892
Schapink, F. W. 278-2272
Steensma, H. Y. 278-2412
van den Ham, A. G. J. 278-4374

van Dijken, J. P. 278-2412
Verheijen, P 278-4326

Assistant Professors

Arends, I.W.C.E.Arends
I.W.C.E.Arends@tnw.tudelft.nl
Gotsis, A.D.Gotsis
A.D.Gotsis@tnw.tudelft.nl
Hanefeld, U.Hanefeld
Kleeff van, B.H.A. van K.
B.H.A.vanKleeff@tnw.tudelft.nl
Ommen van, J.R.vanOmmen
J.R.vanOmmen@tnw.tudelft.nl

† Direct dial with prefix 31 15.

Accredited by: IChE, ABET, VSNU

Undergraduate advisor: Mrs ir A.Schaap

Student organization: Technologisch Gezelschap

Eindhoven University of Technology

Department of Chemical Engineering
Den Dolech 2
5600 MB Eindhoven

University phone 31 (40) 247911
Department phone 31 (40) 2473000
Department FAX 31 (40) 2444321

Professors

Benthem, R. van
Bernard, E.M.G.H. (Director)
Broer, D.J.
Bruin, S.
Cramers, C.A. (Emeritus)
Dautzenberg, F.M.
de Jeu, W.H.
de With, G.
Drinkenburg, A.A.H.
German, A. L. (Emeritus)
Gruter, G.J.
Herk, A. M. van
Hulshof, L.A.
Janssen, F.J.J.G.
Janssen, R. A. J.
Kerkhof, P. J. A. M.
Keurentjes, J.T.F.
Koning, C.E.
Kramer, G.J.
Leegwater, H.
Leeuwen, P.W.N.M. van
Lemstra, P. J.
Linde, R. van der (Emeritus)
Loo, F.J.J. van (Emeritus)
Meijer, E. M.
Meijer, E. W.
Metselaar, R. (Emeritus)
Niemantsverdriet, J.W. (Dean)
Nieuwenhuis, B.
Nolte, R.J.M.
Notten, P.
Santen, R. van
Schouten, J.C. (Associate Dean)
Schubert, U.
Veen, J. A. R. van
Vogt, D.

Associate Professors

Bastiaansen, C.W.M.
Bastin, G. F.
Coumans, W. J.
Hintzen, H. T.
Hoebink, J.
Janssen, A.P.J.
Janssen, L. J. J. (Emeritus)
Klumperman, L
Koper, M.T.M.
Kuster, B. F. M
Meuldijk, J.
Nies, E.
Rastogi, S.
Reijenga, J. C.
Sijbesma, R.
Vekemans, J.A.J M.
Wijers, J. G.

Accredited by: ABET

Degrees granted 2003-2004:

B.S.: 350 M.S.: 40 Ph.D.: 125

Department reports to: Board of University

Twente University of Technology

Faculty of Chemical Technology
P.O. Box 217
7500 AE Enschede
The Netherlands
Deliveries: Drienerlolaan 5, 7522 NB Enschede

University phone 053-4899111
Department phone 053-4893000
Department FAX 053-4892479

Professors

Asselbergs, G. J.
G.J.Asselbergs@ct.utwente.nl
Burggraaf, A. J. (Emeritus)
de Haan, A.B.
a.b.dehaan@ct.utwente.nl
de Smit, J.H.A. (Dean)
j.h.a.desmit@ct.utwente.nl
Engbersen, J.F.J. 053-4892926
j.f.j.engbersen@ct.utwente.nl
Feijen, J.
J.Feijen@ct.utwente.nl
Feil, D. (Emeritus)
Gellings, P. J. (Emeritus)
Grootenhuis, P.D.J.
p.d.j.grootenhuis@ct.utwente.nl
Kuipers, J.A.M.
j.a.m.kuipers@ct.utwente.nl
Lefferts, L.
l.lefferts@ct.utwente.nl
Mulder, M. H. V.
M.H.V.Mulder@ct.utwente.nl
Noordermeer, J.W M.
J.W.M.Noordermeer@ct.utwente.nl
Reinhoudt, D.N.
D.N.Reinhoudt@ct.utwente.nl
Reith, Th. (Emeritus)
Rijnsdorp, J. E. (Emeritus)
Roffel, B. 31-53-489-2920
B.Roffel@ct.utwente.nl

Schuijer, J. (Emeritus)
Smolders, C. A. (Emeritus)
Strathmann, H. (Emeritus)
Struik, L. C. E. L.C.E.Struik@ct.utwente.nl
Taylor, R. r.taylor@ct.utwente.nl
van Aken, W. G. (Emeritus) W.G.vanAken@ct.utwente.nl
van Blitterswijk, C.A.
van den Berg, H. H.vandenBerg@ct.utwente.nl
van der Linden, W. (Emeritus) W.E.vanderLinden@ct.utwente.nl
van Swaaij, W. P. M. W.P.M.vanSwaaij@ct.utwente.nl
Vancso, G. J. G.J.Vancso@ct.utwente.nl
Vermes, I. i.vermes@ct.utwente.nl
Versteeg, G. F. G.F.Versteeg@ct.utwente.nl
Warmoeskerken, M.M.C.G. M.M.C.G.Warmoeskerken@ct.utwente.nl
Wassink, J. G. (Emeritus)
Wessling, M. m.wessling@ct.utwente.nl
Westerterp, K. R. (Emeritus)

Associate Professors

Betlem, B. H. L. B.H.L.Betlem@ct.utwente.nl
Beugeling, T. (Emeritus)
Bos, M. (Emeritus) M.Bos@ct.utwente.nl
Bosch, H. H.Bosch@ct.utwente.nl
Boukamp, B. A. B.A.Boukamp@ct.utwente.nl
Bouwmeester, H. J. M. H.J.M.Bouwmeester@ct.utwente.nl
Brilman, D. W. F. D.W.F.Brilman@ct.utwente.nl
Coenders, F.G.M. f.g.m.coenders@ct.utwente.nl
de Vries, K.J. (Emeritus)
Deen, N. n.deen@ct.utwente.nl
den Hertog, H.J. (Emeritus)
Dierkes, W.K. w.k.dierkes@ct.utwente.nl
Dijkstra, P. J. P.J.Dijkstra@ct.utwente.nl
Engbers, G.H.M. G.H.M.Engbers@ct.utwente.nl
Fokkens, R.H. r.h.fokkens@ct.utwente.nl
Gaymans, R. J. R.J.Gaymans@ct.utwente.nl
Grijpma, D.W. d.w.grijpma@ct.utwente.nl
Harkema, S. S.Harkema@ct.utwente.nl
Heesink, A. B. M. A.B.M.Heesink@ct.utwente.nl
Hempenius, M. A. M.A.Hempenius@ct.utwente.nl
Huskens, J. j.huskens@ct.utwente.nl
Keizer, K. K.Keizer@ct.utwente.nl
Koops, G. H. G.H.Koops@ct.utwente.nl
Kramers-Pals, H. H.Kramers-Pals@ct.utwente.nl
Kuipers, N.J.M. N.J.M.Kuipers@ct.utwente.nl
Meindersma, G.W. g.w.meindersma@ct.utwente.nl
Mojet-Mol, B. b.mojet-mol@ct.utwente.nl
Nierstrasz, V.A. v.a.nierstrasz@ct.utwente.nl
Nieuwenhuijse, B. (Emeritus)
Prins, W. W.Prins@ct.utwente.nl
Rácz, I. G. I.G.Rácz@ct.utwente.nl
Sager, W.F.C. w.f.c.sager@ct.utwente.nl
Schönherr, H. h.schönherr@ct.utwente.nl
Seshan, K. K.Seshan@ct.utwente.nl
ten Elshof, J.E. j.e.tenelshof@ct.utwente.nl
van Dam, T. t.vandam@ct.utwente.nl
van de Velde, G. M. H. (Emeritus)
van den Berg, J. W. A. J.W.A.vandenBerg@ct.utwente.nl
van der Does, L. (Emeritus)
van der Ham, A. G. J. A.G.J.vanderHam@ct.utwente.nl
van der Hoef, M. A. M.A.vanderHoef@ct.utwente.nl
van der Vegt, N. F. A. n.f.a.vandervegt@ct.utwente.nl
van Ommen, J.G. J.G.vanOmmen@ct.utwente.nl
van Sint Annaland, M. m.vansintannaland@ct.utwente.nl
van Veggel, F. C. J. M. F.C.J.M.vanVeggel@ct.utwente.nl
Verboom, W. W.Verboom@ct.utwente.nl
Veugelers, P. P. P.P.Veugelers@ct.utwente.nl
Vrijland, M. S. A. M.S.A.Vrijland@ct.utwente.nl
Weickert, G. g.weickert@ct.utwente.nl
Winnubst, A. J. A. A.J.A.Winnubst@ct.utwente.nl

Student organization: C.T.S.G. Alembic

Department reports to: ing. M. A. C. G. van de Graaf

New Zealand

University of Auckland

Department of Chemical and Materials Engineering
Private Bag 92019
Auckland
Deliveries: Gate 3, Grafton Road, Auckland City

University phone .... (64) (9) 373 7599
Department phone .... (64) (9) 373 7599x88135
Department FAX .... (64) (9) 373 7463

Professors

Broom, Neil D. .... x88974 †
nd.broom@auckland.ac.nz
Chen, John J. J. 88137
j.chen@auckland.ac.nz
Chen, X. D. x87004
d.chen@auckland.ac.nz
Duffy, Geoffrey G. (Associate Dean) x87805
gg.duffy@auckland.ac.nz
Farid, Mohamed M x84807
m.farid@auckland.ac.nz
Ferguson, W. G. (Head) x88133
wg.ferguson@auckland.ac.nz
Gao, Wei (Associate Dean) x88175
w.gao@auckland.ac.nz

Associate Professors

Hossain, Monwar x83962
m.hossain@auckland.ac.nz
Hyland, Margaret M. x87865
m.hyland@auckland.ac.nz

Assistant Professors

Hodgson, Michael A. x88218
ma.hodgson@auckand.ac.nz
James, Bryony J. x85813
b.james@auckland.ac.nz
Liaw, Su-Ling 87003
ms.liaw@auckland.ac.nz
Oakley, Simon x87307
oakley@ihug.co.nz

† Dial (64) (9) 373 7599 and ask for extension.

Accredited by: IChemE (UK), IPENZ

Degrees granted 2003-2004:

B.S.: 28 M.S.: 11 Ph.D.: 5

Undergraduate advisor: Dr M. A. Hodgson

Department reports to: Dean of Engineering

Placement service: Jim Bohm (7488), Careers Advisory Service

University of Canterbury

Chemical and Process Engineering
Private Bag 4800
Christchurch 8001
Deliveries: Corner Creyke & Forestry Rd, Ilam

University phone 64 3 366-7001
Department phone 64 3 364-2543
Department FAX 64 3 364-2063

Professors

Abrahamson, John .... 364-2318 †
john.abrahamson@canterbury.ac.nz
Earl, W. B. 364-2142
brian.earl@canterbury.ac.nz
Keey, Roger B. (Emeritus) 364-2543
r.keey@cape.canterbury.ac.nz
Marsh, Kenneth N. 364-2140
ken.marsh@canterbury.ac.nz
Pang, Shusheng 364-2538
shusheng.pang@canterbury.ac.nz
Weatherley, Laurence R. (Head) 364-2139
laurence.weatherley@canterbury.ac.nz
Williamson, Arthur G. (Emeritus) 364-2943
a.williamson@cape.canterbury.ac.nz

Associate Professors

Gilmour, Ian A. 364-2137
ian.gilmour@canterbury.ac.nz
Gostomski, Peter A. 364-2141
peter.gostomski@canterbury.ac.nz
Jordan, Patrick J. 364-2864
pat.jordan@canterbury.ac.nz
Morison, Ken R. 364-2578
ken.morison@canterbury.ac.nz
Peet, N. J. 364-2142
john.peet@canterbury.ac.nz
Williamson, Chris J. 364-2865
chris.williamson@canterbury.ac.nz

Assistant Professors

Chu, Khim H 364-2217
khim.chu@canterbury.ac.nz

† Direct dial with prefix 64 3.

Accredited by: IChemE (UK), Institution of Professional Engineers

Graduate advisor K. N. Marsh

Undergraduate advisor: L. R. Weatherley

Department reports to: R Sharp, Vice Chancellor

Placement service Lynda Clark, Grad. Appointments Officer

Nigeria

University of Lagos

Department of Chemical Engineering
Akoka, Yaba

University phone 234 1 5454891-5
Department phone 234 1 545 4891-5x1830

Professors

Abdul-Kareem, Hussain K.
Ako, Churchil
Bello, Rahamon A (Head) 234 1 2610809
rab-nnec@alpha.linkserve.com
Denloye, Adetokunbo O.
Kehinde, John A.

Ogboja, Olu (Dean) 234 1 823277
oluogboja@yahoo.com

Olatunji, Francis O.
Oyekunle, Layi O.
Susu, Alfred A. 234 1 822419

Associate Professors

Adeniyi, Victor O.
Aribike, David S.
Dosunmu, O.
Nwalor, J. U.
Ogunbayo, A. O. 234 1 4926958
atoke.og@metrong.com

Assistant Professors

Ajayi, T.
tolu_ajayi@usa.net

Evwierhoma, E. T.
evwierhomaet@yahoo.com

Gimba, A. S. B.
Olafadehan, V. A.
drvaolafadehan@usa.net

Usman, M. A.

Accredited by: Council for the Regulation of Engineering, AIChE

Graduate advisor: F. O. Olatunji

Student organization: Nigerian Society Of Chemical Engineers (UNILAG Chapter)

Department reports to: Olu Ogboja, Dean of Engineering

Placement service: R. A. Bello, Head of Department

Obafemi Awolowo University

Ile-Ife
Osun State
Department phone 234 (036) 230294-6x4111

Professors

Layokun, K. S. 234 (036) 230294-6x4151
sklayok@oauife.edu.ng

Sanni, S. A. .... 234(036)233833
exec-dir@nacetem.oauife.edu.ng

Solomon, O. B.
bsolomon@oauife.edu.ng

Taiwo, Oluwafemi
otaiwo@oauife.edu.ng

Associate Professors

Akeredolu, A. F. 234 (036) 230338
fakered@oauife.edu.ng

Fasesan, O. S.
sfasesan@oauife.edu.ng

Assistant Professors

Ademodi, T. B.
bademodi@oauife.edu.ng

Anozie, N. A. (Head) 234 (036) 230294-6x4111
aanozie@oauife.edu.ng

Macaulay, A. R. S.
ebetiku@oauife.edu.ng

Accredited by: COREN, NSChE

Graduate advisor: Dr. Funso Akeredolu

Undergraduate advisor: Dr. A.N. Anozie, Head of Dept.

Student organization: NSChE

Department reports to: Professor G.A. Adegboyega, Dean Faculty of Technology

Norway

Norwegian University of Science and Technology

Department of Chemical Engineering
Institutt for kjemisk prosessteknologi
NO-7491 Trondheim
Deliveries: Sem Saelands vei 4

University phone 47 73 5950000
Department phone 47 73 594030
Department FAX 47 73 594080

Professors

Berge, Arvid (Emeritus) 94138
arvid.berge@chemeng.ntnu.no

Blekkan, Edd A. 94157
edd.anders.blekkan@chemeng.ntnu.no

Chen, De 93149
de.chen@chemeng.ntnu.no

Christensen, Per K. (Emeritus) 94029
pkochchr@chemeng.ntnu.no

Erga, Olav (Emeritus) 94120
Olav.Erga@chemeng.ntnu.no

Gregersen, Øyvind 94029
gregersen@chemeng.ntnu.no

Grislingas, Arne (Adjunct) 94119
argri@statoil.com

Helle, Torbjorn (Emeritus) 94031
Torbjoern.Helle@chemeng.ntnu.no

Hertzberg, Terje 94113
Terje.Hertzberg@chemeng.ntnu.no

Holmen, Anders 94151
anders.holmen@chemeng.ntnu.no

Hägg, May-Britt 94033
May-Britt.Hagg@chemeng.ntnu.no

Jakobsen, Hugo A. 94132
Hugo.Jakobsen@chemeng.ntnu.no

Lovland, Jorgen (Emeritus) 94124
jolovland@hotmail.com

Malthe-Sørenssen, Didrik (Adjunct) 94119
dms@chemeng.ntnu.no

Moljord, Kjell (Adjunct) 94147
kmol@statoil.com

Mørk, Preben 94148
preben.cato.moerk@chemeng.ntnu.no

Nesse, Norvald 94133
Norvald.Nesse@chemeng.ntnu.no

Preisig, Heinz 92807
preisig@chemeng.ntnu.no

Rytter, Erling (Adjunct) 94147
err@statoil.no

Sjoblom, Johan 95505
johsj@chemeng.ntnu.no

Skogestad, Sigurd (Head) 94154
Sigurd.Skogestad@chemeng.ntnu.no

Stenius, Per (Adjunct) 94030
per.stenius@chemeng.ntnu.no
Svendsen, Hallvard 94100
Hallvard.Svendsen@chemeng.ntnu.no
Thorsen, Gunnar (Emeritus) 94115
Gunnar.Thorsen@chemeng.ntnu.no

Associate Professors

Andreassen, Jens-Petter 94209
jensp@chemeng.ntnu.no
Haanaes, Egil (Emeritus) 94137
egil.haanas@chemeng.ntnu.no
Haanæs, Egil (Emeritus) 94137
egil.haanas@chemeng.ntnu.no
Haug-Warberg, Tore 94108
Tore.Haug-Warberg@chemeng.ntnu.no
Moe, Stoerker T. 94032
Stoerker.Moe@chemeng.ntnu.no
Rønning, Magnus 94121
Magnus.Ronning@chemeng.ntnu.no

Accredited by:

Degrees granted 2003-2004:

M.S.: 41 Ph.D.: 9

Department reports to: David Nicholson (47 73596204)

Philippines

Bicol University

College of Engineering
Chemical Engineering Department
Campus 1
EM's Bo.
Legazpi City 4500 Philippines

University phone ....
University FAX ....
Department phone ....
Department FAX ....

Professors

Padre, Natividad (Chair)

Associate Professors

Macam, Anelia
annmacam@yahoo.com

Accredited by:

Graduate advisor:

Undergraduate advisor: Natividad M.Padre,Ph.D.

De La Salle University

Chemical Engineering Department
2nd Floor,Velasco Building
De La Salle University
2401 Taft Avenue, Manila

University phone .... (632) 524-4611
University FAX .... (632)526-1403
Department phone .... (632) 536-0257
Department FAX .... (632) 524-0563

Professors

Gallardo, Susan M.
gallardos@dlsu.edu.ph
Gaspillo, Pag-Asa D.
gaspillop@dlsu.edu.ph
Olano, Servillano, Jr. S.B.
olanos@dlsu.edu.ph
Roces, Susan A
rocess@dlsu.edu.ph
Salazar, Carlito M.
salazarc@dlsu.edu.ph

Associate Professors

Abella, Leonila C. (Chair)
abella@dlsu.edu.ph
Auresenia, Joseph
aureseniaj@dlsu.edu.ph
Bacani, Florinda T.
bacanif@dlsu.edu.ph
Brondial, Yolanda P. (Associate Chair)
brondialy@dlsu.edu.ph
Cabigon, Noel P
cabigonn@dlsu.edu.ph
Maridable, Julius B. (Dean)
maridablej@dlsu.edu.ph
Razon, Luis F.
razonl@dlsu.edu.ph
Tan, Raymond R.
tanr_a@dlsu.edu.ph

Assistant Professors

Baraoidan, Wilheliza A.
baraoidanw@dlsu.edu.ph
Uy, Marylou M.
uym@dlsu.edu.ph

Accredited by: PAASCU, CHED

Degrees granted 2003-2004:

B.S.: 29 M.S.: 6 Ph.D.: 1

Graduate advisor: Dr. Pag-asa Gaspillo

Undergraduate advisor: Dr. Leonila Abella

Student organization: CHEN Society
Advisor: Mr. Teo Monroy

Department reports to: Dean of Engineering

University of Philippines at Los Banos

Chemical Engineering Department
College of Engineering and Agro-Industrial Technology
College, Laguna 4031, Philippines

University phone 63-49-536-2508
Department phone 63-49-536-2315
Department FAX 63-49-536-2315

Professors

Abrigo, Casiano J. S. 63-49-536-2315
csaj@mudspring.uplb.edu.ph
Acda, Reynaldo I. (Dean) 63-49-536-2315
ria@mudspring.uplb.edu.ph
Movillon, Jovita L. 63-49-536-2315
jlm@mudspring.uplb.edu.ph

Assistant Professors

Alcanzare, Robyn J. C. 63-49-536-2315
dcalcanzare@hotmail.com
Arquiza, Apollo C. 63-49-536-2315
aca@mudspring.uplb.edu.ph
Borines, Myra G. 6349-536-2315
Demafelis, Rex B. (Chair) 63-49-536-2315
rbd@mudspring.uplb.edu.ph
Nabatilan, Marilou M. 63-49-536-2315
mmn@mudspring.uplb.edu.ph
Navarro, Ronald R. 63-49-536-2315
rrn@mudspring.uplb.edu.ph
Promentilla, Michael A. B. 63-49-536-2315
mabp@mudspring.uplb.edu.ph
Saquing, Carl D. 63-49-249-2315
Valencia, Sixto A. 63-49-536-2315
sav@mudspring.uplb.edu.ph

Student organization: Society of Chemical Engineering Students
Advisor: Sixto Valencia, Rex Demafelis

Department reports to: Reynaldo I Acda, Dean of CEAT

Placement service: Rex B. Demafelis

University of Santo Tomas

Department of Chemical Engineering
Roque Ruaqo Buliding
Espaqa, Manila 1008

University phone 00632 7313101
University FAX 00632 406 1611
Department phone .... 00632 7313101 loc. 8275
Department FAX 00632 7314041

Professors

Laurito, Evelyn (Head) .... 00632 7313101 loc. 8275
odgs@ust.edu.ph
Mabini, Marilyn (Dean) .... 00632 7313101 loc. 8275
odgs@ust.edu.ph

Accredited by:

Graduate advisor: Christina Castro Cabral

Undergraduate advisor: Dr. Marilyn C. Mabina

Poland

Technical University of Lodz

Faculty of Process and Environmental Engineering
ul. Wolczanska 213/215
93-005 Lodz

University phone 48 42 6312000
Department phone 48 42 6313700
Department FAX 48 42 6365663

Professors

Gorak, Andrzej (Visiting)
Heim, Andrzej (Dean)
Kaminski, Wladyslaw
Ledakowicz, Stanislaw
Mucha, Maria
Piddubniak, Oleksa
Strumillo, Czeslaw (Emeritus)
Swiatkowski, Witold (Emeritus)
Wodzinski, Piotr
Zarzycki, Roman
Zbicinski, Ireneusz (Associate Dean)

Associate Professors

Chacuk, Andrzej
Dziubinski, Marek
Kuncewicz, Czeslaw (Associate Dean)
Nowicki, Lech
Pakowski, Zdzislaw
Petera, Jerzy
Skrzypski, Jerzy

Department reports to: Dean of the Faculty of Process and Environmental Engineering

Silesian Technical University

Faculty of Chemistry
Department of Chemical and Process Engineering
ul. M. Strzody 7,
44-101 Gliwice
Deliveries: Professor Andrzej Jarzebski

University phone 48-32-2312349
Department phone 48-32-2371461
Department FAX 48-32-2371461

Professors

Jarzebski, Andrzej (Director) 48 32 2371461
a.jarzeb@iich.gliwice.pl

Associate Professors

Palica, Michal (Head) 48 32 2372832
mpalica@zeus.polsl.gliwice.pl
Synowiec, Piotr 48 32 2372704
pm.synowiec@zeus.polsl.gliwice.pl
Thullie, Jan 48 32 2371461
jthullie@zeus.polsl.gliwice.pl

Assistant Professors

Dzido, Grzegorz (Adjunct) 48 32 2371636
dzido@zeus.polsl.gliwice.pl
Gierczycki, Andrzej (Adjunct) 48 32 2372770
andgier@zeus.polsl.gliwice.pl
Kocurek, Jacek 48 32 2371636
jkocurek@zeus.polsl.gliwice.pl
Merta, Henryk (Adjunct) 48 32 2372795
merta@zeus.polsl.gliwice.pl
Raczek, Jerzy (Adjunct) 48 32 2372783
jraczek@zeus.polsl.gliwice.pl
Smyczek, Helena (Adjunct) 48 32 2371937
Wojcik, Janusz (Associate Dean) 48 32 2372783
janusz.wojcik@polsl.gliwice.pl
Ziolo, Jozef 48 32 2372795
ziolo@zeus.polsl.gliwice.pl

Accredited by: PTChem, SiTPChem, NOT

Graduate advisor: Professor Jan Thullie

Undergraduate advisor: Jerzy Raczek, Ph.D.

Department reports to: Jan Zawadiak, Dean of Faculty of Chemistry

Placement service: Krzysztof Piotrowski, Ph.D.

Technical University of Szczecin

Department of Chemical Engineering
and Environmental Protection Processes
Al. Piastow 42
71-065 Szczecin

University phone	91 4346751
Department phone	91 4494472
Department FAX	91 4494642

Professors

Jaworski, Zdzislaw	4494020 †	jaworski@carbon.tuniv.szczecin.pl
Karcz, Joanna (Head)	4494335	joanka@carbon.tuniv.szczecin.pl
Kosmider, Joanna	4494519	jakos@carbon.tuniv.szczecin.pl
Masiuk, Stanislaw	4494847	smasiuk@carbon.tuniv.szczecin.pl
Paderewski, Mscislaw (Emeritus)	4494426	mpader@carbon.tuniv.szczecin.pl
Strek, Fryderyk (Emeritus)	4494560	stef@carbon.tuniv.szczecin.pl

Associate Professors

Dudczak, Jan	4494922	dudus@carbon.tuniv.szczecin.pl
Haba, Alfred (Emeritus)	4494528	ahaba@carbon.tuniv.szczecin.pl
Nastaj, Jozef	4494084	jonas@carbon.tuniv.szczecin.pl
Szaniawska, Daniela	4494688	dszan@carbon.tuniv.szczecin.pl

Assistant Professors

Ambrozek, Bogdan	4494622	ambog@carbon.tuniv.szczecin.pl
Derecki, Wladyslaw	4494023	wder@carbon.tuniv.szczecin.pl
Gabrus, Elzbieta	4494925	elga@carbon.tuniv.szczecin.pl
Kawecka-Typek, Julita	4494183	Julita.Kawecka-Typek@ps.pl
Kuzniewska-Lach, Irena	4494528	drikula@carbon.tuniv.szczecin.pl
Lach, Krzysztof	4494436	klach@carbon.tuniv.szczecin.pl
Lacki, Henryk	4494155	hlacki@carbon.tuniv.szczecin.pl
Majkut, Aleksander	4494437	almat@carbon.tuniv.szczecin.pl
Major, Marta	4494243	mmajor@carbon.tuniv.szczecin.pl
Michalska, Marzena	4494243	mamich@carbon.tuniv.szczecin.pl
Paterkowski, Wojciech	4494399	wpater@carbon.tuniv.szczecin.pl
Suszek, Ewa (Associate Dean)	4494731	evelinas@carbon.tuniv.szczecin.pl
Zakrzewska, Barbara	4494159	zakrzewska@carbon.tuniv.szczecin.pl

† Faculty numbers have prefix 91.

Department reports to: Prof. Eugeniusz Milchert, Dean of College

Technical University of Szczecin, Faculty of Chemical Engineering

Institute of Chemical and Environment Engineering
Pulaskiego 10
70-322 Sczecin

University phone	(4891) 4494105
Department phone	(4891) 4494730
Department FAX	(4891) 4330352

Professors

Arabczyk, Walerian (Head)	(4891) 449-4597	arab@mailbox.ps.pl
Grzmil, Barbara	(4891) 449-4211	bg@mailbox.ps.pl
Kaleñczuk, Ryszard J. (Director)	(4891) 449-4656	rk@mailbox.ps.pl
Kaucki, Kazimierz (Dean)	(4891) 449-4326	kalucki @mailbox.ps.pl
Morawski, A. W. (Headamor@mailbox.ps.pl)	(4891) 449-4474	
Narkiewicz, Urszula	(4891) 449-4687	un@mailbox.ps.pl
Tomaszewska, Maria	(4891) 449-4367	mariatom@mailbox.ps.pl

Assistant Professors

Gryta, Marek	(4891) 449-4682	margryta @mailbox.ps.pl
Grzechulska, Joanna	(4891) 449-4277	joanna@mailbox.ps.pl
Karakulski, Krzysztof	(4891) 449-4277	peteatp1@mailbox.ps.pl
Lendzion-Bieluñ, Zofia	(4891) 449-4397	zosi@mailbox.ps.pl
Lewicki, Marek	(4891) 449-4211	
Michalkiewicz, Beata	(4891) 449-4096	cydzik@mailbox.ps.pl
Moszyñski, Dariusz	(4891) 449-4730	Mosiek@mailbox.ps.pl
Przepiórski, Jacek	(4891) 449-4277	jacek@mailbox.ps.pl
Ziebro, Janusz	(4891) 449-4132	jvp@mailbox.ps.pl

Warsaw University of Technology

Faculty of Chemical and Process Engineering
ul. Warynskiego 1
00-645 Warszawa

University phone	(4822) 660 7220
Department phone	(4822) 660 6369
Department FAX	(4822) 825 1440

Professors

Baldyga, Jerzy (Associate Dean)	660 6376 †	baldyga@ichip.pw.edu.pl
Bin, Andrzej	660 6373	bina@ichip.pw.edu.pl
Chmielewski, Andrzej	660 6249	chmielewski@ichip.pw.edu.pl
Gawronski, Roman	660 6276	gawronski@ichip.pw.edu.pl
Gradon, Leon (Dean)	660 6279	gardon@ichip.pw.edu.pl

Marcinkowski, Ryszard (Emeritus) 660 6419
marcinkowski@ichip.pw.edu.pl
Pohorecki, Ryszard 660 6314
pohorecki@ichip.pw.edu.pl
Sieniutycz, Stanislaw 825 6340
sieniutycz@ichip.pw.edu.pl
Szewczyk, Krzysztof 660 6413
szewczyk@ichip.pw.edu.pl
Szwast, Zbigniew (Associate Dean) 825 6340
szwast@ichip.pw.edu.pl
Urbanek, Andrzej (Emeritus) 660 6352
Warych, Jerzy 660 6372
warych@ichip.pw.edu.pl
Wolny, Andrzej 660 6350
wolny@ichip.pw.edu.pl
Wronski, Stanislaw (Emeritus) 660 6295
wronski@ichip.pw.edu.pl

Associate Professors

Molga, Eugeniusz 660 6293
molga@ichip.pw.edu.pl
Piatkiewicz, Wojciech 660 5352
Podgorski, Albert 660 6351
podgorsa@ichip.pw.edu.pl

† Faculty numbers have prefix 4822.

Accredited by: Conference Rectors Academic Schools, Poland

Graduate advisor: Dr. Molga Eugeniusz

Undergraduate advisor: Prof. Szwast Zbigniew

Department reports to: Mankowski Stanislaw, Rector of Warsaw University of Technology

Wroclaw Technical University

Institute of Chemical Engineering and Heating Equipment
ul. Norwida 4/6
50-373 Wroclaw

University phone 48 71 227336
Department phone 48 71 3280475
Department FAX 48 71 3280475

Professors

Kmiec, Andrzej .... 3202838 †
kmiec@iic.pwr.wroc.pl
Koltuniewicz, B. A. (Head) 3202919
akolt@iic.pwr.wroc.pl
Koziol, Antoni 3203358
koziol@iic.pwr.wroc.pl
Noworyta, Andrzej 3281318
noworyta@iic.pwr.wroc.pl
Pajak, Maksymilian 3203567
pajak@iic.pwr.wroc.pl
Skoczylas, Adam (Emeritus) 3202558

Associate Professors

Kawala, Zdzislaw 3203437
kawala@iic.pwr.wroc.pl
Sadowski, Zygmunt 3203863
sadowski@iic.pwr.wroc.pl
Zaleski, Janusz 3203463
zaleski@iic.pwr.wroc.pl

Assistant Professors

Bryjak, Jolanta 3202677
bryjak@iic.pwr.wroc.pl
Budzianowski, Wojciech 3203814
budzianowski@iic.pwr.wroc.pl
Dziak, Janusz 3202975
dziak@iic.pwr.wroc.pl
Gorska, Lucyna 3203908
gorska@iic.pwr.wroc.pl
Kaplon, Jacek 3203206
kaplon@iic.pwr.wroc.pl
Kolek, Andrzej 3203314
kolek@iic.pwr.wroc.pl
Krolikowski, Lechoslaw 3203334
lkroliko@iic.pwr.wroc.pl
Kuzniar, Janusz 3203425
kuzniar@iic.pwr.wroc.pl
Ludwig, Wojciech 3202975
ludwig@iic.pwr.wroc.pl
Lugowski, Zygmunt 3203814
lugowski@iic.pwr.wroc.pl
Mielczarski, Sylwester 3202838
sylmiel@iic.pwr.wroc.pl
Nowinska, Elzbieta 3203206
Sawinski, Wojciech 3203649
sawinski@iic.pwr.wroc.pl
Skrzypinski, Wojciech 3202406
skrzyp@iic.pwr.wroc.pl
Trusek-Holownia, Anna 3203908
trusek@iic.pwr.wroc.pl

† All phone numbers have prefix 48 71.

Placement service: Irena Zielinska

Portugal

Universidade de Coimbra

Departamento de Engenharia Química
Pólo II
Pinhal de Marrocos
3030-290 Coimbra

Department phone 351 239 798700
Department FAX 351 239 798703

Professors

Figueiredo, M. M. L. 351239798739
mmf@eq.uc.pt
Lobo, Lelio Q. (Chair) 351239798733
Silva, A. T. da 351239798726
ats@eq.uc.pt

Associate Professors

Castro, Jose A. A. M. 351239798740
eq1jmc@eq.uc.pt
Ferreira, Rosa M. O. Q. 351239798723
rqf@eq.uc.pt
Garcia, Fernando A. P. 351239798724
fgarcia@eq.uc.pt
Gil, M. H. M. 351239798743
hgil@eq.uc.pt
Portugal, Antonio A. T. G. 351239798741
atp@eq.uc.pt
Rasteiro, M. G. V. 351239798725
mgr@eq.uc.pt

Assistant Professors

Baptista, Cristina M. G. 351239798738
cristina@eq.uc.pt

Carvalho, M. G. V. S. 351239798746
mgc@eq.uc.pt

Ferreira, Abel G. M. 351239798729
eq1amf@eq.uc.pt

Ferreira, Licinio M. G. A. 351239798745
eq1lmf@eq.uc.pt

Ferreira, Paulo J. T. 351239798747
paulo@eq.uc.pt

Fonseca, Isabel M. A. 351239798728
fonseca@eq.uc.pt

Oliveira, Nuno M. C. 351239798742
nuno@eq.uc.pt

Rocha, Jorge M. S. 351239798731
jrocha@eq.uc.pt

Saraiva, Pedro M. T. L. A. 351239798730
eq1pas@eq.uc.pt

Sousa, Hermínio J. S. 351239798749
hsousa@eq.uc.pt

Accredited by: Portuguese Institution of Engineers, (Ordem dos Engenheiros)

Department reports to: Dean and Science Dean of Faculdade de Ciências e Tecnologia

Universidade do Porto

Departamento de Engenharia Quimica
Faculdade de Engenharia
Rua Dr Roberto Frias s/n
4200-465 Porto

University phone .... (351) 22 508 1400
Department phone .... (351) 22 508 1884
Department FAX .... (351) 22 508 1449

Professors

Azevedo, Sebastiao J.C.F. (Chair) ...... (351) 22 508 1647
sfeyo@fe.up.pt

Costa, Carlos .... (351) 22 508 1670
ccosta@fe.up.pt

de Carvalho, João R. G. .... (351) 22 508 1640
jrguedes@fe.up.pt

Figueiredo, José L. C. .... (351) 22 508 1663
jlfig@fe.up.pt

Medina, A. G. .... (351) 22 508 1656
amedina@fe.up.pt

Melo, Luís .... (351) 22 508 1588
lmelo@fe.up.pt

Rodrigues, Alírio E. .... (351) 22 508 1671
arodrig@fe.up.pt

Salcedo, Romualdo L. R. (Associate Chair) (351) 22 508 1644
rsalcedo@fe.up.pt

Associate Professors

Alves, Arminda .... (351) 22 508 1883
aalves@fe.up.pt

Campos, João M. .... (351) 22 508 1692
jmc@fe.up.pt

Costa, Mário R. .... (351) 22 508 1666
mrcosta@fe.up.pt

Dias, Madalena .... (351) 22 508 1661
dias@fe.up.pt

Faria, Joaquim L. .... (351) 22 508 1645
jlfaria@fe.up.pt

Gonçalves, Maria P. .... (351) 22 508 1684
pilarg@fe.up.pt

Lopes, Jose C. .... (351) 22 508 1667
lopes@fe.up.pt

Loureiro, José M. .... (351) 22 508 1672
loureiro@fe.up.pt

Macedo, Eugénia .... (351) 22 508
eamacedo@fe.up.pt

Martins, J. I. .... (351) 22 508 1643
jipm@fe.up.pt

Mendes, Adélio .... (351) 22 508 1695
mendes@fe.up.pt

Pinto, Alexandra .... (351) 22 508 1675
apinto@fe.up.pt

Sereno, Alberto C. .... (351) 22 508 1655
sereno@fe.up.pt

Órfão, José ... . (351) 22 508 1665
jjmo@fe.up.pt

Assistant Professors

Alves, Manuel .... (351) 22 508 1680
mmalves@fe.up.py

Barbosa, Domingos .... (351) 22 508 1660
dbarbosa@fe.up.pt

Bastos, João .... (351) 22 508 1658
jbastos@fe.up.pt

Botelho, Cidália .... (351) 22 508 1885
cbotelho@fe.up.pt

Coelho, Manuel .... (351) 22 508 1679
mcoelho@fe.up.pt

Ferraz, Maria C. A. .... (351) 22 508 1688
aferraz@fe.up.pt

Ferreira, Palmira O. .... (351) 22 508 1668
pof@fe.up.pt

Madeira, L. M. .... (351) 22 508 1519
mmadeira@fe.up.pt

Magalhães, Fernão D. .... (351) 22 508 1601
fdmagalh@fe.up.pt

Martins, Fernando G. .... (351) 22 508 1974
fgm@fe.up.pt

Mendonça, João .... (351) 22 508 1654
mendonca@fe.up.pt

Morgado, José
jmorgado@citeve.pt

Nunes, Olga P. .... (351) 22 508 1917
opnunes@fe.up.pt

Pereira, M. do C. .... (351) 22 508 1590
mccsp@fe.up.pt

Pereira, M. F. . ... (351) 22 508 1468
fpereira@fe.up.pt

Rocha, Fernando A. N. .... (351) 22 508 1678
frocha@fe.up.pt

Santos, Lúcia .... (351) 22 508 1682
lsantos@fe.up.pt

Soares, Helena .... (351) 22 508 1650
hsoares@fe.up.pt

Tavares, Manuel
manuel.leao@petrogal.pt

Accredited by: Ordem dos Engenheiros (Portugal)

Undergraduate advisor: Sebastião Feyo de Azevedo

Department reports to: Carlos A. V. da Costa, Dean, Fac. of Engineering

Instituto Superior Técnico

Department of Chemical Engineering
Torre Sul
1049-001 LISBOA
PORTUGAL

University phone	00351218417259
Department FAX	00351218417246

Associate Professors

Bordado, João C. (Associate Chair) 00351218419182
jcbordado@ist.utl.pt

Accredited by: Ordem dos Engenheiros(Engineers Profissional Association, FUP/CNAVES Fundação das Univ. Portuguesas

Student organization: Filipe Janela

Department reports to: President of the Institute

Placement service: Gire/ Sofia Cabeleira

Qatar

University of Qatar

College of Engineering
P. O. Box 2713
Doha, Qatar

University phone	974-483-5053
Department phone	974-485-2100
Department FAX	974-485-2101

Professors

Al-Mashat, Ali .... 485-2122 †
almashat@qu.edu.qa

Moshfeghian, Mahmoud 485-2120
mosfeg@qu.edu.qa

Wilson, Andrew J. (Head) 485-2100
awilson@qu.edu.qa

Associate Professors

Abdalla, Babiker K. 485-2119
bkamdr@qu.edu.qa

Assistant Professors

Al-Sayegh, Abdel-Redha 485-2123
ralsaygh@qu.edu.qa

Alfadalah, Hassan 485-2985
alfadala@qu.edu.qa

Mehdizadeh, Hossein 485-2121
DHM@qu.edu.qa

† Faculty numbers have prefix 974.

Department reports to: Dean, College of Engineering

Romania

Petroleum and Gas Institute

Department of Chemical Engineering
Bv. Bucuresti No 39
RO-2000 Ploiesti

University phone	40 44 17 31 71
Department phone	40 44 17 62 11

Professors

Antonescu, Lucia
Cuiban, Flavian
Dobrinescu, Dumitru
Ionescu, Constantin
Marinoiu, Vasile
Paraschiv, Nicolae
Patrascu, Maximiliana
Pavel, Alecsandru
Schiopescu, Alexandra (Head)
Stratula, Costica (Head)
Tescan, Vasile

Associate Professors

Apostol, Daniela (Head)
Bohiltea, Ion
Bolocan, Ion (Dean)
Brebeanu, Gheorghe
Dumitrescu, Vasile
Manea, Floarea
Matei, Vasile
Moraru, Maria
Neagoe, Stelian
Onutu, Ion
Oprea, Florin
florin@oprea.org
Opris, Ion
Patrascioiu, Cristian
Pintea, Octav
Rosca, Paul
Stanica-Ezeanu, Dorin
Sterry, Anca

Assistant Professors

Barbu, Eugen
Borcea, Anca-Florentina
Camenita, Dan A.
Cigolea, Victor
Ciuparu, Dragos
Costache, Liliana
Cristea, Steliana
Cristescu, Carmen
Dinca, Ana
Enache, Dan
Matei, Dorina
Panaitescu, Casen
Patrascu, Cornel
Petre, Mihaela
Popovici, Daniela
Sburlea, Gheorghe
Tanasescu, Ion
Trimbitasu, Eva

Accredited by: National Council for Academic Accreditation

Graduate advisor: Bolocan, Ion

Department reports to: Bolocan, Ion, Dean of Fac. of Petroleum Refining and Petrochemistry

Placement service: Bolocan Ion

Saudi Arabia

King Abdulaziz University

Chemical and Materials Engineering Department
P. O. Box 80204
Jeddah, 21589

University phone .... (02) 6400000
Department phone .... (02) 6952187
Department FAX .... (02) 6952257

Professors

Al-Zahrani, A. A. .... x68246 †
aaoz@hotmail.com
Alp, T. Y. x4042
Zahed, A. H. x4248
Zaki, M. H. x4213

Associate Professors

Abdulsalam, Mohemmad I. (Chair) x68043
MAbdulsalam@kaau.edu.sa
Alhamed, Yahiya A. x4244
yasalhamed@hotmail.com
Bashir, M. D. x4039
mdbashir@hotmail.com
Daous, M. A. x4525
mdaous@hotmail.com
Dawoud, U. M. (Chair) x4271
zcc3318@kaau.edu.sa
Iskanderani, F. I. x4254
Noorwall, M. M. x4038
noorwali@yahoo.com

Assistant Professors

Al-Tajam, M. A. x4040
maltajam@hotmail.com
Al-Turaif, Hamad A. x4044
halturaif@hotmail.com
Wazzan, A. A. x4249
awazzan@SAKAAU03.Bitnet

† Dial (02) 6402000 and ask for extension.

Accredited by: Ministry of Higher Education, Kingdom of Saudi Arabia

Graduate advisor: Uthman Dawoud

Undergraduate advisor: Abdulaziz A. Wazzan

Placement service: Uthman Dawoud

King Fahd University of Petroleum & Minerals

Chemical Engineering Department
PO Box 5050
Dhahran 31261

University phone 966-3-860-0000
Department phone 966-3-860-2205
Department FAX 966-3-860-4234

Professors

Al-Saleh, Muhammad A. .... 860-2029 †
masaleh@kfupm.edu.sa
Al-Shalabi, Mazen A. 860-2198
mshalabi@kfupm.edu.sa
Al-Zakri, Abdullah S. 860-3803
aszakri@kfupm.edu.sa
Amin, Mohamed B. (Chair) 860-2205
mbamin@kfupm.edu.sa
Inui, Tomoyuki
Maadhah, Ali G. 860-3882
amaadhah@kfupm.edu.sa
Shaikh, Abdullah A. 860-2257
aashaikh@kfupm.edu.sa

Associate Professors

Abbas, Nureddin M. 860-3667
abbasnm@kfupm.edu.sa
Abu-Sharkh, Basel 860-2744
sharkh@kfupm.edu.sa
Abul-Hamayel, Mohammad A. 860-3903
hamayel@kfupm.edu.sa
Al-Ali, Habib H. 860-2599
hhali@kfupm.edu.sa
Al-Harbi, Dulaihan K. 860-2375
dharbi@kfupm.edu.sa
Al-Naafa, Mohammad A. 860-3688
manaafa@kfupm.edu.sa
Kahraman, Ramazan 860-4987
kahraman@kfupm.edu.sa
Loughlin, Kevin F. 860-2049
loughkf@kfupm.edu.sa
Rahman, S.U. 860-2219
srahman@kfupm.edu.sa

Assistant Professors

Al-Arfaj, Muhammad A. 860-1694
maarfaj@kfupm.edu.sa
Al-Baghli, Nadhir A. 860-1476
nalbaghli@kfupm.edu.sa
Al-Khattaf, Sulaiman S. 860-1429
skhattaf@kfupm.edu.sa
Al-Mubaiyedh, Usamah A. 860-1528
usamah@kfupm.edu.sa
Alnaizy, Raafat 860-1492
alnaizy@kfupm.edu.sa
Hussein, Ibnelwaleed A. 860-2235
ihussein@kfupm.edu.sa
Zaidi, S.M. J. 860-1242
zaidismj@kfupm.edu.sa
Zughbi, Habib D. 860-4729
hdzughbi@kfupm.edu.sa

† Direct dial with prefix 03.

Accredited by: ABET

Degrees granted 2003-2004:

B.S.: 88 M.S.: 6

Graduate advisor: Kevin F. Loughlin

Student organization: ChE Club
Advisor: Dr. Nadhir A. Al-Baghli

Department reports to: Dr. Samir A. Al-Baiyat, Dean, College of Engineering

King Saud University

Chemical Engineering Department
P.O. Box 800
Riyadh 11421

Department phone 966-1-4676851
Department FAX 966-1-4678770

Professors

Abasaeed, E. A. .... 76856 †
abasaeed@ksu.edu.sa

Ajbar, M. A. 76843
aajbar@ksu.ed.sa

Al-Fariss, F. T. 76875
fariss@ksu.edu.sa

Al-Mutaz, S. I. (Chair) 76870
almutaz@ksu.edu.sa

Fakeeha, H. A. 76847
anishf@ksu.edu.sa

Mustafa, M. H. 76854
hmohm@ksu.edu.sa

Wagialla, M. K. 76846
wagialla@ksu.edu.sa

Associate Professors

Abashar, E. E. M. 75843
mabashar@ksu.edu.sa

Al-Ahmed, I. M. 76874
malahmad@ksu.edu.sa

Al-Hazzaa, I. M. 76845
masai@ksu.edu.sa

Al-Zahrani, M. S. 76873
szahrani@ksu.edu.sa

Alhumaizi, l. K. 76813
humaizi@ksu.edu.sa

Ali, M. E. 76871
amkamal@ksu.edu.sa

Almasry, A. W. 76853
walmasry@ksu.edu.sa

Asif, Mohamed 76849
masif@ksu.edu.sa

Assistant Professors

Al-Hussaini, A. A. 76738
amalik@ksu.edu.sa

Al-Mubaddel, S. F. 76848
mubaddel@ksu.edu.sa

Al-Rabiah, A. A. 76844
arabiah@ksu.edu.sa

Al-Sugair, A. K. 76857
alsugair@ksu.edu.sa

Al-Sughair, S. Y. 76855
yszs@ksu.edu.sa

Benaissa, N. M. 76850
benaissa@ksu.edu.sa

Jameel, A. T. 76842
atjameel@ksu.edu.sa

Khaznadar, K. M. 76865
mkkhazn@ksu.edu.sa

† Direct dial with prefix 966-1-46.

Accredited by: ABET

Singapore

National University of Singapore

Department of Chemical and Biomolecular Engineering
Blk E5 #02-09, 4 Engineering Drive 4
Singapore 117576

University phone .... (65) 6775 6666
Department phone .... (65) 6874 2186
Department FAX .... (65) 6779 1936

Professors

Chung, T. S. (Neal) .... (65) 6874-6645
chencts@nus.edu.sg

Kang, En T. .... (65) 6874-2189
cheket@nus.edu.sg

Lee, Jim Y. .... (65) 6874-2899
cheleejy@nus.edu.sg

Neoh, Koon G. .... (65) 6874-2176
chenkg@nus.edu.sg

Rajagopalan, Raj (Head) .... (65) 6874-2186
chehead@nus.edu.sg

Tan, Thiam C. .... (65) 6874-2192
chetantc@nus.edu.sg

Yap, Miranda G. S. .... (65) 6874-6369
cheyapm@nus.edu.sg

Associate Professors

Bai, Renbi .... (65) 6874-4532
chebairb@nus.edu.sg

Balasubramanian, Rajasekhar .... (65) 6874-5135
cherbala@nus.edu.sg

Chen, Shing B. .... (65) 6874-5237
checsb@nus.edu.sg

Chiu, Min S. .... (65) 6874-2223
checms@nus.edu.sg

Farooq, Shamsuzzaman .... (65) 6874-6545
chesf@nus.edu.sg

Feng, Si S. .... (65) 6874-3835
chefss@nus.edu.sg

Foo, Swee C. .... (65) 6874-8721
chefoosc@nus.edu.sg

Garland, Marc .... (65) 6874-6617
chemvg@nus.edu.sg

Hidajat, Kus .... (65) 6874-2191
chehidak@nus.edu.sg

Hong, Liang .... (65) 6874-5029
chehongl@nus.edu.sg

Karimi, Iftekhar A. .... (65) 6874-6359
cheiak@nus.edu.sg

Kawi, Sibudjing .... (65) 6874-6312
chekawis@nus.edu.sg

Kocherginsky, Nikolai M. .... (65) 6874-5083
chenk@nus.edu.sg

Krishnaswamy, P. R. .... (65) 6874-2177
chekrish@nus.edu.sg

Loh, Kai C. .... (65) 6874-2174
chelohkc@nus.edu.sg

Obbard, Jeffrey P .... (65) 6874-2884
chejpo@nus.edu.sg

Pehkonen, Simo O. .... (65) 6874-4729
chesop@nus.edu.sg

Rangaiah, Gade P. .... (65) 6874-2187
chegpr@nus.edu.sg

Ray, Ajay K. .... (65) 6874-8049
cheakr@nus.edu.sg

Ray, Madhumita B. .... (65) 6874-2885
cheraym@nus.edu.sg
Srinivasan, Madapusi P. .... (65) 6874-2171
chesmp@nus.edu.sg
Srinivasan, Rajagopalan .... (65) 6874-8041
chergs@nus.edu.sg
Stanforth, Robert R. .... (65) 6874-8045
chers@nus.edu.sg
Tan, Reginald B. H. .... (65) 6874-6360
chetanbh@nus.edu.sg
Teo, Wah K. .... (65) 6874-6614
cheteowk@nus.edu.sg
Ti, Hwei C. .... (65) 6874-2188
chetihc@nus.edu.sg
Ting, Yen P. .... (65) 6874-2190
chetyp@nus.edu.sg
Uddin, M. S. .... (65) 6874-2886
cheshahb@nus.edu.sg
Wang, Chi H. .... (65) 6874-5079
chewch@nus.edu.sg
Zeng, Hua C. .... (65) 6874-2896
chezhc@nus.edu.sg

Assistant Professors

Chen, Jiaping .... (65) 6874-8092
checjp@nus.edu.sg
Choe, Woo S. .... (65) 6874-5939
checws@nus.edu.sg
Favelukis, Moshe .... (65) 6874-5133
chemf@nus.edu.sg
Saeys, Mark .... (65) 6874-5826
chesm@nus.edu.sg
Samavedham, Lakshminarayanan .... (65) 6874-8484
chels@nus.edu.sg
Tong, Yen W. .... (65) 6874-8467
chetyw@nus.edu.sg
Trau, Dieter W. .... (65) 6874-8052
chetrau@nus.edu.sg
Yu, Liya .... (65) 6874-6474
cheley@nus.edu.sg
Yung, Lanry L. Y. .... (65) 6874-1699
cheyly@nus.edu.sg
Zhao, George X. S. .... (65) 6874-4727
chezxs@nus.edu.sg

Accredited by: IChemE(UK), ABET Substantial Equivalency

Degrees granted 2003-2004:

B.S.: 231 M.S.: 173 Ph.D.: 16

Undergraduate advisor: G. P. Rangaiah

Student organization: Chemical and Biomolecular Engineering Students Society

Department reports to: Seeram Ramakrishna, Dean of Engineering

Slovenia

University of Ljubljana

Faculty of Chem and Chem. Technology
Askerceva 5
1000 Ljubljana
Deliveries: P.O.B. 537, SLO-1001 Ljubljana

University phone 386 1 2419 100
Department phone 386 1 2419 520
Department FAX 386 1 2419 530

Professors

Golob, Janvit (Associate Dean) .... x522 †
janvit.golob@uni-lj.si
Koloini, Tine x500
tine.koloini@uni-lj.si
Levec, Janez x502
janez.levec@uni-lj.si
Modic, Roman (Emeritus) x520
Pejovnik, Stanislav x202
stane.pejovnik@uni-lj.si
Pohar, Ciril x408
ciril.pohar@uni-lj.si
Zumer, Miha x504
miha.zumer@uni-lj.si

Associate Professors

Berovic, Marin x510
marin.berovic@uni-lj.si
Macek, Jadran x200
jadran.macek@uni-lj.si
Pavko, Aleksander x506
saso.pavko@uni-lj.si
Zagorc-Koncan, Jana x537
jana.zagorc@ki.si

Assistant Professors

Krajnc, Matjaz x524
matjaz.krajnc@uni-lj.si
Plazl, Igor x512
igor.plazl@uni-lj.si

† Dial 386 1 2419 and extension.

Department reports to: Tine Koloini

University of Maribor

Fakulteta za kemijo in kemijsko tehnologijo
Faculty of Chemistry and Chemical Engineering
Smetanova 17
SI, 2000 Maribor
Deliveries: Prof.Dr. Peter Glavic

University phone .... (386 2) 23 55 800
Department phone .... (386 2) 22 94 400
Department FAX .... (386 2) 25 27 774

Professors

Dolecek, Valter (Dean) .... (386 2) 22 94 441
valter.dolecek@uni-mb.si
Drofenik, Mihael .... (386 2) 22 94 416
miha.drofenik@uni-mb.si
Glavic, Peter .... (386 2) 22 94 451
glavic@uni-mb.si

Knez, Zeljko .... (386 2) 22 94 461
zeljko.knez@uni-mb.si
Kravanja, Zdravko .... (386 2) 22 94 481
kravanja@uni-mb.si
Krope, Jurij .... (386 2) 22 94 475
jurij.krope@uni-mb.si

Associate Professors

Butinar, Branko .... (386 2) 22 94 489
branko.butinar@uni-mb.si
Stropnik, Crtomir .... (386 2) 22 94 421
crtomir.stropnik@uni-mb.si

Assistant Professors

Ban, Irena .... (386 2) 22 94 417
irena.ban@uni-mb.si
Brodnjak Voncina, Darinka .... (386 2) 22 94 432
darinka.brodnjak@uni-mb.si
Dobcnik, Danilo (Emeritus) .... (386 2) 22 94 431
danilo.dobcnik@uni-mb.si
Goricanec, Darko .... (386 2) 22 94 475
darko.goricanec@uni-mb.si
Gorek, Andreja .... (386 2) 22 94 453
andreja.gorsek@uni-mb.si
Habulin, Maja .... (386 2) 22 94 462
maja.habulin@uni-mb.si
Korpar, Samo .... (386 2) 22 94 490
samo.korpar@ijs.si
Krajnc, Majda .... (386 2) 22 94 452
majda.krajnc@uni-mb.si
Krajnc, Peter .... (386 2) 22 94 422
peter.krajnc@uni-mb.si
Novak Pintaric, Zorka .... (386 2) 22 94 482
zorka.novak@uni-mb.si
Oreki, Severina .... (386 2) 22 94 454
severina.oreski@uni-mb.si
Ozim, Vojko (Emeritus) .... (386 2) 22 94 471
vojko.ozim@uni-mb.si
Petek, Aljana .... (386 2) 22 94 443
aljana.petek@uni-mb.si
Simonic, Marjana .... (386 2) 22 94 472
marjana.simonic@uni-mb.si
Volavek, Bogdan (Emeritus) .... (386 2) 22 94 415
bogdan.volavsek@uni-mb.si
kerget, Mojca .... (386 2) 22 94 463
mojca.skerget@uni-mb.si

Graduate advisor: Zdravko Kravanja, prof. dr.

Undergraduate advisor: Peter Glavic, prof. dr.

Student organization: Vanja Bogadi

Department reports to: Valter Dolecek, Dean

Placement service: Franc Purkeljc, M.Sc.

Republic of South Africa

University of Cape Town

Department of Chemical Engineering
Private Bag
Rondebosch 7701

University phone 27 21 650 9111
Department phone 27 21 650 2518
Department FAX 27 21 650 5501

Professors

Dry, Mark
med@chemeng.uct.ac.za
Hansford, Geoff S.
gsh@chemeng.uct.ac.za
Harrison, Sue T. L.
stlh@chemeng.uct.ac.za
van Steen, Eric (Head) 27 21-6502509
evs@chemeng.uct.ac.za

Associate Professors

Burton, Stephanie
sburton@chemeng.uct.ac.za
Fletcher, Jack C.Q.
jcqf@chemeng.uct.ac.za
Fraser, Duncan M.
dmf@chemeng.uct.ac.za
Gaylard, Peter
pgaylard@chemeng.uct.ac.za
Harris, Peter
pharris@chemeng.uct.ac.za
Lewis, Alison E.
alison@chemeng.uct.ac.za
Moller, Klaus
km@chemeng.uct.ac.za
Petrie, Jim G.
petrie@chemeng.uct.ac.za

Assistant Professors

Bradshaw, Dee J.
db@chemeng.uct.ac.za
Case, Jenni M
jcase@chemeng.uct.ac.za
Chakraborty, Aninda
aninda@chemeng.uct.ac.za
Deglon, Dave A.
dad@chemeng.uct.ac.za
Harris, Martin C
mch@chemeng.uct.ac.za
Musonge, Paul
pmusonge@chemeng.uct.ac.za
von Blottnitz, Harro
hvb@chemeng.uct.ac.za

Accredited by: IChemE (UK), IChemE (SA), ECSA

Degrees granted 2003-2004:
B.S.: 36 M.S.: 14 Ph.D.: 5

Graduate advisor: S Burton

Undergraduate advisor: H von Blottnitz

Department reports to: C. T. O'Connor, Dean

University of Durban-Westville

Department of Chemical Engineering
Private Bag X54001
Durban, 4000
Deliveries: Prof. M. Carsky

University phone 2731 2044111
Department phone 2731 2044187
Department FAX 2731 2044021

Professors

Judd, R. (Emeritus) .... (2731) 2044022
robinjudd@cybertrade.co.za

Associate Professors

Carsky, M. .... (2731) 2044668
mcarsky@pixie.udw.ac.za
Zhelev, T. K. .... (2731) 2044693
zhelev@pixie.udw.ac.za

Assistant Professors

Abba-Omar, M .... (2731) 2044483
Bantcheva, I. .... (2731) 2044270
bantcheva@pixie.udw.ac.za
Bassa, A. F. (Chair) .... (2731) 2044183
afcbassa@pixie.udw.ac.za
Kormuth, E (Joint) .... (2731) 3603194
kormuth@med.und.ac.za
Obwaka, E. .... (2731) 2044748
obwaka@pixie.udw.ac.za

Accredited by: South African Institute of Chemical Engineers, Enginneering Council of South Africa

Student organization: Engineering Student Society

Department reports to: Dean, Faculty of Science and Engineering

University of Natal

School of Chemical Engineering
University of KwaZulu-Natal
Howard College Campus
Durban, 4041
Deliveries: University of Kwazulu-Natal, Howard College Campus, King George V Avenue, DURBAN, 4001

University phone 27-31-2609111
Department phone 27-31-2603115
Department FAX 27-31-2601118

Professors

Buckley, Christopher A. 27-31-2603131
Buckley@ukzn.ac.za
Loveday, Brian K. 27-31-2603121
Loveday@ukzn.ac.za
Mulholland, Mike 27-31-2603123
Mulholland@ukzn.ac.za
Raal, J. D. (Emeritus) 27-31-2603124
Raal@ukzn.ac.za
Ramjugernath, Deresh 27-31-2603128
Ramjuger@ukzn.ac.za

Associate Professors

Arnold, David R. (Head) 27-31-2601228
Arnoldd@ukzn.ac.za
Starzak, Maciej J. 27-31-2603117
Starzak@ukzn.ac.za

Assistant Professors

Pocock, Jon 27-31-2603377
Pocockj@ukzn.ac.za

Accredited by: IChemE (UK), ECSA

Degrees granted 2003-2004:

B.S.: 58 M.S.: 12 Ph.D.: 3

Graduate advisor: Dave Arnold

Department reports to: H.D. Schreiner, Dean

Placement service: Dave Arnold, Head of School

North-West University, Potchefstroom Campus

School of Chemical and Minerals Engineering
Private Bag X6001
Potchefstroom, 2520

University phone 27 (18) 299-1111
Department phone 27 (18) 299-1656
Department FAX 27 (18) 299-1535

Professors

Bruinsma, O.S.L. D. .... 299-1669 †
chioslb@puk.ac.za
Everson, Raymond C. (Emeritus) 299-1986
chirce@puk.ac.za
Kasaini, Henry 299-1664
chihk@puk.ac.za
Vorster, Schalk W. (Emeritus) 299-1660
chiswv@puk.ac.za
Waanders, Frans B. (Director) 299-1994
chifbw@puk.ac.za

Associate Professors

Marx, Sanette 299-1995
chism@puk.ac.za

Assistant Professors

Campbell, Quentin P. 299-1993
chiqpc@puk.ac.za
Koekemoer, Leon R. 299-1953
chilrk@puk.ac.za
le Roux, Marco 299-1990
chimlr@puk.ac.za
van der Gryp, Percy 299-1997
chipvdg@puk.ac.za

† Direct dial with prefix 27 (18).

Accredited by: IChemE(SA), ECSA, IChemE(UK)

Degrees granted 2003-2004:

B.S.: 18 M.S.: 6 Ph.D.: 3

Graduate advisor: Frans Waanders

Undergraduate advisor: Retha Potgieter

Student organization: Engineering Student Society
Advisor: Retha Potgieter

Department reports to: Dean of Faculty of Engineering

University of Pretoria

Chemical Engineering Department
Pretoria 0002

University phone 27 (12) 420-4111
Department phone 27 (12) 420-2197
Department FAX 27 (12) 362-5173

Professors

Chirwa, Evans .... 420-5894 †
evans.chirwa@up.ac.za
deVaal, Philip (Head) 420-2197
philip.devaal@up.ac.za

Focke, Walter 420-2588
xyris@mweb.co.za

Grimsehl, Uys 420-2475
uys.grimsehl@up.ac.za

Heydenrych, Michael 420-2199
mike.heydenrych@up.ac.za

Majozi, Thokozani 420-4130
thoko.majozi@up.ac.za

Mandersloot, Wim (Honorary) 803-6863
wmanders@eng.up.ac.za

Morgan, Dave (Honorary) 420-2856
dave.morgan@up.ac.za

Pretorius, W. A. (Emeritus) 804-3070

Schoeman, Japie 420-3569
japie.schoeman@up.ac.za

Schutte, Frik 420-3571
cschutte@eng.up.ac.za

Skinner, W (Emeritus) 665-0200

Associate Professors

duPlessis, Barend 420-3740
barend.duplessis@up.ac.za

Friend, Francois 420-3741
francois.friend@up.ac.za

Jaffer, Ashraf 420-4173
ashraf.jaffer@up.ac.za

Nicol, Willie 420-3796
willie.nicol@up.ac.za

Tolmay, Andries 420-3020
dries.tolmay@up.ac.za

Assistant Professors

du Toit, Elizbe L. 420-3641
elizbe.dutoit@up.ac.za

Sandrock, Carl 420-3568
carl.sandrock@up.ac.za

† The prefix for all numbers is 27 (12).

Accredited by: IChemE (SA), IChemE (UK), ECSA

Degrees granted 2003-2004:

B.S.: 30 M.S.: 7 Ph.D.: 3

Department reports to: R.F. Sandenbergh (Dean)

University of the Witwatersrand, Johannesburg

School of Process and Materials Engineering
Private Bag 3
WITS
2050
Deliveries: Richard Ward Building, Room 321

University phone .... (011) 717-1000
Department phone .... (011) 717-7510
Department FAX .... (011) 339-7213

Professors

Bryson, A.W. .... (011) 717 7528
awb@chemeng.chmt.wits.ac.za

Eric, R. H. .... (011) 717 7537
rhe@chemeng.chmt.wits.ac.za

Falcon, Rosemary (Visiting) .... (011) 717 7535
falcons@icon.co.za

Glasser, David (Emeritus) .... (011) 717 7512
dg@chemeng.chmt.wits.ac.za

Hildebrandt, Diane .... (011) 717 7527
Dihil@chemeng.chmt.wits.ac.za

Luyckx, Silvana (Visiting) .... (011) 717 7524
sluyckx@chemeng.chmt.wits.ac.za

Moys, Michael .... (011) 717 7518
mhm@chemeng.chmt.wits.ac.za

Te Riele, Wolter (Head) .... (011) 717 7546
wteriele@chemeng.chmt.wits.ac.za

Associate Professors

Koursaris, Andreas .... (011) 717 7530
akoursar@chemeng.chmt.wits.ac.za

Williams, Donald .... (011) 717 7531
dfw@chemeng.chmt.wits.ac.za

Assistant Professors

Akdogan, Guven
Fernandez-Torres, Maria
Sephton, Michelle

Accredited by: ECSA (Engineering Council of SA), IChemE (UK)

Undergraduate advisor: D. Williams

Department reports to: R Nkado, Dean

Spain

University of Alcala de Henares Ingenieria Quimica

Chemical Engineering
Faculty of Sciences
28871 Alcala de Henares
Madrid
Deliveries: Crta. Barcelona, km 33,600

University phone 34 91 885 49 41
Department phone 34 91 885 49 41
Department FAX 34 91 885 50 88

Professors

García Calvo, Eloy 918 854939
eloy.garcia@uah.es

Associate Professors

Guardiola, Jesus 918 854976
jesus.guardiola@uah.es

Leton, Pedro 918 854974
pedro.leton@uah.es

Ramos, Guadalupe 918 855099
guadalupe.ramos@uah.es

Rodriguez, Antonio 918 854974
antonio.rodriguez@uah.es

Rosal, Roberto
roberto.rosal@uah.es

Assistant Professors

Arranz, Miguel A. 918 854950
miguelan.arranz@uah.es

Boltes, Karina 918 854974
karina.boltes@uah.es

Cortes, Candido 918 855099
candido.cortes@uah.es

Elvira, Rosario 918 854950
rosario.elvira@uah.es

Department reports to: Eloy Garcia Calvo

University of Alicante

Ingenieria Quimica
Facultad de Ciencias
Apartado 99
03080 Alicante

University phone 34-96-5903400
Department phone 34-96-5903867
Department FAX 34-96-5903826

Professors

Fernandez-Sempere, Julio
julio.fernandez@ua.es

Font-Montesinos, Rafael
Rafael.Font@ua.es

Gomis-Yagües, Vicente
vgomis@ua.es

Marcilla-Gomis, Antonio
antonio.marcilla@ua.es

Prats-Rico, Daniel
prats@ua.es

Ruiz-Bevia, Francisco (Head)
ruiz.bevia@ua.es

Associate Professors

Asensi-Steegmann, J. C.
JC.Asensi@ua.es

Beltran-Rico, Maribel
maribel.beltran@ua.es

Boluda-Botella, Nuria
Nuria.Boluda@ua.es

Caballero-Suárez, José A.
caballer@ua.es

Conesa-Ferrer, Juan A.
JA.Conesa@ua.es

Fernández-Torres, María J.
fernandez@ua.es

Garcia-Cortes, Angela N.
angela.garcia@ua.es

García-Quesada, J. C.
JC.García@ua.es

Gomez-Siurana, Amparo
amparo.gomez@ua.es

Martin-Gullon, Ignacio
gullon@ua.es

Olaya-Lopez, M. M.
maria.olaya@ua.es

Reyes-Labarta, Juan A.
ja.reyes@ua.es

Varó-Galvañ, Pedro J.
Pedro.Varo@ua.es

Accredited by:

Degrees granted 2003-2004:

B.S.: 23

Student organization: Aiqua Asociacion de Ingenieros Quimicos de Alicante

Autonoma University of Barcelona

Enginyeria Quimica
Edifici C
08193 Bellaterra
Barcelona

University phone 34-93-5811000
Department phone 34-93-5811018
Department FAX 34-93-5812013

Professors

Gòdia, Francesc .... 5812692 †
francesc.godia@uab.es

Lafuente, Javier 5812143
javier.lafuente@uab.es

Lopez, Josep (Director) 5811806
josep.lopez@uab.es

Solà, Carles 5812002
carles.sola@uab.es

Associate Professors

Benaiges, Dolors 5812144
mariadolors.benaiges@uab.es

Cairó, Jordi 5812694
jordijoan.cairo@uab.es

Casas, Carles 5811809
carles.casas@uab.es

de Mas, Carles 5811019
carles.demas@uab.es

Gabarrell, Xavier 5812789
xavier.gabarrell@uab.es

González, Glòria 5812791
gloria.gonzalez@uab.es

Montesinos, José L. 5812142
joseluis.montesinos@uab.es

Sarra, Montserrat 5812789
montserrat.sarra@uab.es

Valero, Francisco 5811809
francisco.valero@uab.es

Vicent, M. T. 5812142
teresa.vicent@uab.es

Assistant Professors

Baeza, Juan 5812695
juanantonio.baeza@uab.es

Blánquez, Francisca 5812141
paqui.blanquez@uab.es

Cabello, Fernando 5811879
fernando.cabello@uab.es

Carrera, Julian 5811808
julian.carrera@uab.es

Cos, Oriol 5812695
oriol.cos@uab.es

Gabriel, David 5812141
david.gabriel@uab.es

Huix, Josep 5811018

Lecina, Martí 5811808
marti.lecina@uab.es

Mach, Josep M. 5811018

Marín, Antoni 5811018
antonio.marin@uab.es

Montràs, Anna 5811808
anna.montras@uab.es

Pérez, Julio 5812141
julio.perez@uab.es

Rieradevall, Joan 5811018

Sendra, Cristina 5811808
cristina.sendra@uab.es

Suau, Trinitat 5811879
trinitat.suau@uab.es

† Faculty numbers have prefix 34 93.

Accredited by:

Degrees granted 2003-2004:

B.S.: 35 M.S.: 6 Ph.D.: 3

Department reports to: Joan Sorribes, Director, Escola Tècnica Superior d'Enginyeria

University of Cadiz

Departamento de Ingeniería Química
Tecnología de Alimentos y Tecnologías del Medio Ambiente
Facultad de Ciencias
Pol. Rio San Pedro, s/n
11510 Puerto Real, Cadiz

University phone .... (56) 015000
Department phone .... (56) 016375
Department FAX .... (56) 016411

Professors

Cantero Moreno, Domingo .... (56) 016381
domingo.cantero@uca.es
Galan Vallejo, Manuel .... (56) 016380
manuel.galan@uca.es
Martinez de la Ossa, Enrique .... (56) 016378
enrique.martinezdelaossa@uca.es
Perez Rodriguez, Luis (Head) .... (56) 016380
luisperez.rodriguez@uca.es
Sales Marquez, Diego .... (56) 016161
diego.sales@uca.es

Associate Professors

Caro Pina, Ildefonso .... (56) 016376
ildefonso.caro@uca.es
Cohen Mesonero, Leon .... (56) 018073
leon.cohen@uca.es
Macias Garcia, Manuel .... (56) 016383
manuel.macias@uca.es
Martin Minchero, Ricardo .... (56) 016377
ricardo.martin@uca.es
Martin Rodriguez, M. L. .... (56) 018071
luz.martin@uca.es
Muñoz Cueto, Maria J. .... (56) 016377
mariajose.munoz@uca.es
Nebot Sanz, Enrique .... (56) 01 6157
enrique.nebot@uca.es
Perez Garcia, Montserrat .... (56) 016158
montserrat.perez@uca.es
Quiroga Alonso, Jose M. .... (56) 016161
josemaria.quiroga@uca.es
Rodriguez Rodriguez, Miguel .... (56) 016379
miguel.rodri@uca.es
Romero Garcia, Luis I. .... (56) 016379
luisisidoro.romero@uca.es
Romero Zuñiga, Luis E. .... (56) 016382
luisenrique.romero@uca.es
Santiago Fernandez, Inmaculada .... (56) 018059
inmaculada.santiago@uca.es
Soto Fernandez, Fernando .... (56)018056
fernando.soto@uca.es
Trujillo Espinosa, Francisco .... (56) 018074
francisco.trujillo@uca.es

Assistant Professors

Blandino Garrido, Ana .... (56)016381
ana.blandino@uca.es
Coello Oviedo, M.Dolores .... (56) 016158
dolores.coello@uca.es
de Ory Arriaga, Ignacio .... (56) 016300x6831
ignacio.deory@uca.es
Garcia de Lujan, Alberto .... (56) 016380
Garcia Morales, Jose L. .... (56) 016157
joseluis.garcia@uca.es
Garcia Ocaña, Emilia .... (56) 018058
emilia.ocana@uca.es
Gomez Benitez, Juan .... (56) 016380
juangomez.benitez@uca.es
Gomez Montes de Oca, Jose M. .... (56) 01636382
josemanuel.gomez@uca.es
Gordillo Romero, M. D. .... (56) 016300x6833
dolores.gordillo@uca.es
Lopez Ramirez, Juan A. .... (56) 016157
juanantonio.lopez@uca.es
Lopez Sanchez, Diego .... (56) 018057
diego.lopez@uca.es
Mantell Serrano, Casimiro .... (56) 016300x6833
casimiro.mantell@uca.es
Manzano Quiñones, Manuel .... (56) 016157
manuel.manzano@uca.es
Mesa Diaz, M.Mar .... (56) 01636383
mariadelmar.mesa@uca.es
Molero Gomez, Andres .... (56) 016300x6833
andres.molero@uca.es
Palacios Macias, Victor .... (56) 016376
victor.palacios@uca.es
Perales Vargas-Machuca, Jose A. .... (56) 016157
joseantonio.perales@uca.es
Pereyra Lopez, Clara .... (56) 016378
clara.pereyra@uca.es
Portela Miguelez, Juan R. .... (56) 016300x6833
juanramon.portela@uca.es
Solera del Rio, Rosario .... (56) 016158
rosario.solera@uca.es

University of Cantabria

Dpto. Ingeniería Química y Quimica Inorganica
Universidad de Cantabria ETSIIT
Avenida de Los Castros, s/n.
39005 Santander Spain

University phone 34 942 20 15 00
University FAX 34 942 20 11 03
Department phone 34 942 20 15 90
Department FAX 34 942 20 15 91

Professors

Blanco Delgado, Carmen 34 942 20 14 71
blancoc@unican.es
Ortiz Uribe, Inmaculada (Head) 34 942 20 15 85
ortizi@unican.es
Viguri Fuente, Javier R. 34 942 20 15 89
vigurij@unican.es

Associate Professors

Fernández Ferreras, Josefa 34 942 20 20 26
fernandj@unican.es
Fernández Olmo, Ignacio 34 942 20 15 86
fernandi@unican.es
Galán Corta, Berta 34 942 20 15 98
galanb@unican.es

Garea Vázquez, Aurora 34 942 20 15 88
gareaa@unican.es
González Martínez, Fernando 34 942 20 14 73
gonzalfe@unican.es
Herrero Romero, Juana 34 942 20 15 80
herreroj@unican.es
Ibáñez Mendizábal, Raquel 34 942 20 14 70
ibanezr@unican.es
Oliván Martínez, Olga 34 942 20 15 96
olivano@unican.es
Otero Hermida, Jose A. 34 942 20 15 95
oteroj@unican.es
Pesquera González, Carmen 34 942 20 17 72
pesquerc@unican.es
Renedo Omaechevarría, Josefina 34 942 20 15 80
renedomj@unican.es
Rico Gutierrez, Jose L. 34 942 20 15 99
ricoj@unican.es
San Roman San Emeterio, M F. 34 942 20 15 83
sanromm@unican.es
Urtiaga Mendía, Ana M. (Associate Dean) 34 942 20 15 87
urtiaga@unican.es

Assistant Professors

Aldaco, Rubén 34 942 201588
aldacor@unican.es
Arce Recio, Roberto 34 942 20 15 93
arcer@unican.es
Ayerbe de Aragón Aguilera, Asuncion 34 942 84 65 45
ayerbea@unican.es
Colina Pérez, Jose M. 34 942 20 15 94
colinajm@unican.es
Coz Fernández, Alberto 34 942 20 13 59
coza@unican.es
García Posadas, Hipolito 34 942 20 14 70
garciah@unican.es
Lasa Díaz, Cristina 34 942 20 15 88
lasac@unican.es
Lena López, Gumersindo 34 942 20 15 92
lenag@unican.es
Poza Fernández, Angel 34 942 20 15 92
pozaa@unican.es
Rivero Martínez, María J. 34 942 20 15 82
riveromj@unican.es
Ruiz Gutierrez, Gema 34 942 20 15 83
gema.ruiz@unican.es

Accredited by: Spanish Education and Science Ministry

Degrees granted 2003-2004:

B.S.: 30 Ph.D.: 3

Graduate advisor: Gorri Cirella, Eugenio Daniel

Undergraduate advisor: Garea Vázquez, Aurora

Placement service: COIE

Universidad de Castilla-La Mancha

Dpto. de Ingenieria Quimica
Av. Camilo Jose Cela s/n
13004 Ciudad Real

University phone .... (34) 902.204.100
Department phone .... (34) 902/204.100 x 3416
Department FAX .... (34) 926-295318

Professors

Canizares, Pablo (Head) .... x3412 †
Pablo.Canizares@uclm.es
de Lucas, Antonio (Associate Chair) x3426
alucas@inqu-cr.uclm.es
Valverde, Jose L. (Director) x3415
JoseLuis.Valverde@uclm.es

Associate Professors

Alonso, Miguel A .... (34) 926-710577
MiguelAngel.Alonso@uclm.es
Carnicer, Angel .... (34) 926-710577
Angel.Carnicer@uclm.es
Dorado, Fernando x3411
Fernando.Dorado
Duran, Antonio (Associate Dean) x3814
Antonio.Duran@uclm.es
Frades, Jesus M. .... (34) 926-710577
Jesús.Frades@uclm.es
Fuertes, Juan x5414
Juan.Fuertes@uclm.es
Gomez, Rocio (Associate Dean) x3768
Rocio.Gomez@uclm.es
Gracia, Ignacio x3418
Ignacio.Gracia@uclm.es
Lobato, Justo x3418
Justo.Lobato@uclm.es
Monteagudo, Jose M. .... (34) 926-710577
JoseMaria.Monteagudo@uclm.es
Perez, Angel x3416
Angel.Perez@uclm.es
Ricon, Jesusa x5412
Jesusa.Rincon@uclm.es
Rodrigo, Manuel A. x3419
Manuel.Rodrigo@uclm.es
Rodriguez, Joaquin x5414
Joaquin.Rodriguez@uclm.es
Rodriguez, Juan F x3416
juanfran@inqu-cr.uclm.es
Rodriguez, Lourdes (Director) x3413
Lourdes.Rodriguez@uclm.es
Rodriguez, Luis .... x 6021
Luis.RRomero@uclm.es
Sanchez, Paula x3418
Paula.Sanchez@uclm.es
Villasenor, Jose x3419
Jose.Villasenor

Assistant Professors

Asencio, Isaac x3417
Isaac.Asencio@uclm.es
Cabra, Luis x3411
lcabrad@repsol-ypf.es
Carmona, Manuel S. x3814
carmonaf@inqu-cr.uclm.es
Fernández, Fco. J. x3417
FcoJesus.FMorales@uclm.es
Garcia, Jesus x3506
Jesus.GGomez@uclm.es
Ramos, Manuel x3416
manuel.ramos.go@teleline.es
Ramos, Maria J. x3417
mjramos@inqu-cr.uclm.es

† Faculty numbers are extensions of (34) 26-295300.

Graduate advisor: Juan F. Rodríguez

Undergraduate advisor: A. de Lucas

Department reports to: Pablo Cañizares

Complutense University of Madrid

Departamento de Ingeniería Química
Facultad de Ciencias Químicas
Universidad Complutense
E-28040 Madrid

Department phone .... (91) 394 4115
Department FAX .... (91) 394 4114

Professors

Aracil, José .... (91) 394-4175
jam1@quim.ucm.es
Corella, José .... (91) 394 4164
narvaez@quim.ucm.es
García-Ochoa, Félix .... (91) 394 4176
fgochoa@quim.ucm.es
López, Federico .... (91) 394 4249
Ovejero, Gabriel .... (91) 394 4111
govejero@quim.ucm.es
Rodríguez, Francisco (Associate Dean) ...... (91) 394 4246
frsomo@quim.ucm.es
Romero, Arturo .... (91) 394 4170
arturo@emducms1.sis.ucm.es
Sotelo, José L. (Head) .... (91) 394 4117
jose.sotelo@quim.ucm.es
Tijero, Julio .... (91) 394 4250
jtijero@quim.ucm.es
Uguina, M. Ángeles .... (91) 394 4113
uguinama@quim.ucm.es

Associate Professors

Aragón, José M. .... (91) 394-4173
jomar@quim.ucm.es
Blanco, Ángeles .... (91) 394 4247
ablanco@quim.ucm.es
Delgado, José A. .... (91) 394 4119
jadeldob@quim.ucm.es
García, Julián .... (91) 394 5119
jgarcia@quim.ucm.es
Guardiola, Elita .... (91) 394 4248
guforedo@quim.ucm.es
Martínez, Mercedes .... (91) 394 4167
mmr1@quim.ucm.es
Mirada, Fernando .... (91) 394 4251
fmirada@quim.ucm.es
Negro, Carlos .... (91) 394 4242
cnegro@quim.ucm.es
Oliet, Mercedes .... (91) 394 4241
moliet@quim.ucm.es
Palancar, M. C. .... (91) 394 4169
mcpalanc@quim.ucm.es
Rodríguez, Araceli .... (91) 394 4182
arodri@quim.ucm.es
Romero, M. D. .... (91) 394 4118
mdolores@quim.ucm.es
Santos, Aurora .... (91) 394 4171
aursan@quim.ucm.es

Assistant Professors

Alba, Cristina
crisalba@quim.ucm.es
Alonso, Virginia .... (91)394 42 47
valonso@quim.ucm.es
Bautista, Fernando .... (91) 394 4167
lfbausan@quim.ucm.es
Calvo, Lourdes .... (91) 394 4185
lcalvo@quim.ucm.es
Espínola, Ascensión .... (91)394 4112
Fuente, Elena 91 394 4245
helenafg@quim.ucm.es
Garcia, Juan .... (91)394 41 12
juangcia@quim.ucm.es
Guijarro, Isabel
migg@quim.ucm.es
Gómez, Emilio
emgomez@quim.ucm.es
Gómez, José M. .... (91) 394 4185
segojmgm@quim.ucm.es
Ladero, Miguel A.
mladero@quim.ucm.es
Lodares, Carmen .... (91)394 4240
clodares@quim.ucm.es
Monte, Concepción .... (91) 394 4245
cmonte@quim.ucm.es
Perez, Alejandro 91 394 4115
alperez@quim.ucm.es
Pérez, Ponciano .... (91) 394 4112
Santos, Victoria .... (91) 394 4179
vesantos@quim.ucm.es
Tijero, Antonio
atijerc@quim.ucm.es
Torrecilla, Santiago .... (91) 394 4240
jstorre@quim.ucm.es
Águeda, Ismael .... (91) 394 4115
vian@quim.ucm.es

Accredited by:

Degrees granted 2003-2004:

B.S.: 83 M.S.: 25 Ph.D.: 13

University of Granada

Department of Chemical Engineering
Faculty of Sciences
c/ Fuentenueva s/n
18071, Granada

University phone 34 958 243025
Department phone 34 958 243308
Department FAX 34 958 248992

Professors

Bravo, V. .... 3310 †
vbravo@ugr.es
Camacho, F. 3309
fcamacho@ugr.es
Gonzalez, P. 3310
pgtello@ugr.es
Jurado, E. (Chair) 3307
ejurado@ugr.es

Associate Professors

Blazquez, G. 0770
gblazque@ugr.es
Calero, M. 3311
mcaleroh@ugr.es
Fernandez, M. 3311
mferse@ugr.es
Galvez, A. 0770
agalvez@ugr.es
Garcia, A. I. 9018
anaigl@ugr.es
Gomez, M. 3313
mggarzon@ugr.es

Guadix, E. M. 2925
eguadix@ugr.es
Hernainz, F. 3315
hernainz@ugr.es
Luzon, G. (Associate Dean) 2925
german@ugr.es
Martinez, L. 8991
lmartin@ugr.es
Martinez, Ma. E. (Associate Chair) 3315
meugenia@ugr.es
Paez, Ma. P. 3307
mppaez@ugr.es
Reyes, A. 9018
areyesr@ugr.es
Rodriguez, S. 3313
srodrig@ugr.es

Assistant Professors

Bailon, R. 4075
bailonm@ugr.es
Jimenez, J. M. 8991
mjimenez@ugr.es
Luque, A. 4075
antluque@ugr.es
Nunez, J. 4075
jnolea@ugr.es
Vicaria, J.M. 0445
vicaria@ugr.es

† Faculty numbers have extension 34 958-24.

Accredited by: Spanish Ministry of Education

Degrees granted 2003-2004:

B.S.: 60 M.S.: 2 Ph.D.: 3

Department reports to: E. Jurado, Departamento de Ingenieria Quimica

University of Illes Balears

Quimica Tecnica
College of Sciences
Palma de Mallorca

Department phone .... (971) 20-7111x283

Professors

Bergueiro, Jose R. .... (971) 17-3240
Rossello, Carmen .... (971) 17-3239

Associate Professors

Canellas, J. .... (971) 17-3241

Department reports to: Felix Grases, Dean of Sciences College

Universidad Politecnica de Madrid-ETSI Industriales

Departamento de Ingenieria Quimica Industrial y Medio Ambiente
c/Jose Gutierrez Abascal, 2
28006 Madrid

University phone 34 913366035
Department phone 34 913363026
Department FAX 34 913363009

Professors

Larena, A. 34 913363181
alarena@iqi.etsii.upm.es
Laso, M. 34 913363015
laso@diquima.upm.es
Losada, J. 34 913363184
jlosada@iqi.etsii.upm.es
Ramos, M. A. 34 913365344
mramos@diquima.upm.es
Rodriguez, E. (Head) 34 913363188
erodriguez@iqi.etsii.upm.es

Associate Professors

De Maria, M R. 34 913363189
rdemaria@diquima.upm.es
Fernandez, A. 34 913363183
afernandez@iqi.etsii.upm.es
Fuente, M. M. de la 34 913363185
mmfuente@iqi.etsii.upm.es
Galan, S. 34 913363185
sgalan@diquima.upm.es
Garcia, P. 34 913363185
pigarcia@iqi.etsii.upm.es
Gonzalez, I. 34 913365341
aaee.ig@gised.com
Martinez, J. 34 913363183
jmartinez@iqi.etsii.upm.es
Matias, M. C. 34 913363182
mcmatias@iqi.etsii.upm.es
Molina, M. J. 34 913363182
mjmolina@iqi.etsii.upm.es
Narros, A. 34 913363186
anarros@iqi.etsii.upm.es
Paz, I. 34 913365343
i.paz@iqi.etsii.upm.es
Peso, M. I. del 34 913363185
ipeso@iqi.etsii.upm.es
Pinto, G. 34 913363183
gpinto@iqi.etsii.upm.es
Quintana, F. J. 34 913363188
fquintanam@clh.es
Santos, A. 34 913365341
asantos@diquima.upm.es
Zubizarreta, J. I. 34 913363182
jizubi@iqi.etsii.upm.es

Assistant Professors

Díez de Garay, E. 34913365341
ediez@diquima.upm.es
Gayoso, J. A. 34 913365341
jagayoso@tecnicasreunidas.es
Jimeno, N. 34 913363015
njimeno@diquima.upm.es
Llorente, V. 34 915451192
vllorente@cyii.es
Quintanilla, J.E. 34 913363185
je.quintanilla@iqfr.csic.es
Rodríguez, M. 34 913363189
mrod@diquima.upm.es
Rojas, R. 34 913365343
rrojas@iqi.etsii.upm.es
Soto, A. 34 913363189
asoto@uee.es

University of Murcia

Departamento de Ingenieria Quimica
Facultad de Quimica
Campus de Espinardo
Aptdo. Correos. 4.021
30071 Murcia

University phone .... (968) 363600
Department phone .... (968) 367359
Department FAX .... (968) 364148

Professors

Bodalo, Antonio 968 367354
abodalo@um.es

Minana, Agustin 968 367353
minana@um.es

Rubio, Manuel (Director) 968 367357
mrubio@um.es

Saez, Jose 968 367358
saezmer@um.es

Soler, Antonio 968 367355
ansoler@um.es

Associate Professors

Bastida, Josefa 968 367361
jbastida@um.es

Cabanes, A. L. 968 367362
cabanes@um.es

Gomez, Demetrio 968 367356
demetrio@um.es

Gomez, Elisa 968 367352
egomez@um.es

Gomez, Jose L. 968 367351
carrasco@um.es

Gonzalez, Enrique 968 367364
ferradas@um.es

Llorens, Mercedes 968 367349
llorens@um.es

Meseguer, Victor 968 367333
vzapata@um.es

Máximo, Fuensanta 968 367367
fmaximo@um.es

Ortuno, Juan F. 968 367360
jfortuno@um.es

Tomas, Francisca 968 367331
ptomas@um.es

Villora, Gloria (Dean) 968 367363
gvillora@um.es

Assistant Professors

Aguilar, M. I. 968 367091
maguilar@um.es

Hidalgo, Asuncion M. 968 367219
iqasocia@um.es

Montiel, Claudia 968 367219
cmontiel@um.es

Quesada, Joaquín 968 367228
quesamed@um.es

University of Oviedo

Department of Chemical Engineering and Environmental Technology
C/ Julian Claveria, 8
33006 Oviedo (Asturias)

University phone 985-103000
Department phone 985-103441
Department FAX 985-103434

Professors

Alvarez, Ricardo 985-103438
raf@sauron.quimica.uniovi.es

Bueno, Julio L. 985-103444
jbh@sauron.quimica.uniovi.es

Coca, Jose (Chair) 985-103443
jcp@sauron.quimica.uniovi.es

Diaz, Jose M. 985-103439
mdf@sauron.quimica.uniovi.es

Sastre, Herminio 985-103437
hsa@sauron.quimica.uniovi.es

Assistant Professors

Alvarez, Jose R. 985-103028
jras@sauron.quimica.uniovi.es

Berrueta, Jose 985-103442
berrueta@correo.uniovi.es

Diez, Fernando V. 985-103508
fds@sauron.quimica.uniovi.es

Garcia, Luis A. 985-103029
lagd@sauron.quimica.uniovi.es

Gonzalez, Carlos 985-103519
cgs@sauron.quimica.uniovi.es

Gutiérez Lavín, Antonio 985-103518
agl@sauron.quimica.uniovi.es

Iglesias, Olvido 985-103517
oih@sauron.quimica.uniovi.es

Luque, Susana 985 102998
slr@sauron.quimica.uniovi.es

Marañón, Elena 985-103668
emara@sci.cpd.uniovi.es

Medina, Ignacio 985-103510
medina@sauron.quimica.uniovi.es

Pazos, Carmen (Associate Chair) 985-103509
cpm@sauron.quimica.uniovi.es

Pulgar, Andres
pccb1@atenea.etsimo.uniovi.es

Riera, Francisco 985-103436
far@sauron.quimica.uniovi.es

Rosal, Roberto 985-103520
rosal@sauron.quimica.uniovi.es

Vega, Aurelio B. 985-103507
avg@sauron.quimica.uniovi.es

University of Pais Vasco

Chemical Engineering
Faculty of Chemistry
P. O. Box 1072
20080, San Sebastian

University phone 34-943-018000
Department phone 34-943-015328
Department FAX 34-943-212236

Professors

Asua, Jose M. .... 018181[†]
jmasua@sq.ehu.es
Montes, Mario 018183
qppmoram@sq.ehu.es

Associate Professors

Barandiaran, Maria J. 015330
qppbasam@sq.ehu.es
Canton, Lourdes 018169
qppcaorl@sq.ehu.es
de la Cal, José C. 015331
qppdedej@sq.ehu.es
Forcada, Jacqueline 018182
qppfogaj@sq.ehu.es
Legorburu, Iñigo (Associate Dean) 018215
qpplefai@sq.ehu.es
Leiza, Jose R. 015329
qpplerej@sq.ehu.es

Assistant Professors

Antxustegi, Mirari 015325
qppanbem@sq.ehu.es

[†] Faculty numbers have prefix 34 943.

Department reports to: J.M. Asua

Rovira i Virgili University

Departament d'Enginyeria Quimica
Escola Tècnica Superior d'Enginyeria Química
Avinguda dels Països Catalans, 26
Campus Sescelades
43007 Tarragona

Department phone 34 977 559603
Department FAX 34 977 559621

Professors

Castells, F. 34 977 559644
fcastell@etseq.urv.es
Fabregat, A. 34 977 559643
afabrega@etseq.urv.es
Farriol, X. 34 977 559642
xfarriol@etseq.urv.es
Giralt, F. 34 977 559638
fgiralt@etseq.urv.es
Giralt, J. 34 977 559650
jgiralt@etseq.urv.es
Sueiras, J. 34 977 558701
jsueiras@etseq.urv.es

Associate Professors

Alabart, J. R. 34 977 559658
jalabart@etseq.urv.es
Bañares-Alcantara, R 34 977 559673
rbanares@etseq.urv.es
Bonet, J. 34 977 559645
jbonet@etseq.urv.es
Ferrando, M. 34 977 558505
mferrand@etseq.urv.es
Font, J. 34 977 559646
jfont@etseq.urv.es
Garcia, R. 34 977 559611
rgarcia@etseq.urv.es
Gavaldà, J. 34 977 559647
jgavalda@etseq.urv.es
Grifoll, J. 34 977 559639
jgrifoll@etseq.urv.es
Güell, C. 34 977 558504
cguell@etseq.urv.es
Herrero, J. 34 977 559649
jherrero@etseq.urv.es
Katakis, I. 34 977 559655
ikatakis@etseq.urv.es
Lopez, F. 34 977 558503
flopez@etseq.urv.es
Mackie, A. 34 977 559674
amackie@etseq.urv.es
Mateo, J. M. 34 977 559676
jmateo@etseq.urv.es
Medina, F. 34 977 559787
fmedina@etseq.urv.es
Medir, M. 34 977 558177
mmedir@etseq.urv.es
Montané, D. 34 977 559652
dmontane@etseq.urv.es
Renau, J. 34 977 558550
jrenau@etseq.urv.es
Rolán, A. 34 977 559651
arolan@etseq.urv.es
Salvadó, J. 34 977 559641
jsalvado@etseq.urv.es
Schuhmacher, M. 34 977 559653
mschuh@etseq.urv.es
Stüber, F. 34 977 559671
fstuber@etseq.urv.es

Assistant Professors

Bes, E. 34 977 559604
Blanco, M. 34 977 558549
mblanco@etseq.urv.es
Blasco, A. 34 977 559604
Boix Rita, B. 34 977 559661
Boix Sabater, R. 34 977 559661
rboix@etseq.urv.es
Borrajo, M.A. 34 977 559661
Cabré, J. 34 977 559661
Chillida Rabada, J 34 977 559661
Claramonte, V. 34 977 559661
vclaram@etseq.urv.es
Cuesta, J. 34 977 558549
Guardiola, S. 34 977 250000
sguardio@etseq.urv.es
Guitérrez, A. 34 977 559661
aguti@etseq.urv.es
Lloret, R. 34 977 559661
rlloret@etseq.urv.es
Mañé, E. 34 977 559661
emanye@etseq.urv.es
Merino, J. 34 977 558549
jmerino@etseq.urv.es
Moreno Cabello, M 34 977 558549
mmoreno@etseq.urv.es
Moreno Garcia, M.E. 34 977 250000
emoreno@etseq.urv.es
Pemán, J.M. 34 977 559604
jpeman@etseq.urv.es
Rabadà, J. 34 977 250000
Ramírez, B. 34 977 250000
bramirez@etseq.urv.es
Robles, M. 34 977 559661
mrobles@etseq.urv.es

Department reports to: A. Fabregat

University of Salamanca

Departamento de Ingenieria Quimica y Textil
Facultad de Ciencias Quimicas
Plaza de los Caidos, 1-5
37008 Salamanca

University phone 34-923294400
Department phone 34-923294479
Department FAX 34-923294574

Professors

Galan, Miguel A. 34-923294479
magalan@usal.es

Associate Professors

Catalan, Jacinto 34-923294479
jcatalan@usal.es

Costa, Carlos 34-923294479
ccosta@usal.es

Cuellar, Jorge (Chair) 34-923294479
cuellar@usal.es

Estevez, Angel M. 34-923294479
estevez@usal.es

Fernandez, Angel 34-923294479
aftena@usal.es

Marquez, M. C. 34-923294479
mcm@usal.es

Martin, Jose L. 34-923294479
jolmasa@usal.es

Montes, Francisco J. 34-923294479
javimon@usal.es

Ramos, Pedro 34-923294479
pramos@usal.es

Rodriguez, Jesus M. 34-923294479
jesusr@usal.es

Assistant Professors

Alvaro, Audelino 34-923294479
audea@usal.es

Arranz, Juan L. 34-923294479
juluar@usal.es

Domingo, Antonio M. 34-923294479
tonidm@usal.es

Ruiz, Cesar A. 34-923294479
car@usal.es

Sánchez, José M. 34-923294479
chemasal@usal.es

Torrente, Carmen 34-923294479
carmina@usal.es

Tovar, Tomas R. 34-923294479
manana@usal.es

Graduate advisor: Francisco J. Montes

Department reports to: Faculty of Chemical Sciences

Universidad de Santiago de Compostela

Departamento de Ingenieria Quimica
Lope Gómez de Marzoa, s/n
15782 Santiago de Compostela

University phone 34 981 563100
Department phone 34 981 563100
Department FAX 34 981 528050

Professors

Antorrena, Gervasio .... 16787 †
eqgantor@usc.es

Arce, Alberto (Head) 16790
eqaaarce@usc.es

Bao, Manuel 16789
eqbao@usc.es

Casares, Juan J. 16794
eqjjcl@usc.es

Lema, Juan M. 16793
eqjmlema@usc.es

Méndez, Ramón 16791
eqrmndzp@usc.es

Nunez, M. J. 16792
eqmajose@usc.es

Vazquez, Gonzalo 16788
eqgvazq@usc.es

Associate Professors

Blanco, Antonio 16796
eqanblas@usc.es

Chenlo, Francisco 16797
eqchenlo@usc.es

Feijoo, Gumersindo 24073
eqfeijoo@lugo.usc.es

Fernandez, Eugenio 24075
usciqefc@cesga.es

Gonzalez, Julia 24102
eqjulia@lugo.usc.es

Navaza, Jose M. 16795
eqnavaza@usc.es

Omil, Francisco 24074
eqomil@usc.es

Roca, Enrique 24074
eqercca@lugo.usc.es

Soto, Ana M. 24134
eqequilf@usc.es

Veiga, Jose L. 24136
jlveiga@lugo.usc.es

Vidal, Isabel 16798
eqivteef@lugo.usc.es

Assistant Professors

Bajo, José A. 16727
jbajo@usc.es

Campos, Jose L. 16777
eqluis@usc.es

Fenoll, Fernando 16727
f.fenoll@usc.es

Freire, Sonia 16758
sfreire@usc.es

Garrido, Juan 16778
equenlla@usc.es

Martinez, Jose M. 24125
eqageito@lugo.usc.es

Moreira, Ramon 16759
eqmoncho@usc.es

Moreira, Teresa 16767
tmoreira@usc.es

Méndez, M. R. 24077
usciqmmg@cesga.es

Pereira, Gerardo 24140
gepegon@usc.es

Rodil, Eva 16760
erodil@usc.es

Sineiro, Jorge 24133
eqxurxo@lugo.usc.es

Vázquez, M. J. 24077
eqmjvv@usc.es

† Dial 34 81 563100 and ask for extension.

Escuela Superior de Ingenieros Industriales de Sevilla

Department of Chemical and Environmental Engineering
Camino de los Descubrimientos s/n
41092 Sevilla
Deliveries: José F. Vale Parapar

University phone 34 95 4486100
Department phone 34 95 4487260
Department FAX 34 95 4461775

Professors

Alcalde Moreno, Manuel 34 95 4487270
alcalde@esi.us.es
Cañadas Serrano, Luis 34 95 4487267
canadas@esi.us.es
Cortes Galeano, Vicente J. 34 95 4487260
vjcortes@esi.us.es
Fernandez Pereira, Constantino 34 95 4487271
pereira@esi.us.es
Garcia Lopez, Angel 34 95 4487271
angelgl@esi.us.es
Gracia Manarillo, Ignacio 34 95 4487273
gracia@esi.us.es
Olivares del Valle, Joaquin 34 95 4487275
jov@esi.us.es
Ollero de Castro, Pedro (Associate Dean) 34 95 4487260
ollero@esi.us.es
Plumed Rubio, Antonio 34 95 4487262
plumed@esi.us.es
Salvador Martinez, Luis 34 95 4487260
salvador@esi.us.es
Sanchez Pena, Eduardo J. 34 95 4487266
eduardoj@esi.us.es
Usero Garcia, Jose 34 95 4487274
usero@esi.us.es
Vale Parapar, José F. (Head) 3494 4487269
vale@esi.us.es

Associate Professors

Arjona Antolin, Ricardo 34 95 4487223
arjona@esi.us.es
Gutierrez Ortiz, Fco. J. 34 95 4487268
fjgo@esi.us.es
Morillo Aguado, José 34 95 4487276
jmorillo@esi.us.es
Navarrete Rubia, Benito 34 95 4487280
navarre@esi.us.es
Rodriguez Piñero, Miguel A. 34 95 4487270
marp@esi.us.es
Vidal Barrero, Fernando 34 95 4487222
vidal@esi.us.es
Vilches Arenas, Luis 34 95 4487282
vilches@esi.us.es
Villegas Sanchez, Rosario 34 95 4487282
villegas@esi.us.es

Accredited by:

Degrees granted 2003-2004:

B.S.: 40 Ph.D.: 2

Department reports to: José F. Vale Parapar

Universidad de Sevilla

Departamento de Ingenieria Quimica
Facultad de Quimica
Universidad de Sevilla
41071 Sevilla

University phone 34 95 4551000
Department phone 34 95 4557180
Department FAX 34 95 4556447

Professors

Alfaro Rodríguez, M. C. 34 95 4557179
alfaro@us.es
Alvarez Mateos, Paloma 34 95 4552846
palvarez@us.es
Berjano Núñez, Manuel 34 95 4557179
mberjano@us.es
Carranza Mora, Francisco 34 95 4557182
fcarranza@us.es
Carrillo de la Fuente, Francisco (Assoc. Head) 34 95 4557178
cfuente@us.es
Cordobés Carmona, Felipe 34 95 4557179
fcordobe@us.es
Cota Galán, Juan 34 95 4557178/3877
cota@us.es
De la Fuente, Julia 34 95 4557183/2845
jfferia@us.es
Flores Luque, Vicente (Director) 34 95 4557179
vfloresluque@us.es
Guerrero Conejo, Antonio 34 95 4557179
aguerrero@us.es
Iglesias González, Nieves 34 95 4557182
mnieves@us.es
Martín Aguilar, Ana 34 95 4557178
amartina@us.es
Mazuelos Rojas, Alfonso 34 95 4557182
mazuelos@us.es
Moraga Borrell, Jose M. 34 95 4557178
moraga@us.es
Muñoz García, Jose 34 95 4557179
jmunoz@us.es
Palencia Pérez, Inmaculada 34 95 4557182
palencia@us.es
Pereda Marín, Juan 34 95 4557183
jpmarin@us.es
Rodriguez Patino, Juan M. 34 95 4556446
jmrodri@us.es
Rodríguez Niño, Rosario 34 95 4556446
mdrrodri@us.es
Romero Guzmán, Fernando 34 95 4557183
fromero@us.es
Roselló Segado, Antonio 34 95 4557178
rosello@us.es
Ruiz Dominguez, Manuela 34 95 4557183
manuela@us.es

Associate Professors

Durán Barrantes, Montaña 34 95 4557183
mmduran@us.es

Assistant Professors

Carrera Sánchez, Ceciclio 34 95 4556446/7147
cecilio@us.es
Carta Escobar, Felisa 34 95 4557183
felisa@us.es

Accredited by:

Degrees granted 2003-2004:

B.S.: 40 M.S.: 3 Ph.D.: 2

Department reports to: dingquimica@us.es

Placement service: dingquimica@us.es

Universitat de Valencia

Departamento de Ingenieria Quimica
Escola Tècnica Superior d'Enginyeria
46100 Burjassot (Valencia)

University phone 34 96 386 4100
Department phone 34 96 354 4325
Department FAX 34 96 354 4898

Professors

Aucejo, Antonio 34 96 354 4318
Antonio.Aucejo@uv.es

Martinez-Andreu, Antoni (Chair) 34 96 354 4319
Antoni.Martinez@uv.es

Monton, Juan B. 34 96 354 4317
Juan.B.Monton@uv.es

Associate Professors

Berna, Angel 34 96 354 4316
Angel.Berna@uv.es

Burguet, M. C. 34 96 354 4318
Cruz.Burguet@uv.es

de la Torre, Javier 34 96 354 4326
Javier.Torre@uv.es

Dejoz, Ana M. 34 96 354 3436
Ana.M.Dejoz@uv.es

Gabaldon, M. C. 34 96 354 4326
Carmen.Gabaldon@uv.es

Gonzalez-Alfaro, M. V. 34 96 354 3436
Vicenta.Gonzalez@uv.es

Llopis, Francisco J. (Associate Chair) 34 96 354 3130
Francisco.Llopis@uv.es

Loras, Sonia 34 96 354 3169
Sonia.Loras@uv.es

Martinez-Soria, Vicente 34 96 354 3169
Vicente.Mtnez-soria@uv.es

Marzal, Paula 34 96 354 4326
Paula.Marzal@uv.es

Miguel, Pablo J. 34 96 354 3130
Pablo.J.Miguel@uv.es

Munoz, Rosa 34 96 354 4319
Rosa.Munoz@uv.es

Orchilles, A. V. 34 96 354 3170
Vicent.Orchilles@uv.es

Pena, M. P. 34 96 354 4316
Maria.P.Pena@uv.es

Penarrocha, Josep M. 34 96 354 3169
Josep.Penarrocha@uv.es

Sanchotello, Margarita 34 96 354 4317
Margarita.Sanchotello@uv.es

Seco, Aurora 34 96 354 4326
Aurora.Seco@uv.es

Vazquez, M. I. (Associate Dean) 34 96 354 3170
Isabel.Vazquez@uv.es

Vercher, Ernesto 34 96 354 3131
Ernesto.Vercher@uv.es

Assistant Professors

Bouzas, Alberto 34 96 354 3434
Alberto.Bouzas@uv.es

Chafer, Amparo 34 96 354 3434
Amparo.Chafer@uv.es

Accredited by: Council of Universities of Spain

Degrees granted 2003-2004:

B.S.: 59 Ph.D.: 2

Department reports to: M.I. Vazquez, Vice-Dean, ETSE

University of Valladolid

Chemical Engineering Department
Faculty of Sciences
C/ Prado de la Magdalena s/n
47011 Valladolid

University phone .... (34) 983 423250
University FAX ....(34)983423234
Department phone .... (34) 983 423166
Department FAX .... (34) 983 423013

Professors

Carton, Angel .... (34) 983 423167
carton@iq.uva.es

Cocero, Maria J. .... (34) 983 423174
mjcocero@iq.uva.es

Fdz-Polanco, Fernando .... (34) 983 423172
ffp@iq.uva.es

Gonzalez, Gerardo (Dean) .... (34) 983 423170
gerardo@iq.uva.es

Mato, Fidel (Emeritus) .... (34) 983 423175
fmato@iq.uva.es

Associate Professors

Alonso, Gloria E. .... (34) 983 423166
ealonso@iq.uva.es

Antolin, Gregorio .... (34) 983 423362
greant@eis.uva.es

Bolado, Silvia .... (34) 983 423958
silvia@iq.uva.es

Espinel, Manuela .... (34) 983 423508
espinel@sid.eup.uva.es

Fdz-Polanco, Maria .... (34) 983 424506
maria@iq.uva.es

Garcia, Maite .... (34) 983 423237
maite@iq.uva.es

Garcia, Pedro (Head) .... (34) 983 423171
pedro@iq.uva.es

Irusta, Ruben .... (34) 983 423693
rubiru@eis.uva.es

Martinez, Bernardo .... (34) 983 423506
bern@iq.uva.es

Mato, Fidel A. .... (34) 983 423169
pepa@iq.uva.es

Mato, Rafael B. .... (34) 983 423177
rbmato@iq.uva.es

Pena, Maria M. .... (34) 983 423176
pena@iq.uva.es

Sobron, Francisco .... (34) 983 423168
sobron@iq.uva.es

Torio, Isabel .... (34) 983 423507
misato@sid.eup.uva.es

Uruena, Miguel A. .... (34) 983 423506
uru@iq.uva.es

Villaverde, Santiago .... (34) 983 423656
svillave@iq.uva.es

Assistant Professors

García Serna, Juan .... (34) 983 423166
jgserna@iq.uva.es

Lucas, Susana .... (34) 983 423166
susana@iq.uva.es

Accredited by: IChemE

Degrees granted 2003-2004:

M.S.: 56 Ph.D.: 3

Department reports to: Jesus M. Sanz Serna, Rector of the University

University of Zaragoza

Department of Chemical and Environmental Engineering
Pedro Cerbuna, 12
E50009, Zaragoza

University phone .... (34) 976 761000
Department phone .... (34) 976 761154
Department FAX .... (34) 976 762142

Professors

Bilbao, Rafael (Head) .... (34) 976 761150
rbilbao@unizar.es

Menendez, Miguel .... (34) 976 761152
qtmiguel@unizar.es

Ovelleiro, J. L. .... (34) 976 761156
oveleiro@unizar.es

Santamaria, Jesus M. (Dean) .... (34) 976 761153
iqcatal@unizar.es

Associate Professors

Alzueta, M. U. .... (34) 976 761876
uxue@unizar.es

Arauzo, Jesus .... (34) 976 761878
qtarauzo@unizar.es

Aznar, M. P. .... (34) 976 762391
paznar@unizar.es

Calvo, Manuel .... (34) 976 762191
calvom@unizar.es

Ceamanos, Jesus .... (34) 976 762160
ceamanos@unizar.es

Coronas, Joaquin .... (34) 976 762471
coronas@unizar.es

Frances, Eva .... (34) 976 762550
efrances@unizar.es

Garcia-Bacaicoa, Pedro .... (34) 976 761880
bacaicoa@unizar.es

Garcia-Nieto, Lucia .... (34) 976 762194
luciag@unizar.es

Herguido, Javier .... (34) 976 762393
jhergui@unizar.es

Iranzo, Carlos .... (34) 976 762190
ciranzo@unizar.es

Lazaro, Luisa .... (34) 976 761877
llazaro@unizar.es

Mastral, Jose F. .... (34) 976 761876
pepe@unizar.es

Matute, Rosa P. .... (34) 976 762550
rmatute@unizar.es

Millera, Angela .... (34) 976 761875
amillera@unizar.es

Monzon, Antonio .... (34) 976 761157
amonzon@unizar.es

Murillo, M. B. .... (34) 976 761880
murillo@unizar.es

Ormad, M. P. .... (34) 976 761877
mpormad@unizar.es

Oro, M. A. .... (34) 976 761151
maoro@unizar.es

Pena, J. A. (Associate Dean) .... (34) 976 762390
jap@unizar.es

Perez, J. I. .... (34) 976 762194
jiperezs@unizar.es

Revuelta, Gillermo .... (34) 976 762190
grevuelt@unizar.es

Royo, Carlos J. .... (34) 974 761334
cjroyo@unizar.es

Salvador, M. L. .... (34) 976 761584
mlsalva@unizar.es

Sanchez, Emilio (Associate Dean) .... (34) 974 761335
blas@unizar.es

Sarasa, Judith .... (34) 976 762392
jsarasa@unizar.es

Tellez, Carlos .... (34) 976 762471
ctellez@unizar.es

Villacampa, J. I. .... (34) 974 761336
villacam@unizar.es

Assistant Professors

Callejas, Alicia .... (34) 976 733977
acalleja@carbon.icb.csic.es

Gea, Gloria .... (34)976 763308
gloria.gea@unizar.es

Mallada, Reyes .... (34)974 761461
rmallada@unizar.es

Oliva, Miriam .... (34) 976 761339
miroliva@unizar.es

Pina, M. P. .... (34) 976 761155
mapina@unizar.es

Romeo, Eva .... (34) 974 761461
evaromeo@unizar.es

Romero, Enrique .... (34) 976 762392
eromero@unizar.es

Ruiz, Joaquin .... (34) 976 761339
jruizp@unizar.es

Accredited by: Ministry of Education

Student organization: AIQA

Department reports to: Prof. R. Bilbao

Placement service: Mrs. P. Munoz

Sweden

University of Lund

Chemical Engineering
Lund Institute of Technology
P.O. Box 124
S-221 00 Lund

University phone .... (46) 46 2220000
Department phone .... (46) 46 2228285
Department FAX .... (46) 46 2224526

Professors

Aly, Gharib (Emeritus) .... (46) 46 2224257
Gharib.Aly@chemeng.lth.se

Andersson, Arne .... (46) 46 2228280
arne.andersson@chemeng.lth.se

Axelsson, Anders .... (46) 46 2228282
Anders.Axelsson@chemeng.lth.se
Bjerle, Ingemar (Emeritus)
ingemar.bjerle@chemeng.lth.se
Jernqvist, Ake (Emeritus) .... (46) 46 2228294
Ake.Jernqvist@chemeng.lth.se
Jonsson, Ann S. .... (46) 46 2228291
Ann-Sofi.Jonsson@chemeng.lth.se
Karlsson, Hans .... (46) 46 2228244
hans.karlsson@chemeng.lth.se
Lidén, Gunnar (Associate Chair) .... (46) 46 2220826
gunnar.liden@chemeng.lth.se
Odenbrand, Ingemar .... (46) 46 2228284
Ingemar.Odenbrand@chemeng.lth.se
Stenstrom, Stig .... (46) 46 2228292
Stig.Stenstrom@chemeng.lth.se
Sverdrup, Harald .... (46) 46 2228274
Harald.Sverdrup@chemeng.lth.se
Warfvinge, Per .... (46) 46 2223626
per.warfvinge@chemeng.lth.se
Wimmerstedt, Roland (Emeritus) .... (46) 46 2228298
Roland.Wimmerstedt@chemeng.lth.se
Zacchi, Guido (Chair) .... (46) 46 2228297
Guido.Zacchi@chemeng.lth.se

Associate Professors

Dolby, Ingemar (Emeritus) .... (46) 46 2228290
Ingemar.Dolby@chemeng.lth.se

Assistant Professors

Alveteg, Mattias .... (46) 46 2223627
Mattias.Alveteg@chemeng.lth.se
Nilsson, Bernt .... (46) 46 2228299
Bernt.Nilsson@chemeng.lth.se

Accredited by:

Degrees granted 2003-2004:

M.S.: 33 Ph.D.: 12

Graduate advisor: Guido Zacchi

Department reports to: Lund Institute of Technology

Royal Institute of Technology

Department of Chemical Engineering and Technology
Teknikringen 28
S-100 44
Stockholm

University phone 46 8 790 6000
Department phone 46 8 790 8229
Department FAX 46 8 105 228

Professors

Björnbom, Pehr (Chair) 46 8 790 8255
pehr@ket.kth.se
Järås, Sven (Chair) 46 8 790 8917
svenj@ket.kth.se
Lindbergh, Göran (Chair) 46 8 790 8143
goeran.lindbergh@ket.kth.se
Neretnieks, Ivars (Head) 46 8 790 8229
niquel@ket.kth.se
Rasmuson, Åke (Chair) 46 8 790 8227
rasmuson@ket.kth.se
Wennersten, Ronald (Chair) 46 8 790 6347
rw@ket.kth.se
Westermark, Mats 46 8 790 6220
mw@ket.kth.se
Yan, Jinyue 46 8 790 6528
yanjy@ket.kth.se

Associate Professors

Alvfors, Per (Chair) 46 8 790 6526
alvfors@ket.kth.se
Berendson, Jaak 46 8 790 7028
jaak@admin.kth.se
Björnbom, Emilia 46 8 790 8256
emilia@ket.kth.se
Boutonnet, Magali 46 8 790 8245
magali@ket.kth.se
Liu, Jinsong 46 8 790 6346
liuv@ket.kth.se
Martinez, Joaquin 46 8 790 6570
jmc@ket.kth.se
Moreno, Luis 46 8 790 6412
lm@ket.kth.se
Pettersson, Lars 46 8 790 8259
larsp@ket.kth.se
Sjöström, Krister 46 8 790 8248
krister@ket.kth.se
Sylwan, Christopher 46 8 790 8258
sylwan@ket.kth.se
Uusi-Penttilä, Marketta 46 8 790 8228
marketta@ket.kth.se

Accredited by: The Swedish Government

Degrees granted 2003-2004:

B.S.: 19 M.S.: 42 Ph.D.: 13

Graduate advisor: Prof Göran Lindbergh

Undergraduate advisor: Ass Prof Per Alvfors

Department reports to: Prof Anders Flodström, President of the Royal Institute of Technology (KTH)

Placement service: Ms Ulla Schött

Switzerland

Swiss Federal Institute of Technology

ETH Hoenggerberg HCI
Institute for Chemical and Bioengineering
CH-8093 Zurich

University phone 41 1 632 1111
Department phone 41 1 632 3048
Department FAX 41 1 632 1509

Professors

Baiker, Alfons 41 1 632 3153
baiker@tech.chem.ethz.ch
Hungerbuehler, K. 41 1 632 6098
hungerbuehler@tech.chem.ethz.ch
Morbidelli, Massimo 41 1 632 3034
morbidelli@tech.chem.ethz.ch
Prins, Roel (Head) 41 1 632 5490
prins@tech.chem.ethz.ch
Rys, Paul 41 1 632 3120
paul.rys@tech.chem.ethz.ch

Assistant Professors

Consiglio, Giambattista 41 1 632 3552
consiglio@tech.chem.ethz.ch

Department reports to: Prof. Roel Prins, Institute for Chemical and Bioengineering

Swiss Federal Institute of Technology, Ecole Polytechnique Federale de Lausanne

Section of Chemistry and Chemical Engineering
Faculty of Sciences
CH, 1015 Lausanne

University phone .... (4121) 693 1111
Department phone .... (4121) 693 9850
Department FAX .... (4121) 693 9855

Professors

Bodenhausen, Geoffrey (Chair) .... 41 21 693 9431 †
geoffrey.bodenhausen@epfl.ch
Dyson, Paul (Chair) 41 21 693 9854
paul.dyson@epfl.ch
Girault, Hubert (Dean) 41 21 693 3151
hubert.girault@epfl.ch
Graetzel, Michael (Chair) 41 21 693 3112
michael.graetzel@epfl.ch
Hubbell, Jeffrey 41 21 693 96 81
jeffrey.hubbell@epfl.ch
Johnsson, Kai (Chair) 41 21 693 9356
kai.johnsson@epfl.ch
Merbach, André E. (Chair) 41 21 693 9871
andre.merbach@epfl.ch
Mutter, Manfred (Chair) 41 21 693 94 11
manfred.mutter@epfl.ch
Pitsch, Stefan (Chair) 41 21 693 9380
stefan.pitsch@epfl.ch
Renken, Albert (Chair) 41 21 693 3181
albert.renken@epfl.ch
Rizzo, Thomas (Chair) 41 21 693 3073
thomas.rizzo@epfl.ch
Roulet, Raymod (Chair) 41 21 693 9861
raymond.roulet@epfl.ch
Röthlisberger, Ursula (Chair) 41 21 693 0321
ursula.roethlisberger@epfl.ch
Schlosser, Manfred (Chair) 41 21 693 9351
manfred.schlosser@epfl.ch
Severin, Kay (Chair) 41 21 693 9302
kay.severin@epfl.ch
Stockar, Urs von (Chair) 41 21 693 3191
urs.vonstockar@epfl.ch
Vogel, Horst (Chair) 41 21 693 3155
horst.vogel@epfl.ch
Vogel, Pierre (Chair) 41 21 693 03 1
pierre.vogel@epfl.ch
Wurm, Florian (Chair) 41 21 693 6141
florian.wurm@epfl.ch

Associate Professors

Comninellis, Christos 41 21 693 3674
christos.comninellis@epfl.ch
Friedli, Claude 41 21 693 3121
claude.friedli@epfl.ch
Stoessel, Francis 41 21 693 3671
francis.stoessel@epfl.ch

† Direct dial with prefix 4121.

Accredited by:

Degrees granted 2003-2004:
M.S.: 30 Ph.D.: 30

Graduate advisor: Prof. H. Girault

Undergraduate advisor: Prof. H. Girault

Student organization: AGEPOLY

Department reports to: Marendaz Jean-Luc ¡jean-luc.marendaz@epfl.ch¿

Taiwan

Chang Gung University

Chemical and Materials Engineering Department
259, Wen-Hwa 1st Road
Kwei-San, Tao-Yuan
Taiwan 333 R.O.C.

University phone 886-3-2118800
Department phone .... 886-3-2118800 ext 5286
Department FAX 886-3-2118668

Professors

Chen, Jyh-Ping .... x5298 †
jpchen@mail.cgu.edu.tw
Cheng, Chen-Yen x5760
Rei, Min-Hon x5379
reimh@mail.cgu.edu.tw
Shiau, Lie-Ding (Chair) x5291
shiau@mail.cgu.edu.tw
Yang, Jen-Ming x5290
jmyang@mail.cgu.edu.tw

Associate Professors

Chiu, Fang-Chyou x5297
maxson@mail.cgu.edu.tw
Hsu, Ruey-Chi x5288
rcHsu@mail.cgu.edu.tw
Hua, Mu-Yi x5289
huamy@mail.cgu.edu.tw
Lin, Andrew x5756
Andrew@mail.cgu.edu.tw
Lu, Hsin-Chun x5292
hsinchun@mail.cgu.edu.tw
Lu, Tsan-Sheng x5675
tsheng@mail.cgu.edu.tw
Lue, Shing-Jiang x5489
jessie@mail.cgu.edu.tw
Wu, Gwo-Mei x5293
wu@mail.cgu.edu.tw
You, Jiann-Hwa x5287
you@mail.cgu.edu.tw

Assistant Professors

Kuo, Hsiu-Po x5488
hpkuo@mail.cgu.edu.tw
Liu, Yu-Kuo x5328
ykliu@mail.cgu.edu.tw
Wang, Gow-Bin x5757
gbwang@mail.cgu.edu.tw

† Faculty numbers are extensions of University phone.

Accredited by: Ministry of Education

Department reports to: Lie-Ding Shiau, Chairman

Placement service: Lie-Ding, Shiau

Chung Yuan Christian University

Department of Chemical Engineering
Chung Li, Taiwan, 320
Republic of China

University phone .... (03) 2659999
Department phone .... (03) 2654101
Department FAX .... (03) 2654199

Professors

Chung, Tsai-Wang
twchung@cycu.edu.tw

Chyang, Chien-Song
cschyang@cycu.edu.tw

Lai, Juin-Yih
jylai@cycu.edu.tw

Li, Meng-Hui
mhli@cycu.edu.tw

Ruaan, Ruoh-Chyu
ruaan@cycu.edu.tw

Associate Professors

Chang, Tsun-Sheng
tschang@che.cycu.edu.tw

Chen, Jung-Hui
jason@wavenet.cycu.edu.tw

Cheng, Chun-Lin
chunlin@cycu.edu.tw

Chuang, Ching-Jung (Chair)
cjchuang@che.cycu.edu.tw

Hsu, Keh-Ying
kehying@cycu.edu.tw

Tung, Kuo-Lun
kuolun@cycu.edu.tw

Wei, Ta-Chin
tcwei@cycu.edu.tw

Assistant Professors

Liu, Ying-Ling
ylliu@cycu.edu.tw

Accredited by: Taiwan Ministry of Education

Undergraduate advisor: Chuang, Ching-Jung

Department reports to: J.Don Den, Dean of Engineering

Feng-Chia University

Chemical Engineering Department
100 Wenhwa Road
Taichung, Taiwan 40724

University phone 886-4-24517250
Department phone .... 886-4-24517250 ext. 3651
Department FAX 886-4-24510890

Professors

Chan, Chih-Chieh .... ext. 3688 †
ccchan@fcu.edu.tw

Chang, Hsin-Fu .... ext. 3687
hfchang@fcu.edu.tw

Chao, Yun-Peng (Chair) .... ext. 3677
ypshao@fcu.edu.tw

Chen, Chyi-Tsong (Associate Dean) .... ext. 3691
ctchen@fcu.edu.tw

Chu, Hou-Hsien .... ext. 3686
hhchu@fcu.edu.tw

Hwang, Jen-Tai .... ext. 3651
jthwang@npf.org.tw

Lin, Ping J. .... ext. 3682
pjlin@fcu.edu.tw

Wu, Der -Chang .... ext. 3689
dcwu@fcu.edu.tw

Wu, Shu-Yii .... ext. 3679
sywu@fcu.edu.tw

Associate Professors

Sheu, Jiang S. .... ext. 3690
chsu@fcu.edu.tw

Ting, Hsing-Yie .... ext. 3079
hyting@fcu.edu.tw

Assistant Professors

Chang, AlexC.-C. .... ext. 3676
acchang@fcu.edu.tw

Lin, Yuang-Sen .... ext. 3659
yslin@fcu.edu.tw

Shih, C. J. .... ext. 3680
chshih@fcu.edu.tw

Yuan, Wei-Li .... ext. 6892
wyuan@fcu.edu.tw

† Faculty numbers are extensions of 886-4-24517250

Accredited by: Ministry of Education

Department reports to: Chao, Yun-Peng

Placement service: Yen-Hsu Kuan

National Central University

Department of Chemical and Materials Engineering
Chung-Li, Taoyuan
Taiwan 32054

University phone .... (03) 4227151
Department phone .... (03) 4227151x4200
Department FAX .... (03) 4252296

Professors

Chang, Feg-Wen x4202
Chen, Hui x4216
huichen@sparc20.ncu.edu.tw
Chen, Teng-Ko x4215
T3100064@ncu865.ncu.edu.tw
Chen, Wen-Yih (Chair) x4222
T313165.ncu865.ncu.edu.tw
Chen, Yin-Zu x4219
YnZuChem@cc.ncu.edu.tw
Chen, Yu-Wen x4203
Chiang, Shiaw-Tseh x4214
STChiang@cc.ncu.edu.tw
Fey, Ting-Kuo x4206
Kao, Chengheng R. x4225
crkao@twncu865.ncu.edu.tw

Kuo, Kung-Tu (Emeritus) x4205
T312310@twncu865.ncu.edu.tw
Lee, Liang-Sun x4208
T3100206@twncu865.ncu.edu.tw
Lin, Hsiao-Tsung x4207
Ruaan, Rouh-Chyu x4232
ruaan@cc.ncu.edu.tw
Shyu, Shin-Shing x4204
Tsao, Heng-Kwong x4226
HKTsao@colloid.Chem.ncu.edu.tw
Yang, Sze-Ming x4217
Szeyang@tpts6.seed.net.tw

Associate Professors

Shu, Chin-Hang x4227
chinshu@cc.ncu.edu.tw
Wang, Ten-Tsai x4211

Assistant Professors

Liu, Cheng-Yi x4228

Accredited by: Ministry of Education

Graduate advisor: Cheng-Tung Chou

Undergraduate advisor: Chengheng R. Kao

Department reports to: Dean, College of Engineering

Placement service: Student Bureau

National Cheng Kung University

Department of Chemical Engineering
1 University Road
Tainan, Taiwan 70148

University phone 886-6-2757575
Department phone 886-6-2757575x62600
Department FAX 886-6-2344496

Professors

Chang, Chien-Hsiang x62671
changch@mail.ncku.edu.tw
Chang, Chuei-Tin x62663
ctchang@mail.ncku.edu.tw
Chang, Jo-Shu x62651
changjs@mail.ncku.edu.tw
Chen, Bor-Kuan (Adjunct)
Chen, Chin-Cheng x62655
ccchen@mail.ncku.edu.tw
Chen, Chuh-Yung (Head) x62643
ccy7@mail.ncku.edu.tw
Chen, Dong-Hwang x62680
chendh@mail.ncku.edu.tw
Chen, Ling-Yuan (Adjunct)
Chen, Teh-Liang x62660
t62660@mail.ncku.edu.tw
Chen, Yun x62657
yunchen@mail.ncku.edu.tw
Chiang, Chien-Lih x62648
clchiang@mail.ncku.edu.tw
Chou, Tse-Chuan x62639
tcchou@mail.ncku.edu.tw
Chu, Tzong-Jeng (Adjunct) x62681x304
z7108017@email.ncku.edu.tw
Chung, Shyan-Lung x62654
slchung@mail.ncku.edu.tw
Hong, Chau-Nan x62662
hong@mail.ncku.edu.tw
Huang, Ting-Chia (Adjunct) x62630
tchuang@mail.ncku.edu.tw
Hwang, Shyh-Hong x62661
shhwang@mail.ncku.edu.tw
Kao, Chen-Feng x62642
cfkao@mail.ncku.edu.tw
Kuo, Jen-Feng (Adjunct) x62638
jenfkuo@mail.ncku.edu.tw
Kuo, Ping-Lin x62658
plkuo@mail.ncku.edu.tw
Lin, Jui-Che x62665
jclin@mail.ncku.edu.tw
Liu, Chung-Chiun (Adjunct)
Liu, Jui-Hsiang x62646
jhliu@mail.ncku.edu.tw
Maa, Jer-Ru (Adjunct) x62632
jerrumaa@mail.ncku.edu.tw
Syu, Mei-Jywan x62631
z8108094@email.ncku.edu.tw
Teng, Hsisheng x62640
hteng@mail.ncku.edu.tw
Tsai, Shau-Wei x62647
t62647@mail.ncku.edu.tw
Tsay, Sun-Yuan (Adjunct) x62681x341
sytsay@mail.ncku.edu.tw
Wang, Chi x62645
chiwang@mail.ncku.edu.tw
Wang, Chun-Shan (Adjunct) x62649
cswang@mail.ncku.edu.tw
Wen, Ten-Chin x62656
tcwen@mail.ncku.edu.tw
Weng, Hung-Shan x62637
z5408008@email.ncku.edu.tw
Woo, Eamor x62670
emwoo@mail.ncku.edu.tw
Yang, Yu-Min x62633
ymyang@mail.ncku.edu.tw

Associate Professors

Chen, Huey-Ing x62667
hueying@mail.ncku.edu.tw
Cheng, Chu-Yuan x62664
cycheng@mail.ncku.edu.tw
Hong, Chia-Hong x62650
z5209001@email.ncku.edu.tw
Juang, Hor-Da x62652
z6710001@email.ncku.edu.tw
Lee, Jeng-Cheng (Adjunct)
Lee, Yuh-Lang x62693
yllee@mail.ncku.edu.tw
Lin, Hung-Jye x62653
z7208020@email.ncku.edu.tw
Ling, Han-Chern x62659
hcling@mail.ncku.edu.tw
Yang, Ming-Chang x62666
mcyang@mail.ncku.edu.tw

Assistant Professors

Chao, Ying-Chen (Adjunct)
Chen, Bing-Hung x62695
bkchen@mail.ncku.edu.tw
Hou, Sheng-Shu x62641
sshou@mail.ncku.edu.tw
Huang, Yao-Hui x62636
yhhuang@mail.ncku.edu.tw
Wei, Hsien-Hung x62681x384
hhwei@mail.ncku.edu.tw

Wu, Jih-Jen x62694
wujj@mail.ncku.edu.tw

Accredited by: Ministry of Education

Department reports to: Jhing-Fa Wang, Dean of Engineering

National Chung Cheng University

Department of Chemical Engineering
San-Hsing, Min-Hsiung
Chaia-Yi, 621-02

University phone 886-5-2720411
Department phone 886-5-2720411x23400
Department FAX 886-5-2721206

Professors

Chang, Jen-Ray x33455
chmjrc@ccu.edu.tw

Chen, Chien-Chong x33462
chmccc@ccu.edu.tw

Hu, Chi-Chang x33411
chmhcc@ccu.edu.tw

Hwang, Chyi x33456
chmch@ccu.edu.tw

Lee, Wen-Chien x33406
chmwcl@ccu.edu.tw

Tsiang, Raymond C. (Chair) x33454
chmcct@ccu.edu.tw

Wang, Feng-Sheng x33404
chmfsw@ccu.edu.tw

Wang, Maw-Ling (Emeritus) x33461
chmmlw@ccu.edu.tw

Associate Professors

Hua, Chi-Chung x33412
chmcch@ccu.edu.tw

Kuo, Yung-Chih x33459
chmyck@ccu.edu.tw

Lin, Tsao-Jen x33405
chmtjl@ccu.edu.tw

Tsai, Jing-Cherng x33460
chmjct@ccu.edu.tw

Assistant Professors

Huang, Kuang-Tse x33407
chmkth@ccu.edu.tw

Li, Yuan-Yao x33403
chmyyl@ccu.edu.tw

National Taiwan University of Science and Technology

Department of Chemical Engineering
43 Keelung Road, Section 4
Taipei, 106-07
Taiwan

University phone 886-2-2733-3141
Department phone 886-2-2737-6610
Department FAX 886-2-2737-6644

Professors

Chern, Chorng-Shyan 886-2-2737-6649
chern@ch.ntust.edu.tw

Chern, Yaw-Terng 886-2-2737-6646
cyt@ch.ntust.edu.tw

Chien, I-Lung 886-2-2737-6652
Chien@ch.ntust.edu.tw

Chou, Yi-Shyong 886-2-2737-6622
yschou@ch.ntust.edu.tw

Hong, Lu-Sheng 886-2-2737-6650
hong@ch.ntust.edu.tw

Hsu, Yung-Chien 886-2-2737-6616
hsu@ch.ntust.edu.tw

Huang, Yan-Jyi 886-2-2737-6625
yjhuang@ch.ntust.edu.tw

Hwang, Bing-Joe 886-2-2737-6624
bjh@ch.ntust.edu.tw

Ju, Yi-Hsu 886-2-2737-6612
ju@ch.ntust.edu.tw

Ku, Young 886-2-2737-6621
Ku@ch.ntust.edu.tw

Lee, Cheng-Kang 886-2-2737-6629
cklee@ch.ntust.edu.tw

Lee, Chiapyng 886-2-2737-6623
cl@ch.ntust.edu.tw

Lee, Ming-Jer (Chair) 886-2-2737-6626
mjl@ch.ntust.edu.tw

Liaw, Der-Jang 886-2-2737-6638
liaw@ch.ntust.edu.tw

Lin, Chun-I 886-2-2737-6614
cilin@ch.ntust.edu.tw

Lin, Ho-Mu 886-2-2737-6643
hml@ch.ntust.edu.tw

Lin, Shi-Yow 886-2-2737-6648
ling@ch.ntust.edu.tw

Liu, Chin-Hsin 886-2-2737-6647
cjl@ch.ntust.edu.tw

Liu, Jhy-Chern 886-2-2737-6627
liu@ch.ntust.edu.tw

Liu, Tuan-Chi 886-2-2737-6617
tuan@ch.ntust.edu.tw

Shiau, Ching-Yeh 886-2-2737-6619
cys@ch.ntust.edu.tw

Tsai, Dah-Shyang 886-2-2737-6618
tsai@ch.ntust.edu.tw

Associate Professors

Chen, Hsiu-Mei 886-2-2737-6651
hsiumei@ch.ntust.edu.tw

Jiang, Jyh-Chiang 886-2-2737-6653
jcjiang@ch.ntust.edu.tw

Liaw, Ben-Ruey 886-2-2737-6628
ruey@ch.ntust.edu.tw

Assistant Professors

Tseng, Wen-Chi 886-2-2730-1078
tsengwc@ch.ntust.edu.tw

Accredited by: Ministry of Education

Degrees granted 2003-2004:

B.S.: 153 M.S.: 67 Ph.D.: 6

Department reports to: Ming-Jer Lee

Placement service: Ming-Jer Lee

National Taiwan University

Department of Chemical Engineering
No.1, Sec. 4, Roosevelt Road, Taipei, Taiwan 106-17

University phone 886-2-2363-0231
Department phone 886-2-2363-5230
Department FAX 886-2-2362-3040

Professors

Chao, Yung-Cheng (Emeritus)
Chen, Cheng-Ching (Emeritus)
Chen, Cheng-Liang 886-2-2363-6194
ccl@ntu.edu.tw
Chen, Leo-Wang (Emeritus) 886-2-2363-9884
Chen, Li-Jen 886-2-2362-3296
ljchen@ntu.edu.tw
Chen, Wen-Chang 886-2-2362-8398
chenwc@ntu.edu.tw
Chen, Yan-Ping 886-2-2366-1661
ypchen@ntu.edu.tw
Chiu, Wen-Yen 886-2-2362-3259
ycchiu@ntu.edu.tw
Ho, Kuo-Chuan 886-2-2366-0739
kcho@ntu.edu.tw
Hsieh, Kuo-Huang 886-2-2362-7688
khhsieh@ntu.edu.tw
Hsu, Jyh-Ping 886-2-2363-7448
jphsu@ntu.edu.tw
Huang, Hsiao-Ping 886-2-2363-8999
husnaghpc@ntu.edu.tw
Huang, Shih-Yow 886-2-2363-2542
hsyo@ntu.edu.tw
Keh, Huan-Jang (Chair) 886-2-2363-5462
huan@ntu.edu.tw
Lan, Chung-Wen 886-2-2363-3917
cwlan@ntu.edu.tw
Lee, Duu-Jong 886-2-2362-5632
djlee@ntu.edu.tw
Lee, Keh-Chyang 886-2-2362-2530
ericlee@ntu.edu.tw
Lee, Min-Dar (Emeritus)
Leu, Lii-Ping 886-2-2365-7200
lleulii@ntu.edu.tw
Liu, Hwai-Shen 886-2-2362-7499
hsliu@ntu.edu.tw
Lu, Chung-Hsin 886-2-2365-1428
chlu@ntu.edu.tw
Lu, Shaw-Mei (Emeritus)
Lu, Wei-Ming (Emeritus) 886-2-2362-2707
wmlu@ntu.edu.tw
Shih, Shin-Min 886-2-2363-3974
smshih@ntu.edu.tw
Tai, Clifford Y. 886-2-2362-0832
cytai@ntu.edu.tw
Wan, Ben-Zu 886-2-2363-0737
benzuwan@ntu.edu.tw
Wang, Da-Ming 886-2-2366-0433
daming@ntu.edu.tw
Wu, Chi-Sheng 886-2-2363-1994
cswu@ntu.edu.tw
Wu, Nae-Lih 886-2-2362-7158
nlw001@ntu.edu.tw
Yen, Shi-Chern 886-2-2363-0397
scyen@ntu.edu.tw
Yu, Cheng-Ching 886-2-3365-1759
ccyu@ntu.edu.tw

Associate Professors

Hsieh, Hsyue-Jen 886-2-2363-3097
hjhsieh@ntu.edu.tw
Sheng, Yu-Jane 886-2-2366-0454
yjsheng@ntu.edu.tw

Assistant Professors

Dai, Chi-An 886-2-2364-3378
polymer@ntu.edu.tw
Tsai, Wei-Bor 886-2-2369-8627
weibortsai@ntu.edu.tw
Wang, Sheng-Shin

Accredited by: Ministry of Education

Degrees granted 2003-2004:

B.S.: 100 M.S.: 65 Ph.D.: 19

Graduate advisor Prof. Leu, Lii-Ping; Prof. Hsieh, Hsyue-Jen

Department reports to: Y. B. Yang, Dean, College of Engineering

Placement service Staff: Tseng,Sui-hwa

National Tsing Hua University

Department of Chemical Engineering
Hsinchu, Taiwan 300

University phone 886-3-5715131
Department phone 886-3-5719036
Department FAX 886-3-5715408

Professors

Chang, Rong-Yu 886-3-5718344
rychang@che.nthu.edu.tw
Chen, Hsin-Lung 886-3-5721714
hlchen@che.nthu.edu.tw
Chen, Kuo-Cheng 886-3-5716249
kcchen@che.nthu.edu.tw
Chen, Show-An 886-3-5710733
sachen@che.nthu.edu.tw
Chen, Sinn-Wen 886-3-5721734
swchen@che.nthu.edu.tw
Chin, Wei-Kuo 886-3-5713721
wkc@che.nthu.edu.tw
Chou, Kan-Sen 886-3-5713691
kschou@che.nthu.edu.tw
Chu, I-Ming 886-3-5713704
imchu@che.nthu.edu.tw
Hsiue, Ging-Ho 886-3-5719956
ghhsiue@che.nthu.edu.tw
Hu, Andrew T. 886-3-5713830
athu@che.nthu.edu.tw
Huang, Ta-Jen 886-3-5731160
chairman@che.nthu.edu.tw
Hwang, Shyh-Jye 886-3-5723221
sjhuang@che.nthu.edu.tw
Jang, Shi-Shang (Chair) 886-3-5713697
ssjang@che.nthu.edu.tw
Ku, Chia-Soon (Visiting) 886-2-23126100
kucs@cpc.com.tw
Lee, Yu-Der 886-3-5713204
ydlee@che.nthu.edu.tw
Liu, Ta-Jo 886-3-5723380
tjliu@che.nthu.edu.tw
Lu, Shih-Yuan 886-3-5714364
sylu@che.nthu.edu.tw

Ma, Chen-Chi 886-3-5713058
ccma@che.nthu.edu.tw

Sung, Hsing-Wen 886-3-5726832
hwsung@che.nthu.edu.tw

Tan, Chung-Sung 886-3-5721189
cstan@che.nthu.edu.tw

Wan, Chi-Chao 886-3-5721664
ccwan@mx.nthu.edu.tw

Wang, I-Kai 886-3-5713763
ikwang@che.nthu.edu.tw

Wang, Yung-Yun 886-3-5713690
yywang@che.nthu.edu.tw

Wong, David S. 886-3-5721694
dshwong@che.nthu.edu.tw

Wu, Wen-Teng 886-3-5742509
wtwu@che.nthu.edu.tw

Associate Professors

Ho, Rong-Ming 886-3-5738349
rmho@mx.nthu.edu.tw

Mei, Marty T. (Visiting) 886-2-27134242-517
marty@harvester.com.tw

Assistant Professors

Hsu, John T. (Adjunct) 886-3-5720287
tsuanhsu@nhri.org.tw

Hu, Yu-Chen 886-3-5718245
yuchen@che.nthu.edu.tw

Accredited by: Ministry of Education R.O.C.

Graduate advisor: Shi-Shang Jang

Undergraduate advisor: Shi-Shang Jang

Student organization: Student Association of Chemical Engineering
Advisor: Shi-Shang Jang

Department reports to: Dean of Engineering

Placement service: Dean of Student Affairs

Tamkang University

Department of Chemical and Materials Engineering
Tamsui, Taipei Hsien, Taiwan, 25137

University phone 886-2-2621-5656
Department phone 886-2-26215656x2614
Department FAX 886-2-2620-9887

Professors

Chang, Yu-Chi .... x2722 †
040791@mail.tku.edu.tw

Chen, Hsi-Jen x2721
c1952814@ms12.hinet.net

Cheng, Liao-Ping x2725
lpcheng@mail.tku.edu.tw

Cheng, Tung-Wen (Chair) x2725
twcheng@mail.tku.edu.tw

Chi, Jung-Chang x2723
jcchi@mail.tku.edu.tw

Han, Kwong-Wing x2601
043467@mail.tku.edu.tw

Ho, Chii-Dong x2724
cdho@mail.tku.edu.tw

Hwang, Kuo-Jen x2726
kjhwang@mail.tku.edu.tw

Yeh, Ho-Ming x2601
hmyeh@mail.tku.edu.tw

Associate Professors

Chang, Cheng-Liang x2723
chlchang@mail.tku.edu.tw

Chang, Hsuan x2721
nhchang@mail.tku.edu.tw

Chen, Ching-Chung x2722
047927@mail.tku.edu.tw

Don, Trong-Ming x2670
tmdon@mail.tku.edu.tw

Lin, Dar-Jong x2724
djlin@mail.tku.edu.tw

Lin, Gwo-Geng x2726
gglin@ms4.hinet.net

Yu, Hseng-Fu x2728
hfyu@mail.tku.edu.tw

† Faculty numbers are extensions of 886-2-26215656.

Accredited by: Ministry of Education

Graduate advisor: Tung-Wen Cheng

Student organization: ChIChE
Advisor: Tung-Wen Cheng

Placement service: Ching-Sheng Shu, Public Relations, Section Secretariate

Tunghai University

Chemical Engineering Department
P.O. Box 833
Taichung, Taiwan R.O.C. 40704

University phone 886-4-23590121
Department phone 886-4-23590262-108
Department FAX 886-4-23590009

Professors

Chang, You-Im 886-4-23590262
yichang@mail.thu.edu.tw

Chiao, Shu-Min 886-4-23590262
smchiao@mail.thu.edu.tw

Do, Jing-Shan 886-4-23590262
jsdo@mail.thu.edu.tw

Huang, Chi-Tsung 886-4-23590262
huangct@mail.thu.edu.tw

Li, Chung-Hsiung 886-4-23590262
chl@mail.thu.edu.tw

Li, Kuo-Tseng 886-4-23590262
ktli@mail.thu.edu.tw

Wang, Yen (Head) 886-4-23590262
yehwang@mail.thu.edu.tw

Yang, Fan-Chiang 886-4-23590262
fcyang@mail.thu.edu.tw

Associate Professors

Cheng, Shueh-Hen 886-4-23590262
shcheng@mail.thu.edu.tw

Hsieh, Shu-Mu 886-4-23590262
smhsieh@mail.thu.edu.tw

Yang, I-Kuan 886-4-23590262
ikyang@mail.thu.edu.tw

Assistant Professors

Gu, Yesong 886-4-23590121-3316
yegu@mail.thu.edu.tw

Accredited by: Ministry of Education

Student organization: ChIChE
Advisor: Wang Yeh

Yuan Ze University

Department of Chemical Engineering and Materials Science
Chung-Li 320, Taiwan

University phone 886-3-463-8800
Department phone 886-3-463-6897
Department FAX 886-3-455-9373

Professors

Hong, Shinn G.
cesghong@ce.yzu.edu.tw

Hwang, Jenn C.
cejhwang@ce.yzu.edu.tw

Juang, Ruey S. (Chair)
rsjuang@ce.yzu.edu.tw

Lin, Sheng H.
ceshlin@ce.yzu.edu.tw

Sun, Yi M.
cesunym@ce.yzu.edu.tw

Wang, Shaw S. (Visiting)
cewshaw@ce.yzu.edu.tw

Wu, Ho S.
cehswu@ce.yzu.edu.tw

Yin, Ken M.
cekenyin@ce.yzu.edu.tw

Yu, Tzyy L.
cetlyu@ce.yzu.edu.tw

Associate Professors

Chang, Yu C.
yjchan@ce.yzu.edu.tw

Hsuen, Hsiao K. D.
skhsuen@ce.yzu.edu.tw

Lin, Hsiu L.
sherry@ce.yzu.edu.tw

Lin, Sheng D.
sdlin@ce.yzu.edu.tw

Shih, Shawn
cess@ce.yzu.edu.tw

Assistant Professors

Liao, Chien S.
csliao@ce.yzu.edu.tw

Liau, Chau K.
lckliau@ce.yzu.edu.tw

Lin, Kuen S.
kslin@ce.yzu.edu.tw

Accredited by: Ministry of Education

Degrees granted 2003-2004:

B.S.: 122 M.S.: 40

Department reports to: R. S. Juang

Thailand

Chulalongkorn University

Faculty of Engineering
Department of Chemical Engineering
Phaya Thai Road
Bangkok 10330

University phone .... (662) 215-0871-73
Department phone .... (662) 218-6878-83
Department FAX .... (662) 218-6877

Professors

Praserthdam, Piyasan (Chair) .... 218-6711 †
piyasan.p@chula.ac.th

Tanthapanichakoon, Wiwut 218-6894
fchwtt@eng.chula.ac.th

Associate Professors

Assabumrungrat, Suttichai 218-6868
fchsas@kankrow.eng.chula.ac.th

Charinpanitkul, Tawatchai 218-6894
ctawat@pioneer.chula.ac.th

Kittisupakorn, Paisan 218-6892
paisan.k@chula.ac.th

Mongkhonsi, Tharathon 218-6867
tharathon.m@chula.ac.th

Muangnapoh, Chirakarn 218-6874
chirakarn.m@chula.ac.th

Pancharoen, Ura 218-6891
ura.p@chula.ac.th

Satayaprasert, Chairit 218-6893
chairit.s@chula.ac.th

Vanichseni, Sutham 218-6407
sutham.v@chula.ac.th

Assistant Professors

Boon-Long, Sasithorn 218-6876
fchsbl@eng.chula.ac.th

Chatsiriwech, Deacha 218-6863
deacha.c@chula.ac.th

Chongvisal, Vichitra 218-6875
vichitra.c@chula.ac.th

Damrongsakkul, Siriporn 218-6862
siriporn.d@chula.ac.th

Duriyabunleng, Hathaichanok 218-6867
hathaic.d@chula.ac.th

Prechanont, Seeroong 218-6860
seeroong.p@chula.ac.th

Thongyai, Supakanok 218-6860
supakanok.t@chula.ac.th

† Direct dial with prefix 662.

Accredited by: Engineering Association of Thailand, Ministry of University Affairs

Department reports to: Somsak Panyakeow (218-6310), Dean of Faculty of Engineering

Kasetsart University

Department of Chemical Engineering
50 Paholyotin Road
POB 1032, Kasetsart Post Office
Ladyao, Jatujak
Bankok 10903

University phone 66-2942-8200
Department phone .... 66-2942-8555 ext. 1204
Department FAX 66-2579-2083

Associate Professors

Kongkachuichay, Paisan .... ext 1207 †
fengpsk@ku.ac.th
Limtrakul, Sunun .... ext 1210
sunun.l@ku.ac.th
Phanawadee, Phungphai (Associate Chair) .... ext 1212
fengphi@ku.ac.th
Srinophakhun, Penjit (Associate Chair) .... ext 1213
fengpjs@ku.ac.th
Srinophakhun, Thongchai .... ext 1214
fengtcs@ku.ac.th
Suvachitanont, Sirikalaya .... ext 1205
fengsks@ku.ac.th

Assistant Professors

Chareonpanich, Metta .... ext 1215
fengmtc@ku.ac.th
Duangchan, Apinya .... ext 1211
fengapd@ku.ac.th
Mungcharoen, Thumrongrut (Chair) .... ext 1201
fengtrm@ku.ac.th
Pitrchart, Krittaya .... ext 1209
fengktp@ku.ac.th
Vatanatham, Terdthai .... ext 1208
terdthai.v@ku.ac.th

† Phone numbers are extensions of 66-2942-8555

Accredited by: Thailand Ministry of University Affairs, Thailand Civil Service Commission

Degrees granted 2003-2004:

B.S.: 46 M.S.: 26

Graduate advisor: Assoc. Prof. Dr.Sunun Limtrakul

Undergraduate advisor: Dr.Kandis Sudsakorn

Student organization: KUChE Student Club
Advisor: Dr. Siripon Anantawaraskul

Department reports to: Professor Nontawat Junjaroen, Dean of Engineering

Placement service: Assoc.Prof.Dr. Penjit Srinophakhun

Prince of Songkla University

Chemical Engineering Department
P.O.B. 2, Kohong
Hat-Yai, Songkla 90112

University phone 66 74 211030-49
Department phone 66 74 212896
Department FAX 66 74 212896

Associate Professors

Tongurai, Chakrit 66 74 212896
tchakrit@ratree.psu.ac.th

Assistant Professors

Innachitra, Paiboon 66 74 212896
jpaiboon@ratree.psu.ac.th
Srisuwan, Galaya 66 74 212896
sgalaya@ratree.psu.ac.th
Thonglimp, Veerasak 66 74 212896
tveerasa@ratree.psu.ac.th
Tirawanichakul, Supawan 66 74 212896

Graduate advisor: Galaya Srisuwan

Department reports to: Pichit Reungsangwattana, Dean of Engineering

Turkey

Ankara University

Department of Chemical Engineering
Tandoan 06100
Ankara

University phone 90-312-2126040
Department phone 90-312-2126720x1321
Department FAX 90-312-2121546

Professors

Aktas, Zeki (Associate Dean) .... x1224 †
zaktas@eng.ankara.edu.tr
Alpbaz, Mustafa x1365
alpbaz@eng.ankara.edu.tr
Berber, Ridvan (Dean) x1351
berber@eng.ankara.edu.tr
Bilgesu, Ali Y. x1349
bilgesu@eng.ankara.edu.tr
Calik, Guzide x1367
calik@eng.ankara.edu.tr
Calimli, Ayla (Chair) x1303
calimli@eng.ankara.edu.tr
Erol, Murat (Associate Chair) x1375
erol@eng.ankara.edu.tr
Hapoglu, Hale x1338
hapoglu@eng.ankara.edu.tr
Mehmetoglu, Ulku x1311
mehmet@eng.ankara.edu.tr
Oguz, Huseyin x1368
oguz@eng.ankara.edu.tr
Ozdamar, H. T. x1354
ozdamar@eng.ankara.edu.tr
Takac, Serpil x1310
takac@eng.ankara.edu.tr
Togrul, Taner x1329
togrul@eng.ankara.edu.tr
Yeniova, Hasip x1381
yeniova@eng.ankara.edu.tr

Associate Professors

Bayraktar, Emine (Associate Chair) x1373
bayrakta@eng.ankara.edu.tr
Cicek, Burhanettin x1127
bcicek@eng.ankara.edu.tr
Guvenc, Afife x1373
guvenc@eng.ankara.edu.tr
Karacan, Suleyman x1302
karacan@eng.ankara.edu.tr

Kocak, Cetin x1330
kocak@eng.ankara.edu.tr
Ozkan, Gulay x1302
gozkan@eng.ankara.edu.tr
Yildiz, Nuray x1314
nyildiz@eng.ankara.edu.tr

Assistant Professors

Karaduman, Ali x1385
akrduman@eng.ankara.edu.tr
Simsek, Emir H. x1374
simsek@eng.ankara.edu.tr
Suyadal, Yahya x1374
suyadal@eng.ankara.edu.tr

† Phone numbers are extensions of 90-312-2126720.

Accredited by:

Degrees granted 2003-2004:

B.S.: 89 M.S.: 10 Ph.D.: 1

Graduate advisor: Dr.Murat Erol

Undergraduate advisor: Dr.Emine Bayraktar

Department reports to: Prof.Dr.Ridvan Berber, Dean of Faculty of Engineering

Bogazici University

Chemical Engineering Department
34342 Bebek, Istanbul

University phone .... (90) 212 358 1540
Department phone .... (90)212 358 1540x1471
Department FAX .... (90) 212 287 2460

Professors

Akman, Ugur .... x1867 †
akman@boun.edu.tr
Arikol, Mahir (Chair) x1870
arikolma@boun.edu.tr
Baysal, Bahattin (Emeritus) x1868
bmbaysal@hotmail.com
Borak, Fahir (Emeritus) x1470
borakf@boun.edu.tr
Camurdan, Mehmet C. x1405
mccamurd@boun.edu.tr
Dincer, Salih x2338
dincer@yildiz.edu.tr
Haliloglu, Turkan x2003
turkan@prc.bme.boun.edu.tr
Hortacsu, Amable x1469
hortacsa@boun.edu.tr
Hortacsu, Oner x1434
hortacsu@boun.edu.tr
Kirdar, Betul (Associate Dean) x2126
kirdar@boun.edu.tr
Onsan, Z.Ilsen x1751
onsan@boun.edu.tr

Associate Professors

Aksoylu, Ahmet E. x2336
aksoylu@boun.edu.tr
Doruker, Pemra (Associate Chair) x2365
pemra@prc.bme.boun.edu.tr
Ulgen, Kutlu x1869
ulgenk@boun.edu.tr

Assistant Professors

Dervisoglu, Murat 1866
mudervis@boun.edu.tr
Ozer, Ercument x1471
biyotek@biyotek.com.tr
Yildirim, Ramazan x2248
yildirra@boun.edu.tr

† Phone numbers are extensions of

Accredited by: ABET

Graduate advisor Mahir Arikol

Department reports to: Ali Riza Kaylan, Dean of Engineering

Placement service BUMED (BU Alumni Association)

University of Ege

Faculty of Engineering
Chemical Engineering Department
35100 Bornova-IZMIR

University phone .... (90) 232 339 0204
University FAX .... (90) 232 339 9090
Department phone .... (90) 232 388 7600
Department FAX .... (90) 232 388 77 76

Professors

Atalay, Ferhan S. .... 2284 †
fsatalay@eng.ege.edu.tr
Atalay, Suheyda (Chair) 2920
satalay@eng.ege.edu.tr
Ayvaz, Zafer 2287
zayvaz@eng.ege.edu.tr
Erbil, Özel 2459
erbil@eng.ege.edu.tr
Gündüz, Gönül 2292
gunduz@eng.ege.edu.tr
Kabay, Nalan 2290
kabay@eng.ege.edu.tr
Kuryel, Beno 2288
bkuryel@ttnet.net.tr
Olgun, Özden 2286
oolgun@eng.ege.edu.tr
Peker, Sümer 2282
pekers@eng.ege.edu.tr
Saglam, Mehmet 2456
msaglam@eng.ege.edu.tr
Yenigül, Mesut 2289
yenigul@eng.ege.edu.tr
Yüksel, Mithat 2456
myuksel@eng.ege.edu.tr
Övez, Bikem 1483
bovez@eng.ege.edu.tr

Associate Professors

Balkan, Firuz 2291
fbalkan@eng.ege.edu.tr
Ballice, Levent 1484
ballice@eng.ege.edu.tr
Besün, Nurgün 1481
besunnur@ttnet.net.tr
Demircioglu, Mustafa 1831
mdemir@eng.ege.edu.tr

Helvaci, Serife 1480
helvaci@eng.ege.edu.tr

Ulutan, Sevgi (Associate Chair) 1492
ulutans@eng.ege.edu.tr

Yapar, Saadet 1486
yapar@eng.ege.edu.tr

Assistant Professors

Cesur, Serap 1481
kmcesur@eng.ege.edu.tr

Özdemir, Günseli 1485
ozdemirg@eng.ege.edu.tr

Özçelik, Yavuz 1488
yozcelik@eng.ege.edu.tr

Özçelik, Zehra (Associate Chair) 1488
zozcelik@bornova.ege.edu.tr

† Direct dial with prefix (90) 232 388 7600.

Accredited by:

Degrees granted 2003-2004:

B.S.: 95 M.S.: 15 Ph.D.: 10

Placement service: Sevgi Ulutan (90) 232 388 7600

Gazi University

Chemical Engineering Department
Faculty of Engineering and Architecture
06570 Maltepe-Ankara

University phone .... (90) 312-2126840
Department phone .... (90) 312-2317400
Department FAX .... (90) 312-2308434

Professors

Alicilar, Ahmet .... x2551 †
alicilar@mikasa.mmf.gazi.edu.tr

Balci, Suna x2506
sbalci@mikasa.mmf.gazi.edu.tr

Bicer, Ahmet (Head) x2548
abicer@mikasa.mmf.gazi.edu.tr

Cabbar, Canan x2557
cabbar@mikasa.mmf.gazi.edu.tr

Culfaz, Mujgan x2503
culfaz@mikasa.mmf.gazi.edu.tr

Dogu, Gulsen x2559
gdogu@mikasa.mmf.gazi.edu.tr

Erdogan, Sabahat x2549
yuceer@mikasa.mmf.gazi.edu.tr

Ergun, Mubeccel x2537
ergun@mikasa.mmf.gazi.edu.tr

Goymen, Guner (Emeritus) x2524

Koc, Timur x2553
tkoç@mikasa.mmf.gazi.edu.tr

Pamuk, Vecihi x2554
vecihi@mikasa.mmf.gazi.edu.tr

Saracoglu, Nurdan x2546
nurdan@mikasa.mmf.gazi.edu.tr

Uysal, B. Z. x2501
bzu@mikasa.mmf.gazi.edu.tr

Yalcin, Hayri (Emeritus) x2552
hayri@mikasa.mmf.gazi.edu.tr

Associate Professors

Ar, Irfan x2517
irfanar@mikasa.mmf.gazi.edu.tr

Arin, Gonenc x2547
arin@mikasa.mmf.gazi.edu.tr

Dogan, Murat x2513
mdogan@mikasa.mmf.gazi.edu.tr

Guldur, Cigdem x2507
cigdem@mikasa.mmf.gazi.edu.tr

Gunduz, Ufuk x2505
uzafer@mikasa.mmf.gazi.edu.tr

Guru, Metin x2555
metin@mikasa.mmf.gazi.edu.tr

Murathan, Atilla x2518
murathan@mikasa.mmf.gazi.edu.tr

Murtezaoglu, Kirali x2523
kirali@mikasa.mmf.gazi.edu.tr

Assistant Professors

Balbasi, Muzaffer x2512
muzo@mikasa.mmf.gazi.edu.tr

Mutlu, S.Ferda x2519
ferda@mikasa.mmf.gazi.edu.tr

Özkan, Göksel x2515
goksel@mikasa.mmf.gazi.edu.tr

† Dial (90) 312-2317400 and ask for extension.

Hacettepe University

Chemical Engineering Department
Beytepe-06532 Ankara

University phone .... (90) 312 3055050
Department phone .... (90) 312 2977400
Department FAX .... (90) 312 2992124

Professors

Aksu, Zumriye .... (90) 3122977434
zaksu@hacettepe.edu.tr

Alper, Erdogan .... (90) 3122977424
ealper@hacettepe.edu.tr

Beskardes, Oktay .... (90) 3122977474
obeskard@hacettepe.edu.tr

Caglar, Arif .... (90) 3122977414
caglar@hacettepe.edu.tr

Celebi, Serdar S. .... (90) 3122977464
celebi@hacettepe.edu.tr

Durusoy, Tulay .... (90) 3122977494
tdurusoy@hacettepe.edu.tr

Kiremitci, Menemse .... (90) 3122977447
menemse@hacettepe.edu.tr

Kutsal, Tulin .... (90) 3122977484
tkutsal@hacettepe.edu.tr

Oncu, Abdurrezak .... (90) 3122977472
rezzak@hacettepe.edu.tr

Ozbas Bozdemir, Tijen .... (90) 3122977455
bozdemir@hacettepe.edu.tr

Ozdural, Ahmet R. .... (90) 3122977475
ozdural@hacettepe.edu.tr

Piskin, Erhan (Chair) .... (90) 3122977473
piskin@hacettepe.edu.tr

Rzayev, Zakir .... (90) 3122977400
zakirr@hacettepe.edu.tr

Sag, Yesim .... (90) 3122977444
yesims@hacettepe.edu.tr

Tanyolac, Abdurrahman .... (90) 3122977404
tanyolac@hacettepe.edu.tr

Tuncel, S. A. .... (90) 3122977433
tuncel@hacettepe.edu.tr

Associate Professors

Evren, Vural .... (90) 3122977422
vural@hacettepe.edu.tr
Mutlu, Selma .... (90) 3122977400
smselma@hacettepe.edu.tr
Tanyolaç, Deniz .... (90) 3122976162
deniztan@hacettepe.edu.tr

Assistant Professors

Ayhan, Hakan .... (90) 3122977400
hayhan@hacettepe.edu.tr

Accredited by: Turkish Higher Educational Council

Graduate advisor: Erhan Piskin

Undergraduate advisor: Deniz Tanyolac

Student organization: Hakan Ayhan
Advisor: S. Ali Tuncel

Department reports to: Selcuk Gecim, Dean

Placement service: Erhan Piskin

Middle East Technical University

Chemical Engineering Department
Inonu Bulvari Ankara, 06531

University phone .... (90) 312 2101000
Department phone .... (90) 312 2102601
Department FAX .... (90) 312 2101264

Professors

Bakir, Ufuk (Associate Chair) .... (90) 312 2102619
ubakir@metu.edu.tr
Culfaz, Ali .... (90) 312 2102611
culfaz@metu.edu.tr
Dogu, Timur (Chair) .... (90) 312 2102631
tdogu@metu.edu.tr
Eroglu, Inci .... (90) 312 2102609
ieroglu@metu.edu.tr
Gunduz, Gungor .... (90) 312 2102616
ggunduz@metu.edu.tr
Guruz, Guniz .... (90) 312 2102634
guniz@metu.edu.tr
Gürkan, Türker .... (90) 312 2102618
tgurkan@metu.edu.tr
Kincal, Suzan .... (90)312 2102617
nskincal@metu.edu.tr
Kisakürek, Bilgin .... (90) 312 2102625
bilgin@metu.edu.tr
Onal, Isik .... (90) 312 2102639
ional@metu.edu.tr
Ozbelge, Onder .... (90) 312 2102628
oozbelge@metu.edu.tr
Ozbelge, Tulay .... (90) 312 2102621
tozbelge@metu.edu.tr
Ozgen, Canan .... (90) 312 2102605
cozgen@metu.edu.tr
Selcuk, Nevin .... (90) 312 2102603
selcuk@metu.edu.tr
Tosun, Ismail .... (90) 312 2102637
itosun@metu.edu.tr
Yilmaz, Levent .... (90) 312 2102607
lyilmaz@metu.edu.tr
Yilmazer, Ulku .... (90) 312 2102615
yilmazer@metu.edu.tr
Yucel, Hayrettin .... (90) 312 2102635
hyucel@metu.edu.tr

Associate Professors

Calik, Pinar .... (90) 312 2104385
pcalik@metu.edu.tr
Karakas, Gurkan Associate Chair) .... (90) 312 2102630
gkarakas@metu.edu.tr
Ozbay, Deniz U. .... (90) 312 2104383
uner@metu.edu.tr

Assistant Professors

Bayram, Goknur .... (90) 3122102632
gbayram@metu.edu.tr
Kalÿpçÿlar, Halil .... (90) 312 2104357
hkalipcilar@metu.edu.tr
Uludag, Yusuf .... (90) 3122104374
yuludag@metu.edu.tr

Accredited by: Recognised by ABET

Graduate advisor: Levent Yilmaz

Student organization: AIChE
Advisor: Deniz Uner

Department reports to: Yildirim Uctug, Dean, Faculty of Engineering

Yildiz Technical University

Faculty of Chemical and Metallurgical Engineering
Chemical Engineering Department
Davutpasa Campus, No.127, 34210 Esenler
Istanbul

University phone 90 212 449 16 67
Department phone 90 212 449 18 94
Department FAX 90 212 449 18 95

Professors

Bolat, Esen
Dincer, Salih
Oner, Mualla
Pala, Mehmet
Piskin, Sabriye (Head)

Associate Professors

Baykara, Sema
Beker, Ulker
Erturan, Seyfettin
Ozbek, Belma

Assistant Professors

Akgun, Mesut
Akgun, Nalan
Gulen, Jale
Ozkan, Semra

Graduate advisor: Sabriye Piskin

Undergraduate advisor: Sabriye Piskin

Department reports to: Nihat Kinikoglu, Dean of Chemical and Metallurgical Engineering

United Arab Emirates

United Arab Emirates University

Department of Chemical and Petroleum Engineering
PO Box 17555
Al Ain

Department phone ....(971-3) 7051-611
Department FAX ....(971-3) 762-4262

Professors

Abou Kassem, Jamal 971-3-7051-599
J.Aboukassem@uaeu.ac.ae

Ashour, Ibrahim 971-3-7051-637
ashour@uaeu.ac.ae

Chebbi, Rachid 971-3-7051-634
RChebbi@uaeu.ac.ae

Zekri, Abdulrazag 971-3-7051-550
A.Zekri@uaeu.ac.ae

Associate Professors

Abu Al-Rub, Fahmi 971-3-7051-740
F.Rub@uaeu.ac.ae

Almehaideb, Reyadh A. (Chair) 971-3-7621-765
Reyadh@uaeu.ac.ae

Benyahia, Farid 971-3-7051-552
farid.benyahia@uaeu.ac.ae

Ghannam, Mamdouh T. 971-3-7051-635
Mamdouh.Ghannam@uaeu.ac.ae

Kocabas, Ibrahim 971-3-7051-546
IKocabas@uaeu.ac.ae

Assistant Professors

Abdulkarim, Mohamed (Director) 971-3-7051-547
Mohamed.AK@uaeu.ac.ae

Al Marzooqi, Ali 971-3-7051-639
HassanA@uaeu.ac.ae

Al Matroushi, Eisa A. 971-3-7051-638
Almatroushi@uaeu.ac.ae

Al-Marzouqi, Mohamed H. (Associate Dean) 971-3-7621-695
MHHassan@uaeu.ac.ae

Al-Muhtaseb, Shaheen 971-3-7051-545
s.almuhtaseb@uaeu.ac.ae

Nounou, Mohamed 971-3-7051-549
mnounou@uaeu.ac.ae

Rahman, Motiur 971-3-7051-551
M.Rahman@uaeu.ac.ae

Shedid, Shedid 971-3-7051-636
S.Shedid@uaeu.ac.ae

Accredited by: ABET

Degrees granted 2003-2004:

B.S.: 36

Department reports to: Dean of College of Engineering

United Kingdom

Aston University

School of Engineering and Applied Science
Division of Chemical Engineering and Applied Chemistry
Aston Triangle
Birmingham B4 7ET, England

University phone ....(44) 121 359-3611
Department phone .... (44) 121 359 3611 x 4632
Department FAX ....(44) 121 359-4094

Professors

Barker, P. E. (Emeritus)
Bridgwater, A. V. (Head)
Porter, K. E. (Emeritus)

Assistant Professors

Brammer, J
Davies, B.
Drahun, J A
Fletcher, J. P.
Generalis, S.
Titiloye, J

Accredited by: IChemE (UK)

Department reports to: Vice Chancellor

Placement service: Dr J A Drahun

University of Bath

Department of Chemical Engineering
University of Bath
Claverton Down
Bath, BA2 7AY, England

University phone .... (1225) 386826
Department phone .... (1225) 386338
Department FAX .... (1225) 385713

Professors

Crittenden, Barry D. (Head) .... 386501 †
B.D.Crittenden@bath.ac.uk

Greaves, Malcolm (Emeritus) 386624
M.Greaves@bath.ac.uk

Guy, Keith W. A. (Visiting) 386338

Howell, John A. (Emeritus) 386338

Kolaczkowski, Stan T. 386440
S.T.Kolaczkowski@bath.ac.uk

Moore, Stephen (Visiting) 386338

Associate Professors

Bird, Michael R 386336
M.R.Bird@bath.ac.uk

Chaudhuri, Julian B. 386349
J.B.Chaudhuri@bath.ac.uk

England, Richard 386516
R.England@bath.ac.uk

Hubble, John 386221
J.Hubble@bath.ac.uk

Lukyanov, Dmitry B 383329
D.B.Lukyanov@bath.ac.uk

Mays, Tim J 386528
T.J.Mays@bath.ac.uk

Perera, Semali 386584
S.Perera@bath.ac.uk
Plucinski, Pawel 386961
P.Plucinski@bath.ac.uk

Assistant Professors

Arnot, Tom C 386707
T.C.Arnot@bath.ac.uk
Hicks, David I 386251
D.I.Hicks@bath.ac.uk
Lapkin, Alexei A. 383369
A.Lapkin@bath.ac.uk
Rathbone, Richard R. 383961
R.R.Rathbone@bath.ac.uk
Rigby, Sean P 384978
S.P.Rigby@bath.ac.uk

† Dial area code 1225

Accredited by: IChemE (UK)

Degrees granted 2003-2004:

B.S.: 10 M.S.: 25 Ph.D.: 7

Graduate advisor: Dr Semali Perera

Undergraduate advisor: Dr. Tim Mays

Student organization: IChemE (UK)
Advisor: Dr. Sean Rigby

Department reports to: Dean, Professor Alan Day (1225) 386469

Placement service: Dr. Mike Bird (1255) 386336

University of Birmingham

School of Chemical Engineering
Edgbaston, Birmingham
B15 2TT, England

University phone 121-414-3344
Department phone 121-414-5323
Department FAX 121-414-5324

Professors

Al-Rubaei, M. .... x5264 †
M.Al-Rubeai@bham.ac.uk
Biddlestone, A.J x5290
A.J.Biddlestone@bham.ac.uk
Bourne, J. R. (Visiting) x5286
Fryer, P. J. x5451
P.J.Fryer@bham.ac.uk
Kendall, K x2739
K.Kendall@bham.ac.uk
Lyddiatt, A. x5278
A.Lyddiatt@bham.ac.uk
Nienow, A. W. x5325
A.W.Nienow@bham.ac.uk
Seville, J. P. K. x5322
J.P.K.Seville@bham.ac.uk
Thomas, C. R. (Head) x5355
C.R.Thomas@bham.ac.uk
Warner, N. A. (Emeritus) x5306
Winterbottom, J. M. x5293
J.M.Winterbottom@bham.ac.uk

Associate Professors

Al-Duri, B. x3969
B.Al-Duri@bham.ac.uk
Arrowsmith, A. A. x5305
A.Arrowsmith@bham.ac.uk
Ashton, N. x5297
N.Ashton@bham.ac.uk
Barigou, M. x5277
M.Barigou@bham.ac.uk
Barker, A. J. x5289
A.J.Barker@bham.ac.uk
Blackburn, S. x5286
S.Blackburn@bham.ac.uk
Emery, A. N. x5281
A.N.Emery@bham.ac.uk
Kent, C. A. x5283
C.A.Kent@bham.ac.uk
McFarlane, C. M. x3887
C.McFarlane@bham.ac.uk
McManamey, W. J. x5294
Pacek, A. x5308
A.W.Pacek@bham.ac.uk
Rowson, N. A. x5298
N.A.Rowson@bham.ac.uk
Thayanithy, K. x5284
K.Thayanithy@bham.ac.uk
Veasey, T. J. x5296
T.J.Veasey@bham.ac.uk
Zhang, Z. x5357
Z.Zhang@bham.ac.uk

Assistant Professors

Bujalski, W. x5279
W.Bujalski@bham.ac.uk
Burbidge, A.S. x5295
A.S.Burbidge@bham.ac.uk
Guy, S. x5282
S.Guy@bham.ac.uk
Hewitt, C. x5360
C.J.Hewitt@bham.ac.uk
Roberts, A.B. x5288
A.B.Roberts@bham.ac.uk
Santos, R.C.D. x5374
R.C.D.Santos@bham.ac.uk
Thornton, R (Visiting) x5363
rt@arbor.dircon.co.uk
Wood, J x5295
J.Wood@bham.ac.uk
Yong, Y-P x5360
Y.P.Yong@bham.ac.uk
Yuan, W x5303
W.Yuan@bham.ac.uk

† Dial 121-414 and extension faculty numbers.

Accredited by: IChemE (UK)

Undergraduate advisor: C. R. Thomas

Department reports to: Vice-Chancellor

Placement service: E. Oakland

University of Cambridge

Chemical Engineering Department
Pembroke Street
Cambridge CB2 3RA, England

Department phone .... (1223) 334777
Department FAX .... (1223) 334796

Professors

Bridgwater, J. (Emeritus)
john_bridgwater@cheng.cam.ac.uk

Chase, H. A. (Head)
chase@cheng.cam.ac.uk

Davidson, J. F. (Emeritus)

Gladden, L. F.
gladden@cheng.cam.ac.uk

Hayhurst, A. N. (Emeritus)
allan_hayhurst@cheng.cam.ac.uk

Mackley, M.R.
malcolm_mackley@cheng.cam.ac.uk

Slater, N K H
nigel_slater@cheng.cam.ac.uk

Associate Professors

Barrie, P. J.
patrick_barrie@cheng.cam.ac.uk

Cardoso, S. S. S.
silvana_cardoso@cheng.cam.ac.uk

Fisher, A.C.
adrian_fisher@cheng.cam.ac.uk

Johns, M.L.
mlj21@cheng.cam.ac.uk

Kaminski, C.F.
clemens_kaminski@cheng.cam.ac.uk

Kraft, M
markus_kraft@cheng.cam.ac.uk

Moggridge, G.M.
geoff_moggridge@cheng.cam.ac.uk

Paterson, W.R.
wrp1@cheng.cam.ac.uk

Scott, D.M.
dms1@cheng.cam.ac.uk

Skelton, R.L.
rls1@cheng.cam.ac.uk

Vassiliadis, V. S.
vsv20@cheng.cam.ac.uk

Wilson, D. I.
ian_wilson@cheng.cam.ac.uk

Accredited by: The Institution of Chemical Engineers (UK)

Degrees granted 2003-2004:

B.S.: 33 M.S.: 34 Ph.D.: 15

Graduate advisor: M L Johns

Undergraduate advisor: H. A. Chase

Student organization: CUCES

Placement service: A. J. Raban

University of Edinburgh

Chemical Engineering
The Kenneth Denbigh Chemical Engineering Building
King's Buildings, Mayfield Rd.
Edinburgh EH9 3JL, Scotland
Deliveries: Engineering & Electronics Stores, Sanderson Building, King's Buildings, Gate 3

University phone 0131-650-1000
Department phone 0131-650-4860
Department FAX 0131-650-6551

Professors

Christy, John R. E. (Head) .... 650-4854 †
John.Christy@ed.ac.uk

Glass, Donald H. 650-4870
Don.Glass@ed.ac.uk

Ponton, Jack W. 650-4858
J.W.Ponton@ed.ac.uk

Pritchard, Colin L. 650-4852
colin@chemeng.ed.ac.uk

Seaton, Nigel A. 650-4867
Nigel.Seaton@ed.ac.uk

Associate Professors

Biggs, Mark J 650-5891
MBiggs@ed.ac.uk

Sefiane, Khellil 650-4873
ksefiane@ed.ac.uk

Serghiou, George 650-8553
George.Serghiou@ed.ac.uk

Skilling, Jennifer M. 650-4863
J.Skilling@ed.ac.uk

Assistant Professors

Duren, Tina 650-4856
tina.duren@ed.ac.uk

Duursma, Gail 650-4868
Gail.Duursma@ed.ac.uk

† Direct dial with prefix 0131.

Accredited by: Institute of Chemical Engineers

Degrees granted 2003-2004:

B.S.: 11 M.S.: 5 Ph.D.: 2

Department reports to: College of Science and Engineering

University of Exeter

Chemical Engineering Group
School of Engineering
North Park Road
Exeter, EX4 4QF, England
Deliveries: John D. Boyle

University phone 01 392 263263
Department phone 01 392 263653
Department FAX 01 392 217965

Professors

Davies, Tom W. 01 392 263683
T.W.Davies@ex.ac.uk

Turner, J. C. R. (Emer.) 01 392 263623
J.C.R.Turner@ex.ac.uk

Wragg, Anthony A. 01 392 263656
A.A.wragg@ex.ac.uk

Associate Professors

Ashcroft, Stan J. 01 392 263728
S.J.Ashcroft@ex.ac.uk

Boyle, John D. (Head) 01 392 263653
J.D.Boyle@ex.ac.uk

Accredited by: IChemE (UK)

Graduate advisor: Prof A.A.Wragg

Undergraduate advisor: Mrs Lorna Howe

Department reports to: The Vice Chancellor

Placement service: Admissions Tutor (Engineering), The Appointments Board

Heriot-Watt University

Chemical Engineering
School of Engineering & Physical Sciences
Riccarton
Edinburgh EH14 4AS, Scotland

University phone .... (44) 131 449 5111
Department phone .... (44) 131 451 3131
Department FAX .... (44) 131 451 3129

Professors

Docherty, R. (Bob) (Visiting)
Geldart, Derek (Visiting)
Ni, Xiong-Wei
Ocone, Raffaella (Head)
Waldie, Brian (Emeritus)

Associate Professors

Murray, Keith R.
Wilkinson, Derek

Assistant Professors

Bell, George
Boron, Stefan
Bustard, Mark T
Goodwin, Julian A.S.
Pekdemir, Turgay
Thomson, Gillian B
Westacott, Robin E
White, Graeme

Accredited by: Institution of Chemical Engineers

Degrees granted 2003-2004:

B.S.: 21 M.S.: 7 Ph.D.: 7

Department reports to: J S Archer, Principal

Placement service: R E Westacott

University of Leeds

School of Process, Environmental and Materials Engineering
Leeds LS2 9JT

University phone 44 (0) 113 243 1751
Department phone 44 (0) 113 343 2404
Department FAX 44 (0) 113 343 2405

Professors

Biggs, Simon R .... 343 2790 †
s.r.biggs@leeds.ac.uk
Ghadiri, Mojtaba 343 2406
m.ghadiri@leeds.ac.uk
Roberts, Kevin J. 343 2400
k.j.roberts@leeds.ac.uk
Williams, Richard A 343 2801
r.a.williams@leeds.ac.uk

† Direct dial with prefix 0113.

Accredited by: IChemE

Graduate advisor: S. Kalliadasis / T. Mahmud

Undergraduate advisor: M.Fairweather

Student organization: Chemical Engineering Society
Advisor: Chair, Chemical Engineering Society

Department reports to: Prof. P.A. Dowd, Head School Process, Environmental and Materials Engineering

Placement service: M.J. Jones

University of London (Imperial College)

Department of Chemical Engineering & Chemical Technology
Prince Consort Road
London, SW7 2AZ UK

University phone 020 7589 5111
Department phone .... (44) 020 7594 5605
Department FAX .... (44) 020 7594 5604

Professors

Briscoe, Brian J. .... (44) 020 7594 5559
b.briscoe@ic.ac.uk
Chadwick, David .... (44) 020 7594 5579
d.chadwick@ic.ac.uk
Dugwell, Denis R. .... (44) 020 7594 5568
d.dugwell@ic.ac.uk
Hewitt, Geoffrey F. (Emeritus) .... (44) 020 7594 5562
g.hewitt@ic.ac.uk
Higgins, Julia S. .... (44) 020 7594 5565
j.higgins@ic.ac.uk
Jackson, George .... (44) 020 7594 5640
g.jackson@ic.ac.uk
Jones, Alan R. .... (44) 020 7594 5577
a.r.jones@ic.ac.uk
Kandiyoti, Rafael .... (44) 020 7594 5581
r.kandiyoti@ic.ac.uk
Kelsall, Geoffrey .... (44) 020 7594 5633
g.kelsall@imperial.ac.uk
Kershenbaum, Lester S. .... (44) 020 7594 5566
l.kershenbaum@ic.ac.uk
Lawrence, Christopher J. .... (44) 020 7594 5622
c.lawrence@ic.ac.uk
Livingston, Andrew .... (44) 020 7594 5582
a.livingston@ic.ac.uk
Luckham, Paul F. .... (44) 020 7594 5583
p.luckham01@ic.ac.uk
Macchietto, Sandro .... (44) 020 7594 6608
s.macchietto@ic.ac.uk
Michels, Hans J. (Emeritus) .... (44) 020 7594 5586
h.michels@ic.ac.uk
Pantelides, Costas C. .... (44) 020 7594 6622
c.pantelides@ic.ac.uk
Perkins, John D. (Dean) .... (44) 020 7594 5556
j.perkins@ic.ac.uk
Pistikopoulos, Efstratios N. .... (44) 020 7594 6620
e.pistikopoulos@ic.ac.uk
Richardson, Stephen M. (Head) .... (44) 020 7594 5626
s.m.richardson@imperial.ac.uk
Sargent, Roger.W.H. (Emeritus) .... (44) 020 7594 6613
r.sargent@ic.ac.uk
Shah, Nilay .... (44) 020 7594 6621
n.shah@ic.ac.uk
Stuckey, David C. .... (44) 020 7594 5591
d.stuckey@ic.ac.uk
Trusler, J. P. M. .... (44) 020 7594 5592
m.trusler@ic.ac.uk
Weinberg, Felix J. (Emeritus) .... (44) 020 7594 5580
f.weinberg@ic.ac.uk

Associate Professors

Bismarck, Alexander .... (44) 020 7594 5578
a.bismarck@imperial.ac.uk

Brandon, Nigel .... (44) 020 7594 5704
n.brandon@imperial.ac.uk

Carabine, Michael D. .... (44) 020 7594 5571
m.carabine@ic.ac.uk

Galindo, Amparo .... (44) 020 7594 5606
a.galindo@imperial.ac.uk

McLeod, Alistair .... (44) 020 7594 6656
alistair.mcleod@imperial.ac.uk

Mendes, M.-Alcina .... (44) 020 7594 5584
a.mendes@ic.ac.uk

Ortiz, E. S. .... (44) 020 7594 5587
s.ortiz@ic.ac.uk

Saville, Graham .... (44) 020 7594 5578
g.saville@ic.ac.uk

Stepanek, Frantisek .... (44) 020 7594 5608
f.stepanek@imperial.ac.uk

Young, David A. (Emeritus) .... (44) 020 7594 5608
d.a.young@ic.ac.uk

Assistant Professors

Adjiman, Claire S. .... (44) 020 7594 6638
c.adjiman@ic.ac.uk

Alpay, Esat .... (44) 020 7594 5625
e.alpay@ic.ac.uk

Kazarian, Sergei G. .... (44) 020 7594 5574
s.kazarian@ic.ac.uk

Kogelbauer, Andreas .... (44) 020 7594 5572
a.kogelbauer@ic.ac.uk

Mantalaris, Athanassios .... (44) 020 7594 5601
a.mantalaris@ic.ac.uk

Matar, Omar K. .... (44) 020 7594 5635
o.matar@ic.ac.uk

Varley, Julie .... (44) 020 7594 5698
j.varley@ic.ac.uk

Xu, X.Yun .... (44) 020 7594 5588
yun.xu@ic.ac.uk

Accredited by: IChemE (UK)

Department reports to: The Rector

Placement service: Professor Geoff Kelsall

University of London (University College London)

Department of Chemical Engineering
Torrington Place
London, WClE 7JE

University phone 44 020 7679-2000
Department phone 44 020 7679-3825
Department FAX 44 020 7383-2348

Professors

Bogle, I. D. L. 3803
d.bogle@ucl.ac.uk

Brandani, S 2315
s.brandani@ucl.ac.uk

Cornish, A. R. H. (Emeritus) 3824
Gibilaro, L. G. (Emeritus) 3838
Gugan, K. (Visiting) 3824
Jones, A. G. (Head) 3828
a.jones@ucl.ac.uk

Mahgerefteh, H. 3835
h.mahgerefteh@ucl.ac.uk

Mullin, J. W. (Emeritus) 3824
Rowe, P. N. (Emeritus) 3824
Simons, S. J. R. 3805
stefan.simons@ucl.ac.uk

Yates, J. G. 3837
john.yates@ucl.ac.uk

Associate Professors

Elson, T. P. 3821
t.elson@ucl.ac.uk

Fraga, E.S. 3817
e.fraga@ucl.ac.uk

Gavriilidis, A. 3811
a.gavriilidis@ucl.ac.uk

Sorensen, E. 3802
e.sorensen@ucl.ac.uk

Assistant Professors

Angeli, P. 3832
p.angeli@ucl.ac.uk

Cains, P. 3806
p.cains@ucl.ac.uk

Lettieri, P. 7867
p.lettieri@ucl.ac.uk

Manos, G 3810
g.manos@ucl.ac.uk

Papageorgiou, L 2563
l.papageorgiou@ucl.ac.uk

Accredited by: IChemE

Graduate advisor: Dr S Brandani

Student organization: Ramsay Society
Advisor: Anna Harrington

Department reports to: Sir Derek Roberts, Provost and President

Placement service: Dr. G. Manos

Loughborough University

Department of Chemical Engineering
Ashby Road
Loughborough, Leicestershire
LE11 3TU, England

University phone 01509-263171
Department phone 01509-222532
Department FAX 01509-223923

Professors

Brooks, Brian W. (Emeritus) 01509-222510
B.W.Brooks@lboro.ac.uk

Buffham, Bryan A. (Emeritus) 01509-222503
B.A.Buffham@lboro.ac.uk

Hall, Alan (Visiting)
Hankinson, Geoff 01509-222540
G.Hankinson@lboro.ac.uk

Kletz, Trevor A. (Visiting)
T.A.Kletz@lboro.ac.uk

Lydon, Richard (Visiting)
Mason, Geoff 01509-222509
G.Mason@lboro.ac.uk

Nassehi, Vahid 01509-222522
V.Nassehi@lboro.ac.uk.

Preston, Malcolm L (Visiting) 01509-222515
M.L.Preston@lboro.ac.uk
Rielly, Chris D 01509-222504
C.D.Rielly@lboro.ac.uk
Starov, Victor M 01509-222508
V.M.Starov@lboro.ac.uk
Streat, Michael (Emeritus) 01509-222506
M.Streat@lboro.ac.uk
Wakeman, Richard J. (Head) 01509-222500
R.J.Wakeman@lboro.ac.uk

Associate Professors

Cumming, Iain W. 01509-222523
I.W.Cumming@lboro.ac.uk
Edwards, David W. (Visiting) 01509-222515
D.W.Edwards@lboro.ac.uk
Holdich, Richard G. 01509-222519
R.G.Holdich@lboro.ac.uk
Pinder, Colin C. (Visiting)
Saha, Basu 01509-222505
B.Saha@lboro.ac.uk
Shama, Gilbert 01509-222514
G.Shama@lboro.ac.uk
Tarleton, E. S. 01509-222535
E.S.Tarleton@lboro.ac.uk

Assistant Professors

Drott, David W. 01509-222516
D.W.Drott@lboro.ac.uk
Malik, Danish J. 01509-222507
D.J.Malik@lboro.ac.uk
Stapley, Andy G.F. 01509-222525
A.G.F.Stapley@lboro.ac.uk
Wilcockson, Robin W. 01509-222521
R.B.Wilcockson@lboro.ac.uk

Accredited by: IChemE

Degrees granted 2003-2004:

B.S.: 35 M.S.: 8 Ph.D.: 12

Undergraduate advisor: ES Tarleton

Student organization: IChemE

Department reports to: Vice Chancellor

Placement service: Wendy Llewellyn, Appts. Officer

University of Manchester Institute of Science and Technology (UMIST)

Department of Chemical Engineering,
P.O. Box 88
Sackville Street
Manchester, M60 1QD, England

University phone 0161 236 3311
Department phone 0161 200 4340
Department FAX 0161 200 4399

Professors

Davey, Roger J 0161 200 4409
roger.davey@umist.ac.uk
Dyakowski, Tomasz 0161 200 4397
tom.dyakowski@umist.ac.uk
Edwards, Michael F (Visiting)
Garrett, Peter (Visiting) 0161 200 4558
p.garrett@umist.ac.uk
Griffiths, Richard F 0161 200 3980
richard.f.griffiths@umist.ac.uk
Heggs, Peter J 0161 200 4370
peter.heggs@umist.ac.uk
Hulse, Joseph (Visiting)
Mann, Reg 0161 200 4378
reginald.mann@umist.ac.uk
Mavituna, Ferda 0161 200 4372
ferda.mavituna@umist.ac.uk
Metcalfe, Ian S 0161 200 4361
ian.metcalfe@umist.ac.uk
Rankin, Hamish (Visiting) 0161 200 3982
hamish.rankin@umist.ac.uk
Sharratt, Paul N 0161 200 4367
paul.sharratt@umist.ac.uk
Slater, David (Visiting)
Tiddy, Gordon J 0161 200 8865
gordon.tiddy@umist.ac.uk
Webb, Colin (Head) 0161 200 4379
colin.webb@umist.ac.uk

Associate Professors

Campbell, Grant M 0161 200 4472
grant.campbell@umist.ac.uk
Cilliers, Jan JL 0161 200 4353
johannes.cilliers@umist.ac.uk
Garforth, Arthur 0161 200 8850
arthur.garforth@umist.ac.uk
Grassia, Paul 0161 200 8851
paul.s.grassia@umist.ac.uk
Korchinsky, Walter J 0161 200 4344
walter.korchinsky@umist.ac.uk
Newsham, David MT 0161 200 4347
david.newsham@umist.ac.uk
Pickett, David J 0161 200 4374
david.pickett@umist.ac.uk
Roberts, Edward PL 0161 200 8849
ted.roberts@umist.ac.uk
Rushton, Eric 0161 200 4402
eric.rushton@umist.ac.uk
Senior, Peter R 0161 200 4403
peter.senior@umist.ac.uk
Stephens, Gill M. 0161 200 4377
gill.stephens@umist.ac.uk

Assistant Professors

Bennett, Michael 0161 200 4388
mike.bennett@umist.ac.uk
Curtis, Robin 0161 200 4401
r.curtis@umist.ac.uk
Holmes, Stuart 0161 200 4376
stuart.holmes@umist.ac.uk
James, Alec 0161 200 4368
alec.e.james@umist.ac.uk
Kookos, Ioannis 0161 200 4346
i.kookos@umist.ac.uk
Lue, Leo 0161 200 4867
leo.lue@umist.ac.uk
Markx, Gerard H 0161 200 4394
gerard.markx@umist.ac.uk
Martin, Alastair 0161 200 4395
alastair.martin@umist.ac.uk
Martin, Philip 0161 200 4340
p.a.martin@umist.ac.uk
Miller, Aline 0161 200 4340
aline.miller@umist.ac.uk
Neethling, Stephen J 0161 200 8854
stephen.neethling@umist.ac.uk

Pandiella, Severino S 0161 200 4429
severino.pandiella@umist.ac.uk
Saiani, Alberto 0161 200 4340
alberto.saiani@umist.ac.uk
Schroeder, Sven 0161 200 4340
sven.schroeder@umist.ac.uk

Accredited by: IChemE

Graduate advisor: Dr E Roberts

Undergraduate advisor: Dr D Newsham

Student organization: Ms A Marron

Department reports to: Prof J Garside (Vice Chancellor)

Placement service: Careers Service (0161 200 4333)

University of Newcastle Upon Tyne

Chemical Engineering and Advanced Materials
University of Newcastle Upon Tyne
Merz Court
Newcastle Upon Tyne
NE1 7RU, England

University phone 0191-222-6000
Department phone 0191-222-7266
Department FAX 0191-222-5292

Professors

Akay, Galip 0191-222-7269
galip.akay@ncl.ac.uk
Bull, Steve 0191 2227913
s.j.bull@ncl.ac.uk
Martin, Elaine B 0191 2226231
e.b.martin@ncl.ac.uk
Montague, Gary 0191-222-7265
gary.montague@ncl.ac.uk
Morris, Julian (Head) 0191-222-7342
julian.morris@ncl.ac.uk
Page, Trevor 0191 2227201
t.f.page@ncl.ac.uk
Ramshaw, Colin 0191-222-7270
colin.ramshaw@ncl.ac.uk
Scott, Keith 0191-222-8771
k.scott@ncl.ac.uk
Snowden, Ken
k.j.snowden
Thompson, Derek 0191 2227202
d.p.thompson@ncl.ac.uk
White, Jim 0191 2227906
jim.white@ncl.ac.uk

Associate Professors

Egerton, Terry 0191 2225618
t.a.egerton@ncl.ac.uk
Norman, Peter W. 0191-222-7278
peter.norman@ncl.ac.uk
Tham, Ming T. 0191-222-7285
ming.tham@ncl.ac.uk
Willis, Mark J. 0191-222-7242
mark.willis@ncl.ac.uk

Assistant Professors

Boodhoo, Kamelia 0191 222 7243
k.v.k.boodhoo@ncl.ac.uk
Charles, Alastair 0191 2227900
e.a.charles@ncl.ac.uk
Glassey, Jarka 0191-222-7275
jarka.glassey@ncl.ac.uk
Haile, Sue 0191-222-7279
s.m.haile@ncl.ac.uk
Jachuck, Roshan 0191-222-5202
r.j.j.jachuck@ncl.ac.uk
Jaros, Milan 0191 2227379
milan.jaros@ncl.ac.uk
Lee, Jon 0191-222-7241
j.g.m.lee@ncl.ac.uk
Roy, Sudipta 0191-222-7274
s.roy@ncl.ac.uk
Siller, Lidija 0191 2227323
lidija.siller@ncl.ac.uk
Zhang, Jie 0191-222-7240
jie.zhang@ncl.ac.uk

Accredited by: IChemE (UK), Institute of Energy (London)

Department reports to: Prof Peter Hills, Dean

Placement service: Cathryn Harvey, Chief Careers Advisor

University of Nottingham

School of Chemical, Environmental and Mining Engineering (SChEME)
University of Nottingham
University Park
Nottingham, NG7 2RD, England

University phone 44 115 951 5151
Department phone 44 115 951 4163
Department FAX 44 115 951 4115

Professors

Azzopardi, B. J. 44 115 951 4160
barry.azzopardi@nottingham.ac.uk
Denby, Bryan 44 115 951 4097
Bryan.Denby@nottingham.ac.uk
Miles, N. J. (Head) 44 115 951 4085
nick.miles@nottingham.ac.uk
Patrick, J. W. 44 115 951 4175
john.patrick@nottingham.ac.uk
Snape, C. E. 44 115 951 4166
colin.snape@nottingham.ac.uk

Associate Professors

Chen, George Z 44 115 951 4171
george.chen@nottingham.ac.uk
Edwards, J. S. 44 115 951 4075
julian.edwards@nottingham.ac.uk
Lowndes, I. S 44 115 951 4086
ian.lowndes@nottingham.ac.uk
Wilson, J. A. 44 115 951 4179
j.a.wilson@nottingham.ac.uk

Assistant Professors

Hankins, N. P. 44 115 951 4173
nick.hankins@nottingham.ac.uk
Hilal, N 44 115 951 4168
nidal.hilal@nottingham.ac.uk
Jones, W. E. 44 115 951 4172
warren.jones@nottingham.ac.uk
Kingman, S. W. 44 115 951 4165
sam.kingman@nottingham.ac.uk
Langston, P. A. 44 115 951 4177
p.langston@nottingham.ac.uk

Large, D. J. 44 115 951 4114
david.large@nottingham.ac.uk
Lester, E 44 115 951 4974
edward.lester@nottingham.ac.uk
Li Puma, G 44 115 951 4170
gianluca.li.puma@nottingham.ac.uk
Steel, K 44 115 951 4078
karen.steel@nottingham.ac.uk
Waller, M. D. 44 115 951 4096
martin.waller@nottingham.ac.uk

Accredited by: IChemE (UK)

Degrees granted 2003-2004:

B.S.: 24 M.S.: 32 Ph.D.: 6

Graduate advisor: Professor B J Azzopardi

Undergraduate advisor: Dr W E Jones

Department reports to: University Senate

Placement service: Careers Advisory Service

Queen's University of Belfast

School of Chemical Engineering
David Keir Building
Stranmillis Road
Belfast BT9 5AG, Northern Ireland
University phone 028 90245133
Department phone 028 90974253
Department FAX 028 90381753

Professors

Allen, Stephen J. 028 90335425
s.allen@qub.ac.uk
Magee, T. R. A. (Head) 028 90274255
r.magee@qub.ac.uk
Murphy, Raymond 028 90274701
w.r.murphy@qub.ac.uk
Richardson, Philip (Visiting)

Associate Professors

Ahmad, Mohammad N. 028 90274389
mnm.ahmad@qub.ac.uk
Gan, Quan 028 90274463
q.gan@qub.ac.uk
Holland, Clive 028 90974256
c.r.holland@qub.ac.uk
McMinn, Wendy A.M. 028 90274065
w.mcminn@qub.ac.uk
McNally, Gerry M. 028 90274705
g.mcnally@qub.ac.uk
Walker, Gavin 028 90274172
g.walker@qub.ac.uk

Assistant Professors

Rooney, David 028 90274050
d.rooney.@qub.ac.uk
Sen Gupta, Bhaskar 028 90274554
b.sengupta@qub.ac.uk
Sun, Guangzhi 028 90274692
g.sun@qub.ac.uk

Accredited by: IChemE (UK)

Degrees granted 2003-2004:

B.S.: 15 M.S.: 14 Ph.D.: 3

Graduate advisor: T.R.A. Magee

Undergraduate advisor: C.R. Holland

Department reports to: The Vice-Chancellor

Placement service: Appointments and Careers, 14 Malone Road, Belfast

University of Sheffield

Department of Chemical and Process Engineering
Mappin Street
Sheffield, S1 3JD, England
University phone 44 114 222 2000
Department phone 44 114 222 7500
Department FAX 44 114 222 7501

Professors

Allen, R.W.K. .... 44-(0)114-222 7601
r.w.k.allen@shef.ac.uk
Hounslow, M.J. (Head) .... 44-(0)114-222 7565
m.j.hounslow@shef.ac.uk
Sharifi, V.N. .... 44-(0)114-222 7518
v.n.sharifi@shef.ac.uk
Swithenbank, J. (Emeritus) .... 44-(0)114-222 7502
j.swithenbank@shef.ac.uk
Tesar, V. (Visiting) .... 44-(0)114-222 7551
v.tesar@shef.ac.uk
Wright, P.C. .... 44-(0)114-222 7577
p.c.wright@shef.ac.uk

Associate Professors

Cliffe, K.R. .... 44-(0)114-222 7505
k.cliffe@shef.ac.uk
Edyvean, R.G.J. (Associate Head) .... 44-(0)114-222 7506
r.edyvean@shef.ac.uk
Pitt, M.J. .... 44-(0)114-222 7513
m.j.pitt@shef.ac.uk
Priestman, G.H. .... 44-(0)114-222 7512
g.priestman@shef.ac.uk
Ristic, R.I. .... 44-(0)114-222 7516
r.i.ristic@shef.ac.uk
Salman, A.D. .... 44-(0)114-222 7560
a.d.salman@shef.ac.uk
Styring, P. .... 44-(0)114-222 7571
p.styring@shef.ac.uk
Zimmerman, W.B.J. .... 44-(0)114-222 7517
w.zimmerman@shef.ac.uk

Assistant Professors

Biggs, C.A. .... 44-(0)114-222 7510
c.biggs@shef.ac.uk
Dickman, M.J. .... 44-(0)114-222 7541
m.dickman@shef.ac.uk
Ewan, B.C.R. .... 44-(0)114-222 7504
b.c.ewan@shef.ac.uk
MacInnes, J.M. .... 44-(0)114-222 7511
j.m.macinnes@shef.ac.uk
Mahoney, A.W. .... 44-(0)114-222 7602
a.w.mahoney@shef.ac.uk
Routh, A.F. .... 44-(0)114-222 7508
a.routh@shef.ac.uk
Russell, N.V. .... 44-(0)114-222 7521
n.russell@shef.ac.uk
Wu, Y. .... 44-(0)114-222 7514
y.wu@shef.ac.uk

Accredited by: IChemE, IEnergy

Degrees granted 2003-2004:

B.S.: 62 M.S.: 25 Ph.D.: 4

Department reports to: The Vice-Chancellor of the University

South Bank University

Division of Applied Chemical Sciences
103 Borough Road, London, SE1 0AA, England

University phone 020-7928-8989
Department phone 020-7815-7952
Department FAX 020-7815-7999

Professors

Best, Robert J. (Head) 020-7815-7910
rjbest@sbu.ac.uk
Cadbury, Richard G. 020-7815-7925
cadburrg@sbu.ac.uk
Nolan, P. F. 020-7815-7901
Reuben, B. G. (Emeritus) 020-7815-7946

Associate Professors

Henderson, J. P. 020-7815-7946
henderjp@sbu.ac.uk
Kassim, Hamida O. 020-7815-7953
kassimho@sbu.ac.uk
Larkai, Samuel 020-7815-7913
larkais@sbu.ac.uk
Orrin, John E. 020-7815-7950
orrinje@sbu.ac.uk
Steele, Clive H. 020-7815-7989
steelech@sbu.ac.uk

Accredited by: IChemE (UK), Institute of Fire Engineering

Graduate advisor: J.P.Henderson

Undergraduate advisor: R. G. Cadbury

Student organization: Institution of Chemical Engineers
Advisor: S. Larkai

Department reports to: C. Clare 020-7815-8181, Dean

Placement service: H. O. Kassim

University of Strathclyde

Department of Chemical and Process Engineering
James Weir Building
Montrose St.
Glasgow, G1 1XJ, Scotland

University phone 141 552 4400
Department phone 141 548 2361
Department FAX 141 552 2302

Professors

Grant, Colin (Head)
Hall, Peter

Associate Professors

Cousins, Rod
Crawley, Frank
Larsen, Vidar
Muir, David
Postlethwaite, Bruce
Schaschke, Carl
Shilton, Simon

Assistant Professors

Bull, Laura-Anne
Campbell, Linda
Dickson, Brian
Heslop, Mark
Morrison, Karen

Accredited by: Institution of Chemical Engineers

Degrees granted 2003-2004:

B.S.: 9 M.S.: 26 Ph.D.: 5

Graduate advisor: Professor P.J. Hall

Undergraduate advisor: Dr L.A. Bull (Yr1), Dr S.J. Shilton(Yrs 2&3), Mr R.B. Cousins (Yr4), Dr C.J. Schaschke(Yr5,MEng)

Student organization: Chemical Engineering Society

Department reports to: Dean of the Faculty of Engineering

Placement service: University Careers Service

University of Surrey

Department of Chemical and Process Engineering,
School of Engineering, Guildford, Surrey GU2 7XH, England

University phone 44 1483 300800
University FAX 44 1483 300803
Department phone 44 1483 686580
Department FAX 44 1483 686581

Professors

Koenders, M.A. (Visiting)
koenders@kingston.ac.uk
Kokossis, A.C. .... 44(0)1483 686573
A.Kokossis@surrey.ac.uk
Lawson, G. (Visiting)
Smith, J. M. (Emeritus) .... 44(0)1483 689276
J.Smith@surrey.ac.uk
Taylor, K. (Visiting)
K.Taylor@surrey.ac.uk
Thorpe, R.B. .... 44(0)1483 689270
Rex.Thorpe@surrey.ac.uk
Tuzun, U. (Head) .... 44(0)1483 686587
U.Tuzun@surrey.ac.uk

Associate Professors

Kirkby, N. F. .... 44(0)1483 686577
N.Kirkby@surrey.ac.uk
Millington, C. A. .... 44(0)1483 686582
A.Millington@surrey.ac.uk

Assistant Professors

Baxter, J. .... 44(0)1483 686594
J.Baxter@surrey.ac.uk
Cleaver, J. A. S. .... 44(0)1483 686598
J.Cleaver@surrey.ac.uk
Linke, P .... 44(0)1483 689116
P.Linke@surrey.ac.uk
Sharif, A.O. .... 44(0)1483 686584
A.Sharif@surrey.ac.uk

Accredited by: IChemE (UK)

Degrees granted 2003-2004:

B.S.: 21 M.S.: 18 Ph.D.: 4

Graduate advisor: Prof U Tuzun

Undergraduate advisor: Dr N Kirkby

Department reports to: Prof Paul Smith, Head of School of Engineering

Placement service: Dr J Baxter

University of Wales Swansea

Department of Chemical and Biological Process Engineering
Singleton Park, Swansea SA2 8PP

University phone 01792-295195
Department phone 01792-295195x5196
Department FAX 01792-295701

Professors

Atkinson, Bernard
Bowen, W. R. .... x5862 †
r.bowen@swansea.ac.uk
Preece, Paul E (Head) x5664
p.e.preece@swansea.ac.uk
Richardson, Jack F (Emeritus) x5194
j.f.richardson@swansea.ac.uk
Williams, P. R. x4081
p.r.williams@swansea.ac.uk

Associate Professors

Bowen, J. H. (Emeritus) x5735
j.h.bowen@swansea.ac.uk
Conder, John R. x4079
j.r.conder@swansea.ac.uk
Lovitt, R. (Bob) W. x5705
r.w.lovitt@swansea.ac.uk
Wardle, A. (Tony) P. x4088
a.p.wardle@swansea.ac.uk

Assistant Professors

Bryant, Rob x5201
r.bryant@swansea.ac.uk
Fulford, David x5646
r.fulford@swansea.ac.uk
Jones, Meirion G. x5206
m.jones@swansea.ac.uk
Knights, A. (Tony) J. x5965
a.j.knights@swansea.ac.uk
Sinnott, Ray K x5198
r.k.sinnott@swansea.ac.uk

† Faculty numbers are extensions of 01792-295195.

Accredited by: IChemE(UK)

Undergraduate advisor: Dr Tony Wardle

Student organization: Chem & Biochem Soc

Department reports to: R. H. Williams, Vice-Chancellor

Placement service: F. Slater

University of Teesside

School of Science and Technology
University of Teesside,
Middlesbrough, TS1 3BA, England

University phone 44 (0) 1642 218121
Department phone 44 (0) 1642-342499
Department FAX 44 (0) 1642-342401

Professors

Matchett, A. J. .... 44 (0) 1642-342438 †
a.j.matchett@tees.ac.uk

Associate Professors

Gerrard, A. M. 44 (0) 1642-342444
a.m.gerrard@tees.ac.uk
Pritchard, D. W. 44 (0) 1642-342448
d.w.pritchard@tees.ac.uk

Assistant Professors

Martin, G. F. 44 (0) 1642-342447
g.f.martin@tees.ac.uk
Nithiamanthan, V. 44 (0) 1642-342445
v.nthiananthan@tees.ac.uk
O'Reilly, A. J. 44 (0) 1642-342447
a.j.oreilly@tees.ac.uk
Peel, C 44 (0) 1642 342436
c.peel@tees.ac.uk

† Phone numbers are preceded by country code 44.

Accredited by: IChemE (UK)

Degrees granted 2003-2004:

B.S.: 11 M.S.: 11 Ph.D.: 2

Graduate advisor: A J Matchett

Undergraduate advisor: D W Pritchard

Department reports to: Director of School of Science and Technology

Placement service: Careers Office

Venezuela

Universidad Central de Venezuela

Escuela de Ing. Quimica
Apartado 50656-50361
Sabana Grande
Caracas 1040-A

University phone 02-619811x2720
Department phone 02-6627682
Department FAX .... (02) 58-6626181

Professors

Bonnet, Julio C.
Marin, Godofredo
Martin, Rafael
Marzuka, Samir (Director)
Papa, Jose
Pimentel, Jose A.
Ramos, Manuel
Mariangeles Tosi, E.

Associate Professors

Albano, Carmen
Alonso, Mary L.
de del Giorgio, Ana M.
de del Giorgio, Jose F.
de Herrera, Josefina T.
de Roccaro, Vincenza

Assistant Professors

Elman, Hinda
Hernandez, Jose D.
Perez, Cesar
Perez, Jorge
Sciamanna, Rosalba
Sorrentino, J.

Zimbabwe

National University of Science and Technology

Department of Chemical Engineering
P.O. Box AC 939 Ascot
Bulawayo, Zimbabwe
Deliveries: ZESA Training Centre, Accrington Road, Off Plumtree Road

University phone 263-9-282842
Department phone 263-9-475192x464304
Department FAX 263-9-464363

Assistant Professors

Ikhu-Omoregbe, Daniel I.O. . . . . 464304 [†]
dikhuomoregbe@hotmail.com
Kuipa, Pardon K. (Head) 464304
pardonkkuipa@hotmail.com
Manoja, Grace B. G. 464304
Muzenda, Edison 464304
Nyirongo, Simon 464304
Shumba, Harrison 464304
Sibanda, Vusumuzi 464304

[†] Faculty numbers are extensions of 263-9-475192

Department reports to: Dean, Faculty of Industrial Technology, NUST

Placement service: Chairman, Department of Chemical Engineering

Omega Chi Epsilon

Omega Chi Epsilon promotes high scholarship, encourages original investigation in all areas of chemical engineering, and recognizes the valuable traits of character, integrity and leadership. The society renders service to chemical engineering department and its students and fosters meaningful faculty-student dialogue within forty-nine chapters.

ALPHA	University of Illinois, 1931 (inactive)
BETA	Iowa State University, 1932, reactivated 1966
GAMMA	University of Minnesota, 1934 (inactive)
DELTA	Clarkson College of Technology, 1941
EPSILON	The University of Texas, 1941
ZETA	Purdue University, 1943
ETA	New Jersey Institute of Technology, 1957
THETA	West Virginia University, 1958
IOTA	University of Pittsburgh, 1959
KAPPA	Polytechnic Institute of New York, 1964
LAMBDA	City University of New York, 1964
MU	Oklahoma State University, 1964
NU	University of Detroit, 1965
XI	Northeastern University, 1965
OMICRON	Lamar University, 1967
PI	Rose-Hulman Institute of Technology, 1969
RHO	Texas A&M University, 1970
SIGMA	University of Arkansas, 1970
TAU	University of Alabama, 1971
UPSILON	University of Kentucky, 1971
PHI	Louisiana Tech University, 1971
CHI	University of Maryland, 1973
PSI	University of Southern California, 1974
OMEGA	University of Missouri-Rolls, 1974
ALPHA ALPHA	Auburn University, 1975
ALPHA BETA	Northwestern University, 1975
ALPHA GAMMA	University of Lowell, 1975
ALPHA DELTA	University of Houston, 1975
ALPHA EPSILON	Kansas State University, 1976
ALPHA ZETA	Michigan State University, 1976
ALPHA ETA	Mississippi State University, 1978
ALPHA THETA	Youngstown State University, 1978
ALPHA IOTA	Tulane University, 1978
ALPHA KAPPA	University of South Carolina, 1979
ALPHA LAMBDA	University of Southwestern Louisiana, 1979
ALPHA MU	New Mexico State University, 1979
ALPHA NU	Tri-State University, 1980
ALPHA XI	University of Colorado, 1980
ALPHA OMICRON	Texas Tech University, 1981
ALPHA PI	Howard University, 1982
ALPHA RHO	Georgia Institute of Technology, 1983
ALPHA SIGMA	University of Connecticut, 1983
ALPHA TAU	Colorado State University, 1984
ALPHA UPSILON	California State Polytechnic University, 1984
ALPHA PHI	University of Iowa, 1984
ALPHA CHI	Virginia Polytechnic Institute & State University, 1984
ALPHA PSI	Manhattan College, 1985
ALPHA OMEGA	Louisiana State University, 1988
BETA ALPHA	Pennsylvania State University, 1988
BETA BETA	University of Missouri-Columbia, 1991
BETA GAMMA	North Carolina A&T State University, 1992
BETA DELTA	Tuskegee University, 1992
BETA EPSILON	University of Cincinnati, 1995
BETA ZETA	University of Minnesota-Duluth, 1995
BETA ETA	Texas A&M University-Kingsville, 1995
BETA THETA	University of Michigan, 1995
BETA IOTA	Prairie View A&M University, 1996
BETA KAPPA	University of South Florida, 1996
BETA LAMBDA	Washington University, 1996
BETA MU	Widener University, 1996

Index of U.S. Faculty

A

Abbasian, Javad 30
Abbott, James 112
Abbott, Michael M. 76
Abbott, Nicholas L. 120
Abegg, Carl F. 34
Abraham, Martin A. 87
Abrams, Cameron 93
Achenie, Luke E. K. 19
Ackerman, John 121
Ackerson, Michael D. 6
Acrivos, Andreas 72
Adams, Craig 60
Aderangi, Nader 30
Adewuyi, Yusuf G. 79
Adeyiga, Adeyinka A. 115
Adler, Stuart 117
Admassu, Wudneh 27
Adomaitis, Raymond A. 45
Aghara, Sukesh 107
Agrawal, Pradeep K. 26
Aguayo, Guillermo 99
Ahmed, Zikri 73
Akers, William W. 108
Aki, Sudhir 32
Akins, Richard G. 38
Aksay, Ilhan A. 66
Akyurtlu, Ates 115
Akyurtlu, Jale F. 115
Al-Dahhan, Muthanna 61
Al-Hallaj, Said 30
Al-Saadoon, Faleh T. 110
Alamo, Rufina 24
Albright, Lyle F. 33
Alcantar, Norma 25
Aldrich, J. Winthrop 14
Alexandridis, Paschalis 78
Alkire, Richard C. 29
Allan, G. Graham 117
Allen, David T. 111
Allen, Emily 14
Allen, Jonathan 4
Allen, Mark G. 26
Ally, Moonis 104
Altman, Eric 20
Altwicker, Elmar R. 76
Aluko, Mobolaji E. 22
Amaral, Luis A. N. 31
Amato, Wayne 79
Amiridis, Michael D. 102
Amundson, Neal R. 106
Andersen, Paul 71
Anderson, Frank 58
Anderson, Harold M. 70
Anderson, John L. 82, 92
Anderson, Kimberly 39
Anderson, Paul 30
Anderson, Thomas F. 19
Anderson, Timothy J. 23
Andreadis, Stelios T. 78
Andres, Ronald P. 33
Angenent, Lars 61
Angus, John C. 82
Anisimov, Mikhail A. 45
Anklam, Mark R. 34
Annapragada, Ananth 106
Anseth, Kristi S. 16
Anthony, Rayford G. 109
Anton, A. Brad 74
Appleby, Anthony J. 109
April, Gary C. 1
Aranovich, Gregory 44
Arastoopour, Hamid 30
Arce, Pedro E. 105
Archer, Lynden A. 74
Argyle, Morris 121
Aris, Rutherford 57
Armaou, Antonios 96
Armenante, Piero 65
Armeniades, Constantine 108
Armstrong, Robert C. 49
Arnold, David W. 1
Arnold, Frances H. 13
Arnold, John R. 104
Arnold, Robert 5
Arnold, Stephen 75
Artigue, Ronald S. 34
Ashbaugh, Henry S. 43
Ashurst, William (Bob) 2
Askew, William C. 3
Assaf-Anid, Nada 75
Asthagiri, Anand R. 13
Aston, Eric 27
Ataai, Mohammad M. 97
Atanassov, Plamen 70
Atwood, Glenn A. 81
Auerbach, Scott M. 47
Aurand, Gary 35
Aven, Russell E. 58
Avilés-Molina, Misael O. 99
Aydil, Eray S. 12
Azad, Abdul-Majeed 87
Azbel, David S. 60

B

Babb, Albert L. 117
Babcock, Robert E. 6
Babu, S. V. 72
Baertsch, Chelsey D. 33
Bagajewicz, Miguel J. 88
Baglin, Frank G. 63
Bailey, Raymond V. 43
Bailie, Richard C. 119
Baird, Donald G. 116
Bajpai, Rakesh K. 59
Baker, Ian 64
Bakshi, Bhavik 85
Balakotaiah, Vemuri 106
Balazs, Anna C. 97
Balbuena, Perla 109
Balcarcel, R. Robert 105
Baldwin, John T. 109
Balk, John 39
Balsara, Nitash P. 7
Baltus, Ruth E. 72
Baltzis, Basil 65
Banerjee, Sanjoy 12
Banerjee, Sujit 26
Baneyx, François 117
Banish, Michael 1
Bankoff, S. George 31
Banta, Scott 73
Bao, Zhenan 15
Barabino, Gilda A. 50
Barat, Robert 65
Baratuci, William B. 117
Barbari, Timothy A. 45
Barile, Ron 24
Barkat, Omar 42
Barker, Dee H. 113
Barkey, Dale P. 64
Barna, Bruce A. 55
Barnett, Stanley M. 100
Barron, Annelise 31
Barron, Charles H. 101
Bart, Ernest 65
Barteau, Mark A. 21
Bartholomew, Calvin H. 113
Barton, Paul I. 49
Baskaran, Harihara 82
Bates, Frank S. 57
Bauer, Larry G. 102
Baum, Gary 26
Baumann, Melissa 54
Baxter, Gregory 74
Baxter, Larry L. 113
Bayer, Charlene W. 26
Baygents, James 5
Bayles, Taryn 46
Bayuzick, Robert J. 105
Beér, János M. 49
Beard, John N. 101
Beatty, Kenneth O. 80
Beaudoin, Stephen 33
Beckman, Eric J. 97
Beckman, James R. 4
Beckstead, Merrill W. 113
Beckwith, William F. 101
Beddow, John K. 35
Beers, Kenneth J. 49
Behbahani, Ahmed 41
Beissinger, Richard 30
Beitle, Robert 6
Belfiore, Laurence A. 18
Belfort, Georges 76
Bell, Alexis T. 7
Bell, David A. 121
Bell, James P. 19
Bell, Kenneth J. 89
Bell, Richard L. 8
Belovich, Joanne M. 84
Ben-Jebria, Aziz 96
Benítez, Jaime 99
Bender, Jonathan W. 102
Bender, Timothy 54
Bennett, Carroll O. 19

Bennett, Gary F. 87
Bent, Stacey 15
Bentley, William E. 45
Benziger, Jay B. 66
Bequette, B. Wayne 76
Berg, John C. 117
Berglund, Kris A. 54
Beris, Antony N. 21
Berkland, Cory J. 37
Berman, Neil S. 4
Bernstein, Barry 30
Beroes, Charles 97
Besser, Ronald 69
Betenbaugh, Michael J. 44
Bethea, Robert M. 112
Bevan, Michael 109
Bhada, Rohinton K. (Ron) 71
Bhatia, Surita R. 47
Bhattacharyya, Dibakar 39
Bhethanabotla, Venkat R. 25
Bidani, Akhil 106
Bidstrup Allen, Sue Ann 26
Biegler, Lorenz T. 92
Bieler, Thomas 54
Bienkowski, Paul R. 104
Bier, Milan 5
Biernacki, Joseph J. 105
Bird, R. Byron 120
Bischoff, Kenneth B. 21
Bishop, Kenneth A. 37
Biswas, Pratim 61
Biver, Carl J. 25
Blanch, Harvey W. 7
Blanks, Robert F. 69
Blankschtein, E. Daniel 49
Blau, Gary E. 33
Block, David E. 8
Block, Robert J. 88
Block, Seymour S. 23
Blowers, Paul 5
Blythe, Philip A. 94
Boder, Eric T. 95
Bogaert-Alvarez, Ricardo 93
Bogere, Moses N. 99
Bommarius, Andreas 26
Bonnecaze, Roger T. 111
Bonner, Francis J. 48
Book, Neil L. 60
Booth, J. Richard 105
Borhan, Ali 96
Borovetz, Harvey S. 97
Bose, Arijit 100
Bothwell, Michelle K. 90
Botsaris, Gregory B. 51
Botte, Gerardine G. 86
Boudart, Michel 7, 15
Boulton, Roger 8
Bousfield, Douglas W. 43
Bowden, Warren W. 34
Bowen, J. Ray 117
Bowman, Christopher N. 16
Bowman, Frank M. 105
Braatz, Richard D. 29
Bradford, John R. 112
Bradshaw, Jerry L. 109
Brady, John F. 13
Brand, Jennifer I. 62
Braun, Walter G. 96
Bray, Robert S. 14
Brazel, Chris 1
Brazinsky, Irv 73
Breedveld, L. Victor 26
Brennecke, Joan F. 32
Brenner, Howard 49
Brenner, James R. 24
Bretz, Robert E. 69
Breuer, Kenneth 100
Briano, Julio G. 99
Bricka, R. Mark 58
Briedis, Daina M. 54
Briggs, Dale E. 53
Briggs, James M. 106
Brinker, C. Jeffrey 70
Broadbelt, Linda 31
Brock, James R. 111
Brodkey, Robert 85
Bromley, LeRoy A. 7
Brown, George M. 31
Brown, Gilbert Jay 48
Brown, Richard A. 100
Brown, Robert A. 49
Brown, Robert C. 36
Browning, Nigel 8
Bruce, David A. 101
Bruns, Duane D. 104
Bryant, Stephanie 16
Buettner, Helen M. 68
Bukur, Dragomir B. 109
Bullard, Lisa 80
Bullin, Jerry A. 109
Bungay, Henry R. III 76
Bunge, Annette L. 17
Buonopane, Ralph A. 50
Burghardt, Wesley R. 31
Burkey, Daniel 50
Burnet, George 36
Burns, Mark A. 53
Burris, Conrad T. 75
Burrows, Veronica 4
Busch, Robert D. 70
Busot, J. Carlos 25
Butera, Robert J. 21
Buthod, Paul 89
Butler, Audrey 35
Butler, Jason 23
Butt, John B. 31
Buttrey, Douglas J. 21
Byrne, Mark E. 2

C

Córdova-Figueroa, Ubaldo M. 99
Caenepeel, Christopher L. 14
Cahela, Don 2
Cairncross, Richard 93
Cairns, Elton J. 7
Cal, Mark 69
Calabrese Barton, Scott 73
Calabrese, Richard V. 45
Cale, Timothy S. 76
Calo, Joseph M. 100
Camarda, Kyle V. 37
Cameron, Michael R. 87
Camesano, Terri A. 52
Campbell, Gregory A. 72
Campbell, Kenneth 118
Campbell, Scott W. 25
Caneba, Gerard T. 55
Cannon, Joseph N. 22
Capitano, Adam T. 106
Caram, Hugo S. 94
Carbeck, Jeffrey D. 66
Carbonell, Ruben G. 80
Cardona-Martínez, Nelson 99
Caretta, Raul 57
Carlson, Alfred 34
Carlson, Eric S. 1
Carmichael, Gregory R. 35
Carnahan, Brice 53
Carr, Robert W. Jr. 57
Carr, Russell T. 64
Carr, Stephen H. 31
Carrier, Rebecca 50
Carta, Giorgio 115
Carter, C. Barry 57
Carter, John N. 74
Caruthers, James M. 33
Case, Eldon 54
Caskey, Jerry A. 34
Castaldi, Marco 75
Castellanos, Mariajose 46
Castner, David G. 117
Caswell, Bruce 100
Cavanagh, Daniel P. 91
Cecchi, Joseph L. 70
Center, Alfred M. 74
Cerro, Ramon L. 1
Cha, Chang Yul 121
Chaikof, Elliot L. 26
Chakraborty, Arup K. 7
Chalmers, Jeffrey 85
Chambers, Robert P. 2
Chan, Christina 54
Chan, Paul C. H. 59
Chang, Chih-hung (Alex) 90
Chang, Hsueh-Chia 32
Chang, Jane P. 10
Chang, W. Victor 15
Chao, K. C. 33
Chapman, Thomas W. 120
Chapman, Walter G. 108
Charles, Marvin 94
Chase, George G. 81
Chatzimavroudis, George P. 84
Chau, Pao C. 11
Chaudhury, Manoj K. 94
Chauhan, Anuj 23
Chawla, Ramesh C. 22
Chelikowsky, James R. 57
Chella, Ravindran 24
Chellam, Shankar 106
Chen, Chien-Pin 1
Chen, Daniel H. 107
Chen, David H. T. 98
Chen, Jingguang 21
Chen, John C. 94
Chen, Kevin 24
Chen, Rachel 26
Chen, Shaw-Horng 77
Chen, Wei-Yin 58
Chen, Wen-Jia 86
Chen, Wilfred 11
Cheng, Shang-I 73
Cheung, H. Michael 81
Chian, Wei 102

Chiang, Shiao-Hung 97
Chiew, Yee C. 68
Chimowitz, Eldred H. 77
Chin, Der-Tau 72
Chin, Wei-Chun 24
Chisholm, John 110
Chitra, Surya 98
Chittur, Krishnan K. 1
Chmelka, Bradley F. 12
Chmielewski, Donald 30
Cho, Eung H 119
Choi, Kyu Yong 45
Chou, Russell 9
Christofides, Panagiotis D. 10
Chuang, Steven 81
Churchill, Stuart W. 95
Cilento, Eugene V. 119
Cinar, Ali 30
Ciric, Amy 84
Cirino, Patrick C. 96
Clancey, M. Sean 55
Clancy, Paulette 74
Clark, Douglas S. 7
Clark, Peter E. 1
Clark, William M. 52
Clausen, E. C. 6
Clements, L. Davis 62
Clements, Wm. C. Jr. 1
Clough, David E. 16
Clump, Curtis W. 94
Co, Albert 43
Co, Carlos 83
Co, Tomas B. 55
Cobb, James T. Jr. 97
Cocke, David L. 107
Cocker, David R. 11
Codd, Sarah 62
Cohen, Claude 74
Cohen, Robert E. 49
Cohen, William C. 31
Cohen, Yoram 10
Cokelet, Giles 62
Colón, Guillermo 99
Cole, Robert 72
Coleman, Maria R. 87
Collier, John 64
Collier, John R. 104
Collins, Dermot J. 40
Collins, William E. 22
Collura, Michael A. 20
Colton, Clark K. 49
Colucci-Rios, José A. 99
Composto, Russell J. 95
Conner, William C. Jr. 47
Constantinides, Alkis 68
Converse, Alvin O. 64
Cooney, Charles L. 49
Cooper, Douglas J. 19
Cooper, Gary W. 70
Cooper, Stuart 85
Corn, John 85
Cornell, David 58
Cornilsen, Bahne C. 55
Coronell, Daniel G. 34
Coronella, Charles J. 63
Corripio, Armando B. 41
Corti, David S. 33
Cosart, William P. 5
Cosgrove, Stanley 83
Couchman, Peter 68
Coughanowr, Donald R. 93
Coughlin, Robert W. 19
Coulman, George A. 84
Counce, Robert M. 104
Couper, James R. 6
Couzis, Alexander 72
Cowman, Mary 75
Cox, David F. 116
Cox, Kenneth 108
Cramer, Steven M. 76
Creed, Joe L. 71
Crimp, Martin 54
Crisalle, Oscar D. 23
Crisman, Everett 100
Crist, Buckley Jr. 31
Crist, Kevin C. 86
Crocker, John C. 95
Cross, Robert 6
Crosser, Orrin K. 60
Crothers, Donald 20
Crowl, Daniel A. 55
Crunkleton, Daniel 89
Crynes, Billy L. 88
Csernica, Jeffrey 91
Cui, Shingting 104
Cullinan, Harry T. 2
Cummings, Peter T. 104, 105
Cunningham, James 103
Curl, Rane 53
Curtis, Christine W. 2
Curtis, Jennifer 23
Curtis, Wayne R. 96
Cushman-Roisin, Benoit 64
Cussler, Edward L. 57
Cutlip, Michael 19
Cygan, Margaret 41

D

d'Itri, Julie L. 97
da Rocha, Sandro 56
Dadyburjor, Dady B. 119
Dahler, John S. 57
Dahm, Kevin D. 67
Dai, Lenore 112
Dale, Bruce E. 54
Dalzell, William H. 49
Dan, Nily 93
Dandy, David S. 18
Daniels, Raymond D. 88
Danner, Ronald P. 96
Daoutidis, Prodromos 57
Darby, Ronald 109
Darcy, Patricia A. 59
DaSilva, Nancy A. 9
Datta, Ravindra 52
Datye, Abhaya 70
Daubert, Thomas E. 96
Daugherty, Patrick 12
Dautenhahn, Pamela C. 42
David, Morton M. 117
Davidson, Burton Z. 68
Davis, Denny C. 118
Davis, E. James 117
Davis, H. Ted 57
Davis, James 10
Davis, Jeffrey M. 47
Davis, Mark E. 13
Davis, Richard 56
Davis, Richey M. 116
Davis, Robert H. 16
Davis, Robert J. 115
Davis, Sam H. 108
Davison, B.H. 104
Davison, Richard R. 109
Daw, C. Stuart 104
De Kee, Daniel 43
de Nevers, Noel H. 114
de Pablo, Juan J. 120
De Witt, Kenneth J. 87
Dean, Anthony M. 17
Deans, Harry A. 121
Debelak, Kenneth A. 105
Debenedetti, Pablo G. 66
Deen, William A. 49
deGrazia, Janet 16
Deibert, Max C. 62
Del Valle, Francisco 71
DeLancey, George B. 69
Delcamp, Robert M. 83
Delgass, W. Nicholas 33
DeLisa, Matthew P. 74
Demekhin, Evgeny 32
Demirel, Yasar 116
Deng, Shuguang 71
Deng, Yulin 26
Denn, Morton M. 7, 72
Denson, Costel D. 21
Deo, Milind D. 114
DePaoli, David 104
Derby, Jeffrey J. 57
Deshpande, Pradeep B. 40
Deshusses, Marc A. 11
DeSimone, Joseph M. 80
Desio, Peter J. 20
DeSisto, William 43
Detamore, Michael S. 37
Devine, Kevin 75
DeWitte, Renee 56
Dhawan, Jagdish C. 3
Dhurjati, Prasad 21
Diamond, Scott L. 95
DiBenedetto, Anthony T. 19
DiBiasio, David 52
Dickinson, Richard 23
Dillner, Ann 4
DiMarzio, Edmund A. 45
Dimitrakopoulos, Panagiotis 45
Dinos, Nicholas 86
Discher, Dennis 95
Dismukes, John P. 87
Dittman, F. W. 68
Diwekar, Urmila 28
Dixon, Anthony G. 52
Dixon, David J. 102
Dobbins, Richard A. 100
Doherty, Michael F. 12, 47
Domach, Michael M. 92
Donahue, Darrell 43
Donahue, Francis 53
Donahue, Neil M. 92
Donatelli, Alfred Anthony 48
Dong, Junhang 69
Dong, Winny 14
Donnelly, Vincent M. 106

Donohue, Marc D. 44
Dooley, Kerry M. 41
Doraiswamy, Deepak 119
Doraiswamy, L. K. 36
Dordick, Jonathan S. 76
Dorgan, John R. 17
Dorland, Dianne 67
Doskocil, Eric 59
Dougherty, Elmer L. 15
Douglas, James M. 47
Doyle, Frank 12
Doyle, Patrick S. 49
Dranoff, Joshua S. 31
Drown, David C. 27
Drozdov, Aleksey 119
Drzal, Lawrence T. 54
Du, Henry 69
Duda, J. Larry 96
Dudukovic, Milorad 61
Duffie, John A. 120
Duffy, James E. 62
Duke, Steve R. 2
Dumesic, James A. 120
Dunbar, Paul 39
Duncan, T. Michael 74
Dungan, Stephanie R. 8
Dunston, Doug 69
Durbin, Leonel D. 109
Durning, C. J. 73
Durrill, Preston L. 116
Dycus, Patricia J. M. 105
Dyson, Derek C. 108

E

Eakman, James M. 62
Earthman, James C. 9
Eckert, Charles A. 26
Economides, Michael J. 106
Economou, Demetre J. 106
Eddings, Eric 114
Eden, Mario 2
Edgar, James H. 38
Edgar, Thomas F. 111
Edie, Danny D. 101
Edwards, Aurelie 51
Edwards, Brian J. 104
Edwards, Jeremy S. 21
Edwards, Louis L. Jr. 27
Edwards, Richard M. 5
Edwards, Robert V. 82
Ehrman, Sheryl H. 45
Eisenberg, Richard F. 77
Ekechukwu, Kenneth 22
Ekerdt, John G. 111
El-Aasser, Mohamed S. 94
El-Farra, Nael 8
El-Genk, Mohamed S. 70
El-Halwagi, Mahmoud M. 109
Ela, Wendell 5
Elabd, Yossef 93
Eldridge, R. Bruce 111
Eliassen, John D. 105
Elimelech, Menachem 20
Ellington, Rex T. 88
Elliott, J. Richard 81
Elmore, Bill B. 41
Elnashaie, Said Salah 2
Ely, James F. 17
Emery, Alden H. 33
Empie, Howard L. 26
Engel, Alfred J. 96
Engstrom, James R. 74
Enick, Robert M. 97
Erickson, Larry E. 38
Erjavec, John 81
Erkey, Can 19
Ernst, William R. 26
Errington, Jeffrey R. 78
Ershaghi, Iraj 15
Escobar, Isabel C. 87
Escobedo, Fernando 74
Espino, Ramon 115
Estévez, L. Antonio 99
Etchells, Arthur W. III 21
Eubank, Philip T. 109
Evans, Edward A. 81
Evers, John F. 121
Eylon, Daniel 84

F

Fabiano, Leonard A. 95
Fair, James R. 111
Falconer, John L. 16
Faller, Roland 8
Famularo, Jack 75
Fan, L. S. 85
Fan, L. T. 38
Fan, Stephen 64
Fang, Cheng S. 42
Farag, Ihab H. 64
Farmer, Richard 63
Farneth, William E. 21
Farrell, James 5
Farrell, Stephanie 67
Farshad, Fred F. 42
Federspiel, William 97
Fedkiw, Peter S. 80
Feeney, Robert G. 14
Feintuch, Howard 75
Feke, Donald L. 82
Felder, Richard M. 80
Fenton, James M. 19
Fernandez, Erik J. 115
Ferrada, J.J. 104
Ferri, James 93
Ferron, John R. 77
Fichthorn, Kristen A. 96
Field, James 5
Finlayson, Bruce A. 117
Finley, David 34
Finn, Robert K. 74
Finney, Wright C. 24
Fisher, Edward R. 55
Fisher, John P. 45
Flach, Lawrance 84
Flagan, Richard C. 13
Fleischer, Miguel T. 106
Fleischman, Marvin 40
Fletcher, John V. 114
Fletcher, Thomas H. 113
Flood, H. William 48
Floudas, Christodoulos A. 66
Floyd, Tamara 3
Flumerfelt, Raymond W. 106
Flynn, Ann Marie 75
Flynn, George W. 73
Flytzani-Stephanopoulos, Maria ..51
Fogler, H. Scott 53
Foley, Henry C. 96
Fontijn, Arthur 76
Forbes, Neil S. 47
Forciniti, Daniel 60
Ford, David M. 109
Ford, Laura P. 89
Ford, Roseanne M. 115
Forney, Larry J. 26
Forsman, William C. 95
Forsten Williams, Kimberly 116
Fort, Tomlinson 105
Foss, Alan S. 7
Foster, David 77
Fotouh, Kamel H. 107
Foutch, Gary L. 89
Fox, George 106
Fox, Rodney O. 36
Francis, Lorraine F. 57
Frank, Curtis W. 15
Franses, Elias I. 33
Frazier, George C. 104
Frechet, Jean M. J. 7
Frederick, William 26
Fredrickson, Arnold G. 57
Fredrickson, Glenn H. 12
Freeman, Benny D. 111
French, W. Todd 58
Frey, Douglas D. 46
Fricke, Arthur L. 23
Fried, Joel R. 83
Friedlander, Sheldon K. 10
Friedly, John C. 49
Frisbie, C. Daniel 57
Froment, Gilbert 109
Frost, Harold 64
Frymier, Paul 104
Fuchs, Alan 63
Fulghum, Julia E. 70
Fuller, Gerald G. 15
Fuller, Thomas 26
Furst, Eric M. 21

G

Gabitto, Jorge F. 107
Gadala-Maria, Francis A. 102
Gaddy, Glen 44
Gaddy, James L. 6
Gaden, Elmer L. Jr. 115
Gainer, John L. 115
Galkin, Oleg N. 106
Galli, Alfred F. 119
Gallivan, Martha 26
Ganesan, Venkat 111
Garber, James D. 42
Garde, Shekhar 76
Garetz, Bruce 75
Garlid, Kermit L. 117
Garr, Jeanette 87
Gary, James H. 17
Gasem, Khaled A. M. 89
Gast, Alice P. 49
Gates, Bruce C. 8
Gatica, Jorge E. 84
Gatzke, Edward P. 102
Gavalas, George R. 13
Geankoplis, Christie J. 57, 85

Gecol, Hatice 63
Geertsema, Arie 39
Gehrke, Stevin H. 37
Gellman, Andrew J. 92
Genco, Joseph M. 43
Gentry, James W. 45
Genzer, Jan 80
Georgakis, Christos 75
Georgakis,Christos 51
George, Clifford E. 58
George, Steve 16
George, Steven C. 9
Georgiou, George 111
Gephardt, Zenaida Otero 67
Gerberich, William W. 57
Gerhard, Earl R. 40
Gerngross, Tillman 64
Ghorashi, Bahman 84
Giapis, Konstantinos P. 13
Gibbons, Joseph H. 102
Gibbs, Steve 24
Gibeling, Jeffery C. 8
Gibson, Urusla 64
Gidaspow, Dimitri 30
Gilbert, Richard 25
Gilbert, Richard E. 62
Gilchrist, James F. 94
Gilchrist, M. Lane 72
Gilcrease, Patrick 102
Gill, Ryan 16
Gill, William N. 76
Gillham, John K. 66
Gillis, Peter P. 39
Gin, Doug 16
Glandt, Eduardo D. 95
Glasgow, Larry A. 38
Glasser, Benjamin J. 68
Glatz, Charles E. 36
Gleason, Karen K. 49
Gleaves, John T. 61
Glotzer, Sharon C. 53
Glover, Charles J. 109
Godbey, W T. 43
Godbold, Thomas M. 105
Godleski, Edward S. 84
Goetz, Douglas J. 86
Gogos, Costas 65
Golden, John O. 17
Goldstein, Aaron 116
Goldstick, Thomas K. 31
Gomezplata, Albert 45
González-Pérez, Braulio 99
Gonzalez, Ramon 36
Gonzalez, Richard 43
Good, Robert J. 78
Good, Theresa 46
Gooding, Charles H. 101
Goodwin, Bernard M. 50
Goodwin, James G. 101
Goren, Simon L. 7
Gorte, Raymond J. 95
Gossage, John L. 107
Govind, Rakesh 83
Gow, Arthur S. III 20
Graboski, Michael S. 17
Gracias, David 44
Grady, Brian P. 88
Graedel, Thomas E. 20
Graessley, William W. 31, 66
Graff, Robert A. 72
Graham, Michael D. 120
Grant, Christine S. 80
Grant, James J. III 21
Grant, Stanley 9
Graves, David B. 7
Graves, David J. 95
Gray, Donald J. 100
Gray, Jeffrey 44
Green, David L. 115
Green, Don W. 37
Green, Mark 75
Green, Peter F. 111
Green, William H. 49
Greenberg, David B. 83
Greene, Howard 81
Greener, Jehuda 77
Greenfield, Michael L. 100
Greenkorn, Robert A. 33
Greenstein, Teddy 65
Greer, Sandra C. 45
Gregory, Otto J. 100
Grens, Edward A. 7
Grenville, Richard K. 21
Griffin, Gregory L. 41
Griffith, D. John 42
Gross, Joseph F. 5
Gross, Richard 75
Grossberg, A. L. 7
Grossmann, Elihu D. 93
Grossmann, Ignacio E. 92
Groves, Frank R. 41
Groza, Joanna R. 8
Grulke, Eric A. 39
Grummon, David 54
Gryte, Carl C. 73
Grzybowski, Bartosz 31
Gu, Tingyue 86
Gubbins, Keith E. 74, 80
Guin, James A. 2
Gulari, Erdogan 53
Gulari, Esin 56
Guliants, Vadim 83
Gulino, Daniel A. 86
Gupta, Arunava 1
Gupta, Nivedita 64
Gupta, Rakesh K. 119
Gupta, Ram 2
Gustafson, Richard 117
Gutoff, Edgar B. 50
Guymon, C. Allan 35
Guzman, Javier 37
Guzman, Roberto 5
Gyamerah, Michael 107
Gyure, Dale C. 51

H

Hacker, Barbara 14
Hackleman, David 90
Haddon, Robert C. 11
Haering, Edwin R. 85
Hahm, Jong-in 96
Hahn, Dittmar 65
Hahn, Juergen 109
Hales, Hugh B. 113
Hall, Carol K. 80
Hall, Kenneth R. 109
Hall, William B. 58
Haller, Gary L. 20
Hammack, William S. 29
Hammer, Daniel A. 95
Hammond, Paula T. 49
Hamrin, Charles E. Jr. 39
Han, Sang 70
Handy, Lyman L. 15
Hanes, Justin 44
Hanesian, Deran 65
Hanna, Owen T. 12
Hanneman, Larry F. 36
Hannemann, Robert 33
Hanratty, Thomas J. 29
Hanson, Daniel T. 109
Hanson, Donald N. 7
Hanson, Francis V. 114
Hanyak, Michael E. Jr. 91
Hara, Masanori 68
Harb, John N. 113
Harcum, Sarah 101
Harden, James 44
Harding, David R. 77
Harding, W. David 20
Hariri, M. Hossein 34
Harold, Michael P. 106
Harper, Dean O. 40
Harriott, Peter 74
Harris, H. Gordon 121
Harris, Michael T. 33
Harris, Robert E. 82
Harris, Sandra L. 72
Harris, William B. 109
Harris, William M. 14
Harrison, B. Keith 3
Harrison, Douglas P. 41
Harrison, Graham M. 101
Harrison, Roger G. 88
Harwell, Jeffrey H. 88
Hasan, A. Rashid 56
Hasenwinkel, Julie 79
Hassler, John 43
Hatton, T. Alan 49
Hatzimanikatis, Vassily 31, 74
Hauan, Steinar 92
Haug, Warren 31
Haugh, Jason 80
Havens, Jerry A. 6
Hawley, Martin C. 54
Haynes, Henry W. 121
Hebert, Kurt R. 36
Hecker, William C. 113
Heenan, William A. 110
Heideger, William J. 117
Heist, Richard 75
Helble, Joseph J. 19
Helfferich, Friedrich G. 96
Heller, Adam 111
Hellums, J. David 108
Henderson, Clifford 26
Henderson, Louis 98
Hendrix, James L. 62
Henley, Ernest J. 106
Henry, Jim 103
Henry, Joseph D. 119
Henson, Michael A. 47
Henthorn, David 60
Henthorn, Kimberly 60

Hernández-Maldonado, Arturo J. . 99
Hernandez, Rafael 58
Herring, Andy 17
Hersel, Allen 34
Hershey, Daniel 83
Hershey, Harry C. 85
Herz, Richard K. 11
Hesketh, Robert P. 67
Hess, Dennis W. 26
Heydweiller, John C. 79
Heys, Jeffrey 4
Hicks, Robert F. 10
Higdon, Jonathan J. L. 29
Higgins, Brian G. 8
Higgins, Charles J. 48
High, Karen A. 89
High, Martin S. 89
Hightower, Joe W. 108
Hile, Lloyd 13
Hill, Archibald G. 42
Hill, Charles G. Jr. 120
Hill, Davide A. 32
Hill, Donald O. 58
Hill, James C. 36
Hill, Marquita 43
Hill, Priscilla J. 58
Hillhouse, Hugh W. 33
Hillier, Andrew 36
Hilt, J. Zachary 39
Himmelblau, David M. 111
Hinds, Bruce 39
Hirasaki, George J. 108
Hirt, Douglas E. 101
Hirth, Leo J. 2
Hirtzel, Cynthia 87
Hjortso, Martin 41
Hlavacek, Vladimir 78
Ho, Chia-Chi 83
Ho, Thomas C. 107
Ho, W.S. Winston 85
Hoagland, David A. 47
Hodges, Russel 2
Hodgson, Kevin 117
Hoffman, Allan S. 117
Hoffman, Robert 67
Hoflund, Gar B. 23
Hoh, Jan 44
Hohmann, Edward C. 14
Hohn, Keith L. 38
Holder, Gerald D. 97
Holland, Charles D. 109
Holland, Nolan B. 84
Holland, William D. 105
Hollein, Helen C. 75
Holles, Joseph H. 55
Holman, Chad 34
Holmes, John M. 104
Holste, James C. 109
Holt, Bradley R. 117
Holtzapple, Mark T. 109
Hong, Juan 9
Hoo, Karlene 112
Hoopes, John W. Jr. 98
Hopfenberg, Harold B. 80
Hopke, Philip K. 72
Hopper, Jack R. 107
Horbett, Thomas A. 117
Hougen, Joel O. 111
Houze, R. Neal 33
Howard, G. Michael 19
Howard, Jack B. 49
Howat, Colin S. 37
Howat, Julie 37
Howitt, David G. 8
Howley, Maureen 65
Hrenya, Christine M. 16
Hsieh, Jeffery 26
Hsu, James T. 94
Hu, Wei-Shou 57
Huang, Ching-Rong 65
Huang, Michael 65
Huang, Xiwen 2
Huang, Yinlun 56
Huddleston, Ted 3
Hudson, J. L. 115
Huffman, Gerald 39
Huggins, Frank 39
Hurt, Robert H. 100
Husson, Scott M. 101
Hwalek, John J. 43
Hwang, Gyeong S. 111
Hwang, Sun-Tak 83
Hyun, Kun Sup 65

I

Ierapetritou, Marianthi G. 68
Iglesia, Enrique 7
Ignarro, L. J. 10
Ilias, Shamsuddin 79
Isaacs, Leslie L. 72
Isdale, Charles E. 109
Israelachvili, Jacob 12
Ivory, Cornelius F. 118

J

Jabarin, Saleh A. 87
Jabbari, Esmaiel 102
Jablonski, Erin L. 91
Jachuck, Roshan J.J. 72
Jackman, Alan P. 8
Jackson, Melbourne L. 27
Jackson, Roy 66
Jacobs, Stephen D. 77
Jacobson, Allan J. 106
Jacoby, William A. 59
Jaffe, Charles M. 119
Jagota, Anand 94
Jang, Larry 13
Jaroszeski, Mark 25
Jayaraman, Arul 109
Jayaraman, Krishnamurthy 54
Jenekhe, Samson A. 117
Jennings, G. Kane 105
Jennings, Michael B. 14
Jennings, Paul A. 24
Jensen, Klavs F. 49
Jeon, Hyun-Sik (Sam) 69
Jessop, Julie 35
Jhon, Myung S. 92
Jiang, Shaoyi 117
Johannes, Arland H. 89
Johanson, Lennart N. 117
John, Vijay T. 43
Johns, Lewis E. 23
Johnson, Charles L. 71
Johnson, Duane T. 1
Johnson, Ernest F. 66
Johnson, J. Karl 97
Johnston, Barry S. 49
Johnston, Keith P. 111
Jolls, Kenneth R. 36
Jones, Allan E. 21
Jones, Christopher 26
Jones, Frank 103
Joo, Yong Lak 74
Jorne, Jacob 77
Jose-Yacaman, Miguel 111
Joseph, Babu 25
Josephson, William 3
Jothimurugesan, K. 115
Jovanovic, Goran 90
Joye, Donald D. 98
Ju, Jingyue 73
Ju, Lu-Kwang 81
Juda, Walter 51

K

Kabadi, Vinayak N. 79
Kabel, Robert L. 96
Kadlec, Robert H. 53
Kaler, Eric W. 21
Kalika, Douglass 39
Kalu, Eric 24
Kalyon, Dilhan 69
Kamat, Prashant 32
Kane, Ravi 76
Kang, Kyung A. 40
Kannan, Rangaramanujam 56
Kantor, Jeffrey C. 32
Kao, Camilla 15
Kao, Y. K. 83
Kaplan, David 51
Kardos, John L. 61
Karel, Marcus 49
Karkalits, O. C. 42
Karlsson, Sture K. F. 100
Karweit, Michael 44
Katz, Alexander 7
Katz, Joseph L. 44
Kauffman, David 70
Kauffman, Ken J. 11
Kavianian, Hamid 13
Kawatra, S. Komar 55
Kazantzis, Nikolaos K. 52
Kaznessis, Yiannis 57
Keane, Mark 39
Keasling, Jay D. 7
Keffer, David 104
Keith, Jason M. 55
Keller, George 119
Keller, Kenneth H. 57
Kelley, Brian 51
Kelly, C. Michael 98
Kelly, Robert M. 80
Kelly, William J. 98
Kenis, Paul J. A. 29
Kennedy, Francis 64
Kermode, Richard I. 39
Kerr, Clayton P. 105
Kevrekidis, Yannis G. 66
Kezirian, Michael 15
Khalil, Yehia 19
Khalil, Yehia F. 20
Khalili, Nasrin R. 30

Khan, Saad A. 80
Khang, Soon-Jai 83
Khinast, Johannes 68
Khomami, Bamin 61
Khoroosi, Hossain 56
Khosla, Chaitan S. 15
Kidnay, Arthur J. 17
Kiefer, John H. 28
Kilbey, S. Michael 101
Kilpatrick, Peter K. 80
Kim, Dong-Shik 87
Kim, Jingsang 53
Kim, Nam K. 55
Kim, Sangtae 33
Kim, Seong Han 96
Kimmel, Howard 65
Kimura, Shoichi 90
King, C. Judson 7
King, Franklin G. 79
King, Julia A. 55
King, Susan A. 62
King, Terry S. 38
King, William E. Jr. 91
Kiran, Erdogan 116
Kirmse, Dale W. 23
Kirwan, Donald J. 115
Kiselev, Sergei B. 17
Kisutcza, Joseph 65
Klapa, Maria I. 45
Klein, Andrew 94
Klein, Michael T. 68
Klein, Tonya M. 1
Klemm, William A. 102
Klingenberg, Daniel J. 120
Klinzing, George E. 97
Knecht, Robert D. 17
Knickle, Harold N. 100
Knopf, F. Carl 41
Knox, Dana 65
Knudsen, James G. 90
Knuth, Eldon L. 10
Knutson, Barbara L. 39
Kobayashi, Riki 108
Koberstein, J. 73
Koch, Donald L. 74
Koelling, Harold A. 58
Koelling, Kurt W. 85
Koffas, Mattheos 78
Kofinas, Peter 45
Kofke, David A. 78
Kofke, Tamara 78
Koh, Carolyn 17
Kohl, Paul 26
Kokkoli, Efrosini 57
Kolkka, Robert W. 55
Kolodka, Edward B. 81
Kolodney, Morris 72
Komives, Claire 14
Kompala, Dhinakar S. 16
Kono, Hisashi O. 119
Konstantopoulos, Konstantinos ... 44
Kopelevich, Dmitry 23
Koretsky, Milo D. 90
Korgel, Brian A. 111
Kornfield, Julia A. 13
Koros, William J. 26
Korus, Roger A. 27
Kostin, Morton D. 66
Kothare, Mayuresh V. 94
Kotov, Nicholas 53
Kovalchuk, Vladimir 97
Kovenklioglu, Suphan 69
Krahenbuhl, Melinda 114
Krambeck, Fredrick 44
Kramer, Edward J. 12
Krantz, William 83
Krentsel, Evguenii B. 59
Krieger-Brockett, Barbara 117
Krishnagopalan, Gopal A. 2
Krishnagopalan, Jaya 3
Krishnamoorti, Ramanan 106
Kroenke, William 70
Kuech, Thomas F. 120
Kuehl, Guenter H. 95
Kuester, James L. 4
Kugler, Edwin L. 119
Kuhl, Tonya L. 8
Kumar, Sanat 76
Kumar, Satish 57
Kummler, Ralph H. 56
Kung, Harold H. 31
Kung, Mayfair 31
Kunz, H. Russell 19
Kuo, Chiang-Hai 58
Kuo, Yue 109
Kusaka, Isamu 85
Kwon, K. C. 3
Kwon, Soonjo 24
Kyle, Benjamin G. 38

L

Ladd, Anthony J.C. 23
Lahann, Joerg 53
Laibinis, Paul 108
Lam-Anderson, Marca 73
Lamb, H. Henry 80
Lamm, Monica 36
Landau, Uziel 82
Lane, Alan M. 1
Lange, Fred 12
Langer, Robert S. 49
Langer, Stanley H. 120
LaRoche, Richard K. 21
Larsen, Gustavo 62
Larsen, Ronald W. 62
Larson, Ronald G. 53
Lauderback, Lee L. 62
Lauffenburger, Douglas A. 49
Laukhuf, W. L. S. 40
Laurence, Jennifer S. 37
Laurence, Robert L. 47
Laurencin, Cato 115
Laurenzi, Ian J. 94
Lauterbach, Jochen A. 21
Lavernia, Enrique 8
Law, Victor J. 43
Lawal, Adeniyi 69
Lawton, Carl W. 48
Leal, L. Gary 12
LeBlanc, Steven E. 87
Leckband, Deborah E. 29
Lee, Andre 54
Lee, C. Ted 15
Lee, C. William 84
Lee, Carolyn W. T. 50
Lee, Gil U. 33
Lee, Ilsoon 54
Lee, Jae W. 72
Lee, James Ly 85
Lee, James M. 118
Lee, Jay 26
Lee, Kelvin H. 74
Lee, Kyongbum 51
Lee, Lloyd L. 88
Lee, Robert L. 69
Lee, Sang-Yong 110
Lee, Stephen 85
Lee, Sunggyu 59
Lee, T. Randall 106
Lee, William E. III 25
Lee, Woo Y. 69
Lee, Y. Y. 2
Lee-DeSautels, Rhonda 39
Lefebvre, Brian G. 67
Leggoe, Jeremy 112
Leighton, Christopher 57
Leighton, David T. Jr. 32
Leipziger, Stuart 34
Lemlich, Robert 83
Lenhoff, Abraham M. 21
Lenoir, John M. 13
Leonard, Edward F. 73
Leong, Kam W. 44
LeVan, M. Douglas 105
Levenspiel, Octave 90
Levicky, Rasti 73
Levien, Keith L. 90
Levins, John M. 95
Levon, Kalle 75
Lewandowski, Gordon 65
Lewis, Randy S. 89
Li, Ku-Yen 107
Li, Kun 92
Liang, Jenn-Tai 37
Liang, Ruifeng 119
Liao, James C. 10
Liapis, Athanasios I. 60
Libera, Matthew R. 69
Licht, William 83
Liddell, KNona C. 118
Lidstrom, Mary E. 117
Lightfoot, Edwin N. 120
Lighty, Jo Ann S. 114
Lilleleht, Lembit 115
Lim, Henry C. 9
Lim, Phooi K. 80
Lim, Soon-Sik 87
Lin, David 118
Lin, Jerry Y. S. 4
Lin, Jerry Y.S. 83
Linden, Henry 30
Linden, James C. 18
Linderman, Jennifer J. 53
Lindt, Jan T. 97
Linic, Suljo 53
Linninger, Andreas 28
Liotta, Charles L. 26
Lipscomb, G. Glenn 87
Lira, Carl T. 54
List, Harvey L. 72
Littman, Howard 76
Liu, Chung-Chiun 82
Liu, Joseph T. C. 100
Liu, Juncheng 2

Liu, Y. A. 116
Llewellyn, J. Anthony 25
Lloyd, Douglas R. 111
Lobban, Lance 88
Lobo, Raul F. 21
Locke, Bruce R. 24
Locke, Carl E. Jr. 37
Lodge, Keith B. 56
Lodge, Timothy P. 57
Loehman, Ronald E. 70
Loewenberg, Michael 20
Lombardo, Stephen J. 59
Loney, Norman 65
Long, Richard L. Jr. 71
Longo, Marjorie L. 8
Loo, Yueh Lin 111
Lopez, Gabriel P. 70
Lopina, Stephanie T. 81
Lou, Helen H. 107
Lou, Jianzhong 79
Louie, Bev 16
Louvar, Joeseph 56
Loveland, Stephanie 36
Lowman, Anthony 93
Loyalka, Sudarshan 59
Lu, Christopher C. 84
Lu, Jia Grace 9
Lu, Yong 2
Lu, Yunfeng 43
Lucas, Glenn E. 12
Lucas, James 54
Lucas, Robert 75
Lucia, Angelo 100
Ludlow, Douglas K. 60
Luebbers, Ralph H. 59
Luecke, Richard H. 59
Luedeking, Robert 118
Luk, Yan Yeung 79
Luks, Kraemer D. 89
Lund, Carl R. F. 78
Luss, Dan 106
Lustig, Rolf 84
Lustig, Steven R. 21
Lutz, Robert J. 22
Luyben, William L. 94
Luzik, Eddie 20
Lynd, Lee R. 64
Lynn, David M. 120
Lynn, Scott 7
Lyon, David N. 7
Lyons, James E. 21

M

Ma, Teng 24
Ma, Yi Hua 52
Maboudian, Roya 7
Mackay, Michael 54
Macosko, Christopher W. 57
Maddox, Robert N. 89
Madihally, Sundararajan V. 89
Madix, Robert J. 15
Maffia, Gennaro J. 98
Magda, Jules J. 114
Maginn, Edward J. 32
Mahajan, Subhash 4
Mahalingam, R. 118
Mahoney, Melissa 16
Mainardi, Daniela 41
Maiorella, Brian 7
Maldarelli, Charles 72
Mallapragada, Surya K. 36
Mallinson, Richard G. 88
Malone, Michael F. 47
Maloney, James O. 37
Mandavilli, Satya N. 99
Mandell, John F. 62
Maneval, James E. 91
Manke, Charles 56
Manley, David B. 60
Mann, J. Adin Jr. 82
Mann, Michael D. 81
Mann, Uzi 112
Mannan, Sam 109
Manning, Francis S. 89
Manning, Michael 50
Manogue, William H. 21
Manousiouthakis, Vasilios 10
Mansoori, G. Ali 28
Mantzaris, Nikolas 108
Mao, Guangzhao 56
Maples, Glennon 2
Maranas, Costas 96
Maranas, Janna K. 96
Marand, Eva 116
Maravelias, Christos T. 120
Marc de Chazal, L. E. 59
Marnell, Paul 75
Maroudas, Dimitrios 47
Marple, Stanley Jr. 106
Marr, David W. M. 17
Marrero, Thomas R. 59
Marshall, John 69
Martínez-Iñesta, María M. 99
Marten, Mark 46
Martin, George C. 79
Martin, Heidi B. 82
Martin, J. Ronald 93
Martin, Lealon 76
Masel, Richard I. 29
Mason, Donald R. 24
Mason, Geoffrey 121
Massoth, Franklin E. 114
Mather, Patrick 19
Mathur, Virendra K. 64
Matsen, John M. 94
Matsoukas, Themistoklis 96
Matsumoto, Mark R. 11
Matthew, Howard 56
Matthews, John C. 38
Matthews, Michael A. 102
Mavrikakis, Manos 120
Mayer, Michael 53
Maynard, Jennifer 57
McAvoy, Thomas J. 45
McCabe, Clare 105
McCandless, Frank P. 62
McCarthy, Joseph 97
McClurg, Richard B. 57
McCluskey, Richard J. 72
McCormick, Alon V. 57
McCoy, Benjamin J. 8, 41
McCoy, John 69
McCready, Mark J. 32
McDonald, Karen A. 8
McFarland, Eric 12
McFetridge, Peter S. 88
McGee, John C. 105
McGinn, Paul J. 32
McGuire, Joseph 90
McHugh, Anthony J. 29, 94
McIntire, Larry V. 108
McKean, Wm. T. 117
McKelvey, James M. 61
McKenna, Greg 112
McKenzie, Steven E. 95
McKetta, John J. 111
McKinley, Marvin D. 1
McKinnon, J. Thomas 17
McLaughlin, Edward 41
McLaughlin, John B. 72
McMicking, James H. 56
McNeil, Kenneth M. 98
McNeil, Melanie 14
McRae, Gregory J. 49
McWilliams, Thomas G. Jr. 98
Mead, Richard W. 70
Meagher, Michael M. 62
Mecartney, Martha 9
Medlin, Will 16
Mehta, Narinder K. 99
Meldon, Jerry 51
Mellichamp, Duncan A. 12
Melsheimer, Stephen S. 101
Merrill, Edward W. 49
Mesler, Russell B. 37
Metters, Andrew 101
Metzner, Arthur B. 21
Meyer, Stephen P. 93
Michelsen, Donald L. 116
Mihalko, William M. 78
Mijovic, Jovan S. 75
Mikos, Antonios G. 108
Miletic, Marina 29
Millard, Paul 43
Miller, Albert E. 32
Miller, Clarence A. 108
Miller, David C. 34
Miller, David R. 11
Miller, Dennis J. 54
Miller, Kelly 17
Miller, Reid C. 118
Miller, Ronald L. 17
Miller, William M. 31
Mills, David R. 2
Mills, Robert G. 66
Mischke, Roland A. 116
Misra, Devesh 42
Misra, Manish 3
Mitch, William 20
Mitchell, Brian S. 43
Mitchell, Martha C. 71
Mitragotri, Samir 12
Mockros, Lyle F. 31
Moghe, Prabhas 68
Mohamed, Farghalli 9
Mohanty, Kishore K. 106
Mohr, C. Michael 49
Mollenkamp, Robert A. 60
Monbouquette, Harold G. 10
Monson, Peter A. 47
Montagna, Peter 20
Montgomery, Susan M. 53
Moore, Charles F. 104
Moore, Noel E. 34

Moore, O. C. 2
Moreira, Antonio R. 46
Morgan, John A. 33
Morgan, Morris H. III 115
Morikis, Dimitrios 11
Morisani, Stephen J. 3
Morosoff, Nicholas 60
Morris, James G. 39
Morrison, Faith A. 55
Morrow, Norman R. 121
Morse, David C. 57
Morsi, Badie I. 97
Morton, Samuel 93
Moschandreas, Demetrios 30
Moser, William 52
Moss, Melissa A. 102
Motard, Rodolphe L. 61
Motzkin, Shirley 75
Mountziaris, T. J. 78
Mowrey, Daniel B. 41
Muggli, Darrin S. 81
Mukasyan, Alexander 32
Mukherjee, Amiya K. 8
Mulchandani, Ashok 11
Mulholland, Kenneth L. 21
Muller, Susan J. 7
Mullins, C. Buddie 111
Mullins, Joseph C. 101
Mullins, Michael E. 55
Mumm, Daniel 9
Mumme, Kenneth 43
Munir, Zuhair A. 8
Munro, James M. 102
Munson-McGee, Stuart 71
Murad, Sohail 28
Murhammer, David W. 35
Murphy, David 110
Murphy, Regina M. 120
Murphy, Vincent G. 18
Muscat, Anthony 5
Musgrave, Charles 15
Muske, Kenneth R. 98
Mutharasan, Rajakannu 93
Muzzio, Fernando J. 68
Muzzy, John D. 26
Myers, Alan L. 95
Myers, John A. 98
Myers, Kevin J. 84
Myers, Wm. A. 6
Myerson, Allan 30
Myung, Nosang V. 11

N

Nadarajah, Arunan 87
Nagarajan, R. 96
Naimpally, Ashok 13
Nair, Sankar 26
Narang, Atul 23
Narasimhan, Balaji 36
Narayan, Ramani 54
Narayanan, Ranga 23
Naser, Jamil 3
Natoli, John 67
Nauman, E. Bruce 76
Nave, Felecia M. 107
Navrotsky, Alexandra 8
Nazario-Acevedo, Aldo 99
Nealey, Paul F. 120
Nedwick, Robert 96
Neelamegham, Sriram 78
Neivandt, David 43
Nelson, Donald 62
Nenes, Athanasios 26
Neogi, Partho 60
Nerem, Robert M. 26
Nesbitt, Carl 55
Nesic, Srdjan 86
Neuman, Ronald D. 2
Neurock, Matthew 115
Newby, Bi-min Zhang 81
Newell, James A. 67
Newman, Bernard 42
Newman, John S. 7
Ng, K. Y. Simon 56
Nguyen, Thuan K. 14
Nguyen, Trung Van 37
Nicholson, Bruce J. 78
Nickelson, Robert L. 62
Nikolaou, Michael 106
Nikolov, Alex 30
Nippert, Charles R. 98
Nitsche, Johannes M. 78
Nitsche, Ludwig 28
Nobe, Ken 10
Noble, Richard D. 16
Noll, Kenneth 30
Nollert, Matthias U. 88
Norbeck, Joseph M. 11
Nordheden, Karen J. 37
Norris, David 57
Nosa, Egiebor 3
Noureddini, Hossein 62
Nuttall, H. Eric 70

O

O'Connell, John P. 115
O'Connor, Kim C. 43
O'Haver, John Howard 58
O'Rear, Edgar A. III 88
O'Shaughnessy, B. 73
Oberhauser, James P. 115
Ofoli, Robert Y. 54
Ogale, Amod A. 101
Ogden, Kimberly 5
Ogunnaike, Babatunde A. 21
Okazaki, Kenji 39
Okorator, Charles 73
Okos, M. 33
Olbricht, William L. 74
Ollis, David F. 80
Olsen, Thor 77
Olson, David H. 95
Olson, Fred 119
Olson, Jon H. 21
Ontiveros, Cordelia 14
Orazem, Mark E. 23
Oriani, Richard A. 57
Orkoulas, Gerassimos 10
Osborne-Lee, Irvin W. 107
Oscarson, John L. 113
Ostafin, Agnes E. 32
Ostermann, Russell D. 37
Ostermeier, Marc 44
Ottino, Julio M. 31
Overcash, Michael R. 80
Overney, Rene M. 117
Owens, Camelia 46
Owens, Thomas C. 81
Oxford, Charles W. 6
Oyama, S. Ted 116
Ozkan, Mihri 11
Ozkan, Umit 85

P

Pack, Daniel W. 29
Padrón, Federico 99
Pagilla, Krishna 30
Palanki, Srinivas 24
Palazoglu, Ahmet N. 8
Palecek, Sean P. 120
Palmer, Andre F. 32
Palmer, Jim 41
Palmese, Giuseppe 93
Palmstrom, Christopher J. 57
Panagiotopoulos, Athanassios Z. 66
Pandis, Spyros 92
Pang, K. Hing 14
Papadopoulos, Kyriakos D. 43
Papathanasiou, Thanasis D. 102
Papavassiliou, Dimitrios 88
Papoutsakis, E. Terry 31
Park, Chan Beum 4
Park, Chang-Won 23
Park, Eugene 100
Park, Jin Y. 27
Park, SangKwon 2
Park, Yoonkook 3
Parker, Robert S. 97
Parnas, Richard S. 19
Parsons, Gregory N. 80
Parulekar, Satish J. 30
Pasquali, Matteo 108
Patterson, Gary K. 60
Patton, Christi L. 89
Patton, John T. 71
Patzer, John F. II 97
Paul, Donald R. 111
Pavone, Anthony 15
Payne, Gregory F. 46
Peattie, Robert A. 90
Pedersen, Henrik 68
Peeples, Tonya 35
Pekny, Joseph F. 33
Pendse, Hemant 43
Peng, Ching-An 15
Penn, Lynn 39
Penney, W. Roy 6
Peppas, Nicholas A. 111
Peretti, Steven W. 80
Perez, Joseph M. 96
Perez-Luna, Victor 30
Perkins, Tracy 18
Perlmutter, Daniel D. 95
Perna, Angelo 65
Perry, Scott S. 106
Pershing, David W. 114
Peters, Michael H. 24
Petersen, Eugene E. 7
Petersen, James N. 118
Peterson, Richard E. 114
Peterson, Thomas W. 5
Petrenko, Victor 64
Petrou, Athos 78
Petrovan, Simioan 104

Petty, Charles A. 54
Peyton, Brent M. 118
Pfeffer, Robert 65, 72
Pfefferle, Lisa D. 20
Pfeifer, Blaine 51
Pfromm, Peter 38
Phelps, James Parkhurst 48
Philipossian, Ara 5
Phillips, Jonathan 70
Phillips, Paul 83
Phillips, Ronald J. 8
Piergiovanni, Polly 93
Pignatello, Joseph 20
Pike, Ralph W. 41
Pilat, Michael J. 117
Pilehvari, Ali 110
Pine, David 12
Pintar, Anton J. 55
Pintauro, Peter N. 82
Pinto, Jose M. 75
Pipes, Byron 33
Pishko, Michael V. 96
Pitt, William G. 113
Placek, Timothy D. 2
Plank, Charles A. 40
Plawsky, Joel L. 76
Ploehn, Harry J. 102
Plouffe, Paul 7
Podlaha, Elizabeth J. 41
Poehl, Michael 111
Poling, Bruce E. 87
Pommersheim, James M. 91
Ponter, Anthony B. 42
Popov, Branko N. 102
Porter, Marc D. 36
Potoff, Jeffrey J. 56
Pourhashemi, Ali 103
Pourki, Forouza 98
Powell, Robert L. 8
Powers, Gary J. 92
Powers, John E. 53
Pozo de Fernandez, Maria E. 24
Pozrikidis, Constantine 11
Prados, John W. 104
Prakash, Jai 30
Prather, Kristala J. 49
Prausnitz, John M. 7
Prausnitz, Mark R. 26
Preston, Floyd W. 37
Price, Douglas 87
Price, Geoffrey L. 89
Price, Randel M. 103
Prieve, Dennis C. 92
Prince, Michael J. 91
Prinja, Anil K. 70
Proehl, Jeffrey A. 64
Prokop, Ales 105
Prud'homme, Robert K. 66
Prudich, Michael E. 86
Przybycien, Todd 92
Pugmire, Ronald J. 114
Pulliam Holoman, Tracey R. 45
Punzi, Vito L. 98
Puszynski, Jan A. 102
Putatunda, Susil K. 56
Putnam, Andrew 9
Putnam, David A. 74

Q

Qammar, Helen C. 81
Qin, S. Joe 111
Queneau, Paul 64
Quinn, John A. 95
Qutubuddin, Syed 82

R

Radke, Clayton J. 7
Radosz, Maciej 121
Raghavan, Srinivasa R. 45
Rajagopalan, Padma 94
Rajagopalan, Raj 23
Ralston, Patricia A. S. 40
Ramírez, Carlos A. 99
Ramachandran, P. A. 61
Ramanarayanan, Trikur A. 95
Ramarao, Bamdaru 79
Ramirez, W. Fred 16
Ramkrishna, D. 33
Ranade, Madhav B. 45
Randall, D'Arcy 111
Randolph, Theodore W. 16
Rankin, Stephen 39
Rao, Christopher V. 29
Rao, Govind 46
Rao, M. Gopala 22
Raper, Judy A. 60
Rase, Howard F. 111
Rasmussen, Don H. 72
Rathman, James F. 85
Ratner, Buddy D. 117
Raupp, Gregory B. 4
Ravi, Vilupanur 14
Rawlings, James B. 120
Ray, Asit K. 39, 103
Ray, W. Harmon 120
Raymond, Timothy M. 91
Realff, Matthew J. 26
Reardon, Kenneth F. 18
Reed, X B Jr. 60
Regalbuto, John 28
Regan, Thomas M. 45
Register, Richard A. 66
Rehm, Thomas R. 5
Reid, Robert C. 49
Reilly, Peter J. 36
Reimer, Jeffrey A. 7
Reinhardt, James R. 42
Rekhson, Simon 84
Reklaitis, Gintaras V. 33
Ren, Fan 23
Rengaswamy, Raghunathan 72
Resasco, Daniel E. 88
Rethwisch, David G. 35
Retzloff, David G. 59
Reucroft, Philip 39
Reynolds, Joseph 75
Rezac, Mary 38
Rhinehart, R. Russell 89
Ribeiro, Fabio 33
Rice, Philip A. 79
Rice, Richard W. 101
Rice, William J. 98
Richardson, James T. 106
Richardson, Peter D. 100
Richter, Horst 64
Ricker, N. Lawrence 117
Ridgway, Darin 86
Riggs, James B. 112
Rinaldi-Ramos, Carlos Manuel 99
Rinard, Irven H. 72
Ring, Terry A. 114
Rinker, Kristina 18
Rinker, Robert G. 12
Risbud, Subhash 8
Ritchie, Stephen M. 1
Ritter, Arthur 69
Ritter, Edward R. 98
Ritter, James A. 102
Rivero-Hudec, Mercedes 100
Robert, Marc A. 108
Roberts, Christopher 2
Roberts, Christopher J. 21
Roberts, George W. 80
Roberts, Kenneth L. 79
Roberts, Robert W. 81
Roberts, Ronnie S. 26
Roberts, Susan C. 47
Robertson, Channing R. 15
Robinson, Anne Skaja 21
Robinson, L. B. 10
Robinson, Richard L. 42
Robinson, Robert L. Jr. 89
Roche, Edward C. 63
Rochefort, Willie E. 90
Rochelle, Gary T. 111
Rockstraw, David A. 71
Roderick, Norman F. 70
Rodgers, Victor G. J. 35
Rodríguez-Ramírez, Abraham 99
Rodriguez, Ferdinand 74
Rodriguez, Harold V. 3
Rogers, Bridget R. 105
Rogers, Rudy E. 58
Rollins, Derrick K. 36
Rony, Peter R. 116
Root, Thatcher W. 120
Roper, Keith 114
Rorrer, Gregory L. 90
Rose, Vincent C. 100
Rosen, Stephen L. 60
Rosinski, Matthew 121
Rosner, Daniel E. 20
Ross, Julia M. 46
Rossky, Peter J. 111
Rosson, Harold F. 37
Roth, Charles 68
Roth, John A. 105
Rothberg, Gerald 69
Rothberg, Lewis J. 77
Rothe, Erhard W. 56
Rother, Michael 56
Roubicek, Rudolf 71
Rousseau, Ronald W. 26
Rowell, George 93, 98
Rowley, Richard L. 113
Ruckenstein, Eli 78
Rumschitzki, David 72
Russel, William B. 66
Russell, Alan J. 97
Russell, T. W. Fraser 21
Ruthven, Douglas M. 43
Rutkowski, Gregory 56
Rutledge, Gregory C. 49
Ryan, Michael E. 78

Ryder, Daniel 51
Ryu, Dewey 8

S

Saber, Hamed 70
Sacco, Albert Jr. 50
Sadana, Ajit 58
Saddawi, Salma 32
Saez, Eduardo 5
Sahimi, Muhammad 15
Sahinidis, Nikolaos V. 29
Saliba, Tony E. 84
Saliby, Michael J. 20
Saliceti-Piazza, Lorenzo 99
Salim, Majid 34
Salley, Steven O. 56
Salovey, Ronald 15
Saltsburg, Howard 51
Saltzman, Mark 20
Sama, Dominick Anthony 48
Sambanis, Athanassios 26
Sampson, Kendree J. 86
Samuels, Robert J. 26
San, Ka-Yiu 108
Sanchez, Isaac C. 111
Sandall, Orville C. 12
Sandell, John 55
Sandhu, Sarwan S. 84
Sandler, Stanley I. 21
Sandvig, Robert L. 102
Sangani, Ashok 79
Sani, Robert L. 16
Santanam, Suresh 79
Santore, Maria M. 47
Sarikaya, Mehmet 117
Sarofim, Adel 114
Sarofim, Adel F. 49
Sater, Vernon E. 4
Sather, Glenn A. 120
Satterfield, Charles N. 49
Satvat, Behrooz 50
Sauer, John A. 68
Sauer, Sharon G. 34
Savage, Phillip E. 53
Savelski, Mariano J. 67
Saville, Dudley A. 66
Savinell, Robert F. 82
Sawin, Herbert H. 49
Saxena, Satish C. 28
Scamehorn, John F. 88
Scarrah, Warren P. 62
Schaffer, David 7
Schaffer, James P. 93
Schall, Constance A. 87
Schechter, Robert S. 111
Scheinbeim, Jerry 68
Scheldorf, Jay J. 27
Schieber, Jay D. 30
Schiesser, William 94
Schimmel, Keith A. 79
Schlup, John R. 38
Schmidt, Christine 111
Schmidt, Lanny D. 57
Schmidtke, David W. 88
Schmitz, Roger A. 32
Schneider, James 92
Schnelle, Karl B. Jr. 105
Schoenung, Julie M. 8
Schork, F. Joseph 26
Schowalter, William 66
Schowalter, William R. 29
Schrader, Glenn L. 36
Schrieber, Loren 24
Schroder, Klaus 79
Schrodt, J. Thomas 39
Schruben, Dale L. 110
Schulson, Erland 64
Schultz, Jerold M. 21
Schultz, Jerome S. 11
Schulz, Kirk H. 58
Schure, Mark R. 21
Schwaber, James S. 21
Schwank, Johannes W. 53
Schwartz, Daniel 16
Schwartz, Daniel T. 117
Schwartz, Pauline M. 20
Schwarz, James A. 79
Sciance, Carroll T. 111
Scott, Garland E. 14
Scott, Susannah L. 12
Scovazzo, Paul 58
Scranton, Alec 35
Scriven, L. E. 57
Scurto, Aaron M. 37
Seader, J. D. 114
Seagrave, Richard C. 36
Seames, Wayne S. 81
Searson, Peter 44
Seborg, Dale E. 12
Seebauer, Edmund G. 29
Seferis, James C. 117
Segalman, Rachel A. 7
Segura, Tatiana 10
Seidel, Robert W. 57
Seider, Warren D. 95
Seinfeld, John H. 13
Selman, J. Robert 30
Seminario, Jorge 109
Sengers, Jan V. 45
Sengupta, Arup K. 94
Senkan, Selim M. 10
Serbezov, Atanas 34
Serth, Robert W. 110
Seshadri, K. 11
Setzer, C. John 80
Sevick-Muraca, Eva 109
Seymour, Joseph 62
Shackelford, James F. 8
Shadman, Farhang 5
Shaeiwitz, Joseph A. 119
Shaffer, Daniel L. 62
Shah, Dhananjai B. 84
Shah, Dinesh O. 23
Shah, Naresh 39
Shambaugh, Robert L. 88
Shanks, Brent H. 36
Shanks, Jackie V. 36
Shantz, Daniel F. 109
Shapir, Yonathon 77
Shapiro, Andrew 9
Shapley, Nina 73
Shaqfeh, Eric S. G. 15
Sharfstein, Susan 76
Sharma, Mrityunjai P. 121
Sharp, Julie E. 105
Shaw, Montgomery T. 19
Shea, Lonnie D. 31
Sheff, James Robert 48
Sheldon, Brian W. 100
Shen, Youqing 121
Sheng, Henry P. 14
Sheth, Atul 104
Shi, Frank 9
Shieh, Wen K. 95
Shim, Jae-Jin 2
Shine, Annette D. 21
Shing, Katherine S. 15
Shinnar, Reuel 72
Sholl, David 92
Shonnard, David R. 55
Shores, David A. 57
Short, David 21
Shreve, Gina S. 56
Shuler, Michael L. 74
Shuming, Zheng 115
Shusta, Eric V. 120
Shvartsman, Stanislav Y. 66
Siberio-Pérez, Diana Y. 99
Sibulkin, Merwin 100
Sidebotham, George 73
Sides, Paul 92
Sierka, Raymond 5
Sierks, Michael 4
Sierra Alvarez, Maria Reyes 5
Siirola, Jeffrey 104
Sikavitsas, Vassilios I. 88
Silano, A. A. 6
Silcox, Geoffrey D. 114
Silebi, Cesar A. 94
Silverstein, David 39
Simon, Laurent 65
Simon, Sindee 112
Simpkins, Peter G. 94
Sinno, Talid R. 95
Sircar, Shivaji 94
Sirkar, Kamalesh 65
Sitton, Oliver C. 60
Sivertsen, John M. 57
Siviniah, Easan 112
Skaf, Dorothy W. 98
Skelland, A. H. Peter 26
Skliar, Mikhail 114
Slater, C. Stewart 67
Slattery, John C. 109
Sleicher, Charles A. 117
Slider, H. C. 85
Sliepcevich, C. M. 88
Sloan, E. Dendy Jr. 17
Slonaker, Robert E. Jr. 91
Smart, Jimmy 39
Smay, James E. 89
Smirniotis, Panagiotis 83
Smith, Carlos A. 25
Smith, Edwin E. 85
Smith, Grant D. 114
Smith, J.M. 8
Smith, James E. Jr. 1
Smith, Julian C. 74
Smith, Kenneth A. 49
Smith, Philip J. 114
Smith, Stanley 102
Smith, T. Gordon 18
Smith, Theodore G. 45
Smolke, Christina D. 13

Smoot, L. Douglas ... 113
Smotkin, Eugene S. ... 30
Smyrl, William H. ... 57
Snide, James A. ... 84
Snurr, Randall ... 31
Snyder, William J. ... 91
Soane, David S. ... 7
Solen, Kenneth A. ... 113
Solomon, Michael J. ... 53
Solum, Mark ... 114
Sommerfeld, Jude T. ... 26
Sonnerup, Bengt ... 64
Soroush, Masoud ... 93
Sotirchos, Stratis V. ... 77
Southard, Marylee Z. ... 37
Sparrow, Charles A. ... 58
Spencer, Hugh T. ... 40
Spencer, Jordan L. ... 73
Sperling, Leslie H. ... 94
Spicer, Thomas O. ... 6
Spinti, Jennifer ... 114
Spivey, James J. ... 41
Spontak, Richard J. ... 80
Springer, Charles ... 6
Squires, Arthur M. ... 116
Squires, Robert G. ... 33
Sridhar, Lakshmi N. ... 99
Srienc, Friedrich ... 57
Srivastava, Ranjan ... 19
Stadtherr, Mark A. ... 32
Stancell, Arnold ... 26
Stanford, Thomas G. ... 102
Stansbury, Jeff ... 16
Starling, Kenneth E. ... 88
Starr, Thomas L. ... 40
Stasaik, Raymond ... 58
Stebe, Kathleen J. ... 44
Steen, Paul H. ... 74
Stein, Fred P. ... 94
Steiner, Carol ... 72
Steinfink, Hugo ... 111
Stenger, Harvey G. Jr. ... 94
Stephanopoulos, George ... 49
Stephanopoulos, Gregory ... 49
Sterling, Arthur ... 41
Stern, S. Alexander ... 79
Sternberg, Moshe ... 7
Sternberg, Steven P.K. ... 56
Stevens, William F. ... 31
Stewart, Phillip ... 62
Stewart, Richard R. ... 50
Stewart, Warren E. ... 120
Stice, James E. ... 111
Stiel, Leonard I. ... 75
Stiller, Alfred H. ... 119
Stinespring, Charter ... 119
Stock, Richard ... 73
Stoll, A. George ... 14
Storvick, Truman S. ... 59
Strano, Michael S. ... 29
Streett, William B. ... 74
Strieder, William C. ... 32
Stroeve, Pieter ... 8
Stroock, Abraham D. ... 74
Strunk, Mailand R. ... 60
Stuve, Eric M. ... 117
Sublette, Kerry L. ... 89
Subramaniam, Bala ... 37
Subramanian, Anu ... 62
Subramanian, K.N. ... 54
Subramanian, R. Shankar ... 72
Subramanian, Venkat ... 105
Sukanek, Peter C. ... 58
Suleiman-Rosado, David ... 99
Sullivan, Joseph ... 116
Sullivan, Samuel L. Jr. ... 43
Sum, Amadeu ... 116
Sundaresan, Sankaran ... 66
Sundstrom, Donald W. ... 19
Sung, Changmo ... 48
Sung, Nak-Ho ... 51
Suni, Ian I. ... 72
Sunkara, Mahendra K. ... 40
Sunol, Aydin K. ... 25
Sunwoo, Changshin ... 2
Suppes, Galen J. ... 59
Sureshkumar, R. ... 61
Suuberg, Eric M. ... 100
Svanks, Karlis ... 85
Svoronos, Spyros A. ... 23
Swaney, Ross E. ... 120
Swanson, Bernard ... 30
Swartz, James ... 15
Swartz, Melody A. ... 31
Sweeney, Thomas L. ... 85
Sweeny, Robert F. ... 98
Swift, George W. ... 37
Swihart, Mark ... 78
Swinnea, J. Steven ... 111
Sylvester, Nicholas ... 3
Szepe, Stephen ... 28

T

Taconi, Katherine ... 1
Tadmor, Rafael ... 107
Takamura, Koichi ... 121
Takoudis, Christos ... 28
Talbot, Jan B. ... 11
Talu, Orhan ... 84
Tang, Yi ... 10
Tanner, Robert D. ... 105
Tao, Luh C. ... 62
Tardos, Gabriel I. ... 72
Tarrer, A. Ray ... 2
Tassoney, Joseph P. ... 14
Tatarchuk, Bruce J. ... 2
Tatterson, Gary B. ... 79
Tavakoli, Javad ... 93
Tavlarides, Lawrence L. ... 79
Taylor, Ross ... 72
Tedder, Dan W. ... 26
Teja, Amyn ... 26
Tek, M. Rasin ... 53
Tellkamp, Victoria ... 9
Telotte, John C. ... 24
Teraoka, Iwao ... 75
Terry, Ronald E. ... 113
Tester, Jefferson W. ... 49
Tewari, Surendra N. ... 84
Teymour, Fouad A. ... 30
Tharakan, John P. ... 22
Thatcher, Charles M. ... 6
Theodore, Louis ... 75
Theofanous, Theofanis ... 12
Thibodeaux, Louis J. ... 41
Thies, Curt ... 61
Thies, Mark C. ... 101
Thode, Edward F. ... 71
Thoma, Greg ... 6
Thomas, Aaron ... 27
Thompson, Edward V. ... 43
Thompson, Karsten E. ... 41
Thompson, Levi T. ... 53
Thompson, Richard E. ... 89
Thompson, Robert W. ... 52
Thompson, Ronald H. ... 41
Thomson, Kendall T. ... 33
Thomson, William J. ... 118
Thorgerson, Eric ... 50
Thorpe, R. G. ... 74
Tien, Chi ... 79
Tierney, John W. ... 97
Tilton, James N. ... 21
Tilton, Robert ... 92
Timm, Delmar C. ... 62
Tirrell, David A. ... 13
Tirrell, Matthew ... 12
Tock, R. W. ... 112
Todd, David ... 65
Tomadakis, Manolis M. ... 24
Tomasko, David ... 85
Tomassone, M. Silvina ... 68
Tomb, Jean-Francois ... 21
Tomkins, Reginald ... 65
Tomlinson, John L. ... 14
Tooney, Nancy ... 75
Toor, Herbert L. ... 92
Torkelson, John M. ... 31
Torres-Lugo, Madeline ... 99
Toups, Harry ... 41
Tournier, Jean-Michel ... 70
Towler, Brian F. ... 121
Tran, Veronique V. ... 106
Tranquillo, Robert T. ... 57
Tree, D. Alan ... 89
Tripathi, Anubhav ... 100
Troian, Sandra M. ... 66
Trout, Bernhardt L. ... 49
Trujillo, Edward M. ... 114
Truskett, Thomas M. ... 111
Tsai, Shirley ... 13
Tsang, Tate T. H. ... 39
Tsao, George T. ... 33
Tsapatsis, Michael ... 57
Tschoegl, Nicholas W. ... 13
Tseng, Yiider ... 44
Tsianou, Marina ... 78
Tsotsis, Theodore T. ... 15
Tucker, W. Henry ... 34
Turian, Raffi ... 28
Turner, C. Heath ... 1
Turner, Jay R. ... 61
Turpin, Jim L. ... 6
Turro, Nicholas J. ... 73
Turton, Richard ... 119
Tuzla, Kemal ... 94
Tyler, Bonnie ... 114
Tyner, Mack ... 23
Tzanakakis, Emmancuhl ... 78

U

Uebler, E. Alan ... 21
Ueki, Taro ... 70
Ugaz, Victor ... 109

Uitenham, Leonard 79
Ulman, Abraham 75
Ulrich, Gael D. 64
Ulrich, Richard K. 6
Ulrichson, Dean L. 36
Ultman, James S. 96
Ungar, Lyle H. 95
Ungarala, Sridhar 84
Uslenghi, P.L.E. 28
Utgikar, Vivek 27
Uz, Mehmet 93

V

Vaeth, Kathleen 74
Vahdat, Nader 3
Valluzzi, Regina 51
Valsaraj, Kalliat T. 41
Van Brunt, Vincent 102
van Heiningen, Adriaan 43
Van Ness, Hendrick C. 76
van Oss, Carel J. 78
Van Tassel, Paul 20, 56
Van Vorst, W. D. 10
Van Wie, Bernard J. 118
Van Zanten, John 80
Van Zee, John W. 102
VanAuker, Michael 25
Vanderlick, T. Kyle 66
VanKirk, Jesse 67
Vannice, M. Albert 96
VanWormer, Kenneth A. 51
Varanasi, Sasidhar 87
Vasavada, Anita 118
Vasilos, Thomas 48
Vasquez, Victor R. 63
Vasudevan, Palligarnai T. 64
Vaughn, Mark 112
Vekilov, Peter G. 106
Velázquez-Figueroa, Carlos 99
Velankar, Sachin 97
Velegol, Darrell 96
Velev, Orlin 80
Venerus, David C. 30
Venkatasubramanian, V. 33
Venugopalan, Vasan 9
Veser, Goetz 97
Vigeant, Margot A.S. 91
Vigil, R. Dennis 36
Viljoen, Hendrik J. 62
Villafañe-Ruiz, Gilberto 99
Vinals, Jorge 24
Virk, Preetinder S. 49
Visco, Donald P. 105
Viswanath, Dabir S. 59
Vives, Donald L. 2
Vlachos, Dionisios 21
Vlahovska, Petia 100
Vohs, John M. 95
Volloch, Vladimir 51
von Braun, Margrit C. 27
VonBerg, Robert L. 74
Vossoughi, Shapour 37
Vrentas, James S. 96
Vunjak-Novakovic, Gordana 51

W

Wachs, Israel E. 94
Waggoner, Raymond C. 60
Wagner, Eric S. 13
Wagner, Jan 89
Wagner, John E. 34
Wagner, Norman J. 21
Wagner, Robert 52
Wagner, William R. 97
Walas, Stanley M. 37
Walawender, Walter P. 33
Walker, Lynn 92
Walker, Sharon L. .... 1[illegible]
Walkinshaw, John W. .... 4[illegible]
Wallace, William E. 119
Wallis, Graham B. 64
Wallis, James 20
Wallman, P. Henrik 7
Walton, S. Patrick 54
Walz, John Y. 20
Wang, Chunsheng 105
Wang, Daniel I. C. 49
Wang, Henry Y. 53
Wang, Jee-Ching 60
Wang, Joseph 4
Wang, Linda Nien-Hwa 33
Wang, Nam Sun 45
Wang, Pin 15
Wang, Ping 81
Wang, Qiang (David) 18
Wang, Shaw S. 68
Wang, Szu-Wen 9
Wang, Tse-Wei 104
Wang, Zhen-Gang 13
Wankat, Phillip C. 33
Ward, Michael D. 57
Ward, Timothy L. 70
Wasan, Darsh T. 30
Watkins, James J. 47
Watson, A. Ted 18
Watson, Jack S. 104
Watson, John, L. 81
Watters, James C. 40
Way, J. Douglas 17
Wayner, Peter C. 76
Wazzan, A. R. 10
Weatherley, Laurence R. 37
Weaver, Harry 70
Weaver, Jason 23
Weaver, Mark 1
Weber, Frederick E. 104
Weber, James H. 62
Weber, Thomas W. 78
Weber, Walter J. Jr. 53
Wedgewood, Lewis 28
Wei, James 66
Weidner, John W. 102
Weigand, William A. 45
Weimer, Alan W. 16
Weimer, Jeffrey 1
Weinberger, Charles B. 93
Weinkauf, Donald H. 69
Weinstein, Herbert 72
Weinstein, Randy D. 98
Weiss, Alvin 52
Weiss, Robert A. 19
Welker, J. Reed 6
Weller, Sol W. 78
Wender, Irving 97
Wendt, Jost O. L. 5
Wentzcovitch, Renata 57
Wesson, G. Dale 24
Wesson, Laura 41
Wesson, Rosemarie D. 45
West, A. C. 73
West, Jennifer 108
Westerberg, Arthur W. 92
Westmoreland, Phillip R. 47
Westwater, James W. 29
Wetzel, David M. 41
Wheeler, Clayton 43
Wheeler, Dean 113
Wheeler, George L. 20
Wheelock, Thomas D. 36
Whitaker, Stephen 8
White, Don H. 5
White, John Robert 48
White, Lee 92
White, Mark G. 26
White, Ralph E. 102
Whiteley, James R. 89
Whiting, Wallace B. 63
Whitlow, Jonathan E. 24
Whitmire, David R. 105
Wick, Timothy M. 26
Wickramasinghe, Ranil 18
Wicks, Charles E. 90
Wiegandt, H. F. 74
Wiencek, John 35
Wiesner, Theodore F. 112
Wiest, John M. 1
Wikstrom, Leonard Lee 20
Wilcox, Jennifer L. 52
Wilcox, William R. 72
Wildi, Robert 119
Wilding, W. Vincent 113
Wilkens, Robert 84
Wilkes, Garth L. 116
Wilkes, James O. 53
Wilkins, Ebtisam S. 70
Willey, Ronald J. 50
Willhite, G. Paul 37
Williams, Christopher 102
Williams, Edwin T. 55
Williams, John A. 50
Williams, Michael C. 7
Williams, Susan M. 37
Williamson, Kenneth 90
Williford, Clint W. Jr. 58
Willis, Brian G. 21
Wills, George B. 116
Willson, C. Grant 111
Willson, Richard C. 106
Wilson, Bruce 71
Winey, Karen I. 95
Winkler, Stefan 51
Winnick, Jack 26
Winter, Horst H. 47
Winter, Robb M. 102
Wirtz, Denis 44
Wise, Donald L. 50
Wisecarver, Keith D. 89
Wissbrun, Kurt F. 21
Wissler, Eugene H. 111
Wittrup, K. Dane 49
Wnek, Gary 82
Wolden, Colin A. 17
Wolf, Clarence J. 61
Wolf, David 5

Wolf, Eduardo E. 32
Wolf, Edward 75
Won, You-Yeon 33
Wong, Michael 108
Wood, David W. 66
Wood, Thomas K. 19
Woodruff, Gene L. 117
Wool, Richard P. 21
Woolf, Peter 53
Worden, R. Mark 54
Wornat, Mary J. 41
Wrenn, Steven 93
Wu, David T. 17
Wu, Huey 98
Wu, J. H. David 77
Wu, Jianzhong 11
Wu, Jing 65
Wyman, Charles E. 64
Wyslouzil, Barbara 85

X

Xanthos, Marino 65
Xing, Yangchuan 60
Xu, Xinglong 87

Y

Yablonsky, Gregory 61
Yager, Paul 117
Yan, Yushan 11
Yang, Arthur J.-M. 45
Yang, Fuqian 39
Yang, Hong 77
Yang, Ralph T. 53
Yang, Ray Y. K. 119
Yang, Shang-Tian 85
Yang, Yi 70
Yarbrough, David W. 105
Yasuda, H. K. 59
Yates, Matthew 77
Yaws, Carl L. 107
Ydstie, B. Erik 92
Yee, Albert F. 9
Yesavage, Victor F. 17
Yin, John 120
Ying, Jackie Y. 49
Yoganathan, Ajit P. 26
Yokochi, Alexandre 90
York, Otto 65
York, Steven 116
Yortsos, Yanis C. 15
Young, Edwin H. 53
Young, Gregory L. 14
Young, Valerie L. 86
Youngquist, Gordon R. 36
Yu, Qingsong 59
Yuen, Hsiu Huo 98
Yurttas, Lale 109

Z

Zabaloy, Marcelo 63
Zafiriou, Evanghelos 45
Zakin, Jacques L. 85
Zasadzinski, Joseph 12
Zawodzinski, Thomas 82
Zeimer, Katherine S. 50
Zenz, Frederick A. 75
Zhai, Tongguang 39
Zhang, Wu 119
Zhao, Huimin 29
Zhou, Yaoqi 78
Zhou, Youngquan 103
Zhu, Lei 19
Zhu, Wen-Hua 2
Ziegler, Edward N. 75
Ziff, Robert M. 53
Zilm, Kurt 20
Zlatanova, Jordanka 75
Zollars, Richard L. 118
Zondlo, John W. 119
Zukoski, Charles F. 29
Zurawsky, Walter P. 75
Zwiebel, Imre 4
Zydney, Andrew L. 96
Zygourakis, Kyriacos 108

Index of Institutions

A

Aachen Technical University ... 157
Abo Akademi University ... 154
Adelaide, University of ... 125
Akron, University of ... 81
Alabama in Huntsville, University of ... 1
Alabama, University of ... 1
Alberta, University of ... 135
Alcala de Henares Ingenieria Quimica, University of ... 210
Alicante, University of ... 211
Amirkabir University of Technology(Tehran Polytechnic) 173
Amsterdam, University of ... 194
Ankara University ... 230
Anna University ... 164
Annamalai University ... 164
Aristotle University ... 161
Arizona State University ... 4
Arizona, University of ... 5
Arkansas, University of ... 6
Aston University ... 234
Auburn University ... 2
Auckland, University of ... 197
Autonoma University of Barcelona ... 211

B

Bangladesh University of Engineering and Technology ... 129
Bath, University of ... 234
Beijing University of Chemical Technology ... 146
Ben-Gurion University of the Negev ... 175
Berlin, Technical University of ... 157
Bicol University ... 199
Birla Institute of Technology & Science (BITS) - Pilani ... 165
Birmingham, University of ... 235
Bogazici University ... 231
Bologna, Universita di ... 176
Braunschweig, Technische Universitaet ... 157
Brigham Young University ... 113
British Columbia, University of ... 134
Brown University ... 100
Bucknell University ... 91
Budapest University of Technology and Economics ... 163

C

Cadiz, University of ... 212
Cagliari, Universita di ... 176
Calcutta University ... 165
Calgary, University of ... 136
California Institute of Technology ... 13
California State Polytechnic University, Pomona ... 14
California State University, Long Beach ... 13
California, Berkeley, University of ... 7
California, Davis, University of ... 8
California, Irvine, University of ... 9
California, Los Angeles, University of ... 10
California, Riverside, University of ... 11
California, San Diego, University of ... 11
California, Santa Barbara, University of ... 12
Cambridge, University of ... 235
Campinas, Universidade Estadual de ... 130
Cantabria, University of ... 212
Canterbury, University of ... 197
Cape Town, University of ... 208
Carnegie Mellon University ... 92
Case Western Reserve University ... 82
Castilla Universidad ... 213
Chang Guag University ... 223
Chonnam National University ... 187
Christian Brothers University ... 103
Chulalongkorn University ... 229
Chung Yuan Christian University ... 224
Cincinnati, University of ... 83
City College of The City University of New York ... 72
Clarkson University ... 72
Clausthal, Technical University of ... 158
Clemson University ... 101
Cleveland State University ... 84
Coimbra, Universidade de ... 202
Colorado School of Mines ... 17
Colorado State University ... 18
Colorado, University of ... 16
Columbia University ... 73
Complutense University of Madrid ... 214
Concepcion, Universidad de ... 144
Connecticut, University of ... 19
Cooper Union ... 73
Cork Institute of Technology ... 174
Cornell University ... 74
Curtin University of Technology ... 125

D

Dalhousie University ... 135
Darmstadt University of Technology ... 158
Dartmouth College ... 64
Dayton, University of ... 84
De La Salle University ... 199
Delaware, University of ... 21
Delft University of Technology ... 194
Dortmund, University of ... 158
Drexel University ... 93
Durban-Westville, University of ... 208

E

Ecole Polytechnique-University of Montreal ... 137
Edinburgh, University of ... 236
Eduardo Mondlane, Universidade ... 193
Ege, University of ... 231
Eindhoven University of Technology ... 195
Erlangen-Nurnberg, Universitat ... 158
Estudios Superiores de Monterrey, Instituto Tecnologico y de ... 191
Exeter, University of ... 236

F

Fachhochschule Mannheim, University of Applied Sciences ... 161
Feng-Chia University ... 224
Florida Institute of Technology ... 24
Florida State University/Florida A&M University ... 24
Florida, University of ... 23
Fukuoka University ... 180

G

Gazi University ... 232
Gent, University of ... 129

Georg-Simon-Ohm Fachhochschule Nuernberg 161
Georgia Institute of Technology 26
Government Engineering College 165
Granada, University of 214

H

Hacettepe University 232
Hamburg-Harburg, Technical University 159
Hampton University 115
Hannover, ITC, University of 160
Hannover, University of 159
Helsinki University of Technology 154
Heriot-Watt University 237
Himeji Institute of Technology 180
Hiroshima University 180
Hong Kong University of Science and Technology 163
Houston, University of 106
Howard University 22

I

Iberoamericana, Universidad 192
Idaho, University of 27
Illes Balears, University of 215
Illinois at Chicago, University of 28
Illinois at Urbana-Champaign, University of 29
Illinois Institute of Technology 30
Indian Institute of Science 166
Indian Institute of Technology, Bombay 166
Indian Institute of Technology, Delhi 168
Indian Institute of Technology, Kanpur 167
Indian Institute of Technology, Kharagpur 167
Indian Institute of Technology, Madras 168
Indian Institute of Technology, Roorkee 171
Industries Chimiques Institut, Ecole Nationale Superieure des 156
Ingenieurs en Arts Chimiques et Technologiques, Ecole Nationale Superieure des 155
Instituto Superior Técnico 204
Instituto Tecnologico de Buenos Aires 123
Iowa State University of Science and Technology 36
Iowa, University of 35
Isfahan University of Technology 173

J

Johns Hopkins University 44

K

Kaiserslautern, Universitat 160
Kanazawa University 181
Kansai University 181
Kansas State University 38
Kansas, University of 37
Karlsruhe, Universitat 160
Karnataka Regional Engineering College, Surathkal 168
Kasetsart University 230
Keimyung University 187
Kentucky, University of 39
King Abdulaziz University 205
King Fahd University of Petroleum & Minerals 205
King Saud University 206
Korea Advanced Institute of Science and Technology 188
Korea University 188
Kunsan National University 189
Kuwait University 190
Kwame Nkrumah University of Science and Technology . 161
Kyoto University 182
Kyushu University 182

L

L'Aquila, Universita de 177
la Plata, Universidad Nacional de 123
Lafayette College 93
Lagos, University of 197
Lakehead University 137
Lamar University 107
Lappeenranta University of Technology 154
Las Americas, Puebla, Universidad de 192
Laval, Universite 137
Laxminarayan Institute of Technology, Nagpur University 169
Leeds, University of 237
Lehigh University 94
Liege, University of 129
Litoral, Universidad Nacional del 123
Ljubljana, University of 207
Lodz, Technical University of 200
London (Imperial College), University of 237
London (University College London), University of 238
Loughborough University 238
Louisiana at Lafayette, University of 42
Louisiana State University 41
Louisiana Tech University 41
Louisville Speed Scientific School, University of 40
Lund, University of 221

M

Madrid-ETSI Industriales, Universidad Politecnica de ... 215
Maine, University of 43
Malaviya National Institute of Technology, Jaipur 169
Malaya, University of 190
Malaysia, Universiti Sains 191
Manchester Institute of Science and Technology, University of 239
Manhattan College 75
Manipal Institute of Technology 169
Maribor, University of 207
Maringa, Universidade Estadual de 131
Maryland Baltimore County, University of 46
Maryland, College Park, University of 45
Massachusetts Institute of Technology 49
Massachusetts-Amherst, University of 47
Massachusetts-Lowell, University of 48
McGill University 138
McMaster University 138
McNeese State University 42
Melbourne, University of 126
Metropolitana-Iztapalapa, Universidad Autonoma 192
Michigan State University 54
Michigan Technological University 55
Michigan, University of 53
Middle East Technical University 233
Milano, Politecnico di 179
Minas Gerais, Universidade Federal de 131
Minnesota at Duluth, University of 56
Minnesota at Minneapolis St. Paul, University of 57
Mississippi State University 58
Mississippi, University of 58
Missouri-Columbia, University of 59
Missouri-Rolla, University of 60
Monash University 126
Montana State University 62
Mumbai, University of 170
Murcia, University of 216

N

Nagoya University 182

Napoli, Universita di 177
Natal, University of 209
National Central University 224
National Cheng Kung University 225
National Chung Cheng University 226
National Institute of Technology Warangal 170
National Taiwan University 227
National Taiwan University of Science and Technology .. 226
National Tsing Hua University 227
National University of Science and Technology 244
Nebraska, University of 62
Nevada, Reno, University of 63
New Brunswick, University of 139
New Hampshire, University of 64
New Haven, University of 20
New Jersey Institute of Technology 65
New Mexico Institute of Mining and Technology 69
New Mexico State University 71
New Mexico, University of 70
New South Wales, University of 127
Newcastle Upon Tyne, University of 240
Newcastle, University of 127
Niigata University 183
North Carolina Agricultural and Technical State University 79
North Carolina State University 80
North Dakota, University of 81
North-West University, Potchefstroom Campus 209
Northeastern University 50
Northwestern University 31
Norwegian University of Science and Technology 198
Notre Dame, University of 32
Nottingham, University of 240

O

Obafemi Awolowo University 198
Ohio State University 85
Ohio University 86
Oklahoma State University 89
Oklahoma, University of 88
Oregon State University 90
Osaka Prefecture University 183
Osaka University 184
Ottawa, University of 139
Oulu, University of 154
Oviedo, University of 216

P

Pais Vasco, University of 216
Pardubice, University of 153
Patras, University of 162
Pellissippi State Technical Community College 122
Pennsylvania State University 96
Pennsylvania, University of 95
Petroleum and Gas Institute 204
Philippines at Los Banos, University of 199
Pisa, Universita di 178
Pittsburgh, University of 97
Pohang University of Science and Technology 189
Polytechnic University 75
Pontificia Universidad Catolica de Chile 145
Porto, Universidade do 203
Prague Institute of Chemical Technology 153
Prairie View A&M University 107
Pravara Rural Education Society's Pravara Rural Engineering College, Loni 170
Pretoria, University of 209
Prince of Songkla University 230
Princeton University 66
Puerto Rico, University of 99
Purdue University 33

Q

Qatar, University of 204
Queen's University 140
Queen's University of Belfast 241
Queensland, University of 127

R

R. V. College of Engineering, Affiliated with Visveswaraiah Technical University 171
Regional Engineering College, Affiliated with Bharathidasan University 171
Rensselaer Polytechnic Institute 76
Rhode Island, University of 100
Rice University 108
Rio De Janeiro, Universidade Federal Do 132
Rochester, University of 77
Rose-Hulman Institute of Technology 34
Rovira i Virgili University 217
Rowan University 67
Royal Institute of Technology 222
Royal Melbourne Institute of Technology 128
Royal Military College of Canada 140
Rutgers, The State University of New Jersey 68
Ryerson University 141

S

S. V. National Institute of Technology 172
Salamanca, University of 218
Salerno, University of 178
Salta, Universidad Nacional del 124
San Jose State University 14
San Luis Potosi, Universidad Autonoma de 193
Santander, Universidad Industrial de 152
Santiago de Chile, Universidad de 145
Santiago de Compostela, Universidad de 218
Santo Tomas, University of 200
Sao Paulo, Universidade de 133
Saskatchewan, University of 141
Seikei University 184
Seoul National University 189
Sevilla, Escuela Superior de Ingenieros Industriales de .. 219
Sevilla, Universidad de 219
Sharif University of Technology 173
Sheffield, University of 241
Sherbrooke, Universite de 141
Siddaganga Institute of Technology, Affiliated with Viveswaraiah University 172
Silesian Technical University 200
Singapore, National University of 206
South Alabama, University of 3
South Bank University 242
South Carolina, University of 102
South China University of Technology 147
South Dakota School of Mines and Technology 102
South Florida, University of 25
Southern California, University of 15
Stanford University 15
State University of New York at Buffalo 78
Stevens Institute of Technology 69
Strathclyde, University of 242
Sur, Universidad Nacional del 124
Surrey, University of 242
Swiss Federal Institute of Technlolgy 222
Swiss Federal Institute of Technology, Ecole Polytechnique Federale de Lausanne 223

Sydney, University of 128
Syracuse University 79
Szczecin, Technical University of, Faculty of Chemical Engineering 201
Szczecin, Technical University of 201

T

Tamkang University 228
Technion-Israel Institute of Technology 175
Tecnica Federico, Universidad 146
Teesside, University of 243
Tennessee at Chattanooga, University of 103
Tennessee at Knoxville, University of 104
Tennessee Technological University 105
Texas A&M University 109
Texas A&M University-Kingsville 110
Texas at Austin, University of 111
Texas Tech University 112
Thapar Institute of Engineering and Technology 172
Tianjin University 147
Tohoku University 184
Tokyo Institute of Technology 185
Tokyo, University of 185
Toledo, University of 87
Toronto, University of 142
Toyama University 186
Tri-State University 34
Tsinghua University 149
Tufts University 51
Tulane University 43
Tulsa, University of 89
Tunghai University 228
Tuskegee University 3
Twente University of Technology 195

U

Uberlandia, Universidade Federal de 133
United Arab Emirates University 234
University College Dublin 174
Utah, University of 114

V

Valencia, Universitat de 220
Valladolid, University of 220
Valparaiso, Universidad Catolica de 146
Vanderbilt University 105
Vellore Institute of Technology 173
Venezuela, Universidad Central de 243
Veszprem, University of 164
Villanova University 98
Virginia Tech 116
Virginia, University of 115

W

Wales Swansea, University of 243
Warsaw University of Technology 201
Washington State University 118
Washington University 61
Washington, University of 117
Waterloo, University of 143
Wayne State University 56
West Virginia University 119
Western Ontario, University of 144
Widener University 98
Wisconsin-Madison, University of 120
Witwatersrand, Johannesburg, University of the 210
Worcester Polytechnic Institute 52
Wroclaw Technical University 202
Wyoming, University of 121

Y

Yale University 20
Yamagata University 186
Yamaguchi University 186
Yildiz Technical University 233
Yokohama National University 187
Youngstown State University 87
Yuan Ze University 229

Z

Zagreb, University of 152
Zaragoza, University of 221
Zhejiang University 150